QCD@WORK 2005

Proceedings in the Series of International Workshops on Quantum Chromodynamics

Year		Held in	Publisher	ISBN
2005	3rd	Conversano, Bari, Italy	AIP Conf. Proceedings Vol. 806	0-7354-0302-3
2003	2nd	Conversano, Bari, Italy	SLAC eConf C030614	
2001	1st	Martina Franca, Bari, Italy	AIP Conf. Proceedings Vol. 602	0-7354-0046-6

To learn more about these titles, or the AIP Conference Proceedings Series, please visit the webpage **http://proceedings.aip.org**

QCD@WORK 2005

International Workshop on
Quantum Chromodynamics:
Theory and Experiment

Conversano, Bari, Italy   16 – 20 June 2005

EDITORS
Pietro Colangelo
Fulvia De Fazio
Eugenio Nappi
Istituto Nazionale di Fisica Nucleare
Bari, Italy

Giuseppe Nardulli
Università di Bari
Bari, Italy

SPONSORING ORGANIZATIONS
Italian Ministry of Education, University and Research (MIUR)
Istituto Nazionale di Fisica Nucleare (INFN)

Melville, New York, 2006
AIP CONFERENCE PROCEEDINGS ■ VOLUME 806

Editors:

Pietro Colangelo
Fulvia De Fazio
Eugenio Nappi

Istituto Nazionale di Fisica Nucleare
Sezione di Bari
Via Orabona 4
70126 Bari
Italy

E-mail: pietro.colangelo@ba.infn.it
fulvia.defazio@ba.infn.it
eugenio.nappi@ba.infn.it

Giuseppe Nardulli
Dipartimento di Fisica
Università di Bari
Via Orabona 4
70126 Bari
Italy

E-mail: giuseppe.nardulli@ba.infn.it

Image on title page: Castello Aragonese, Conversano (Bari, Italy).
Courtesy of Cooperativa Armida, Conversano.

Authorization to photocopy items for internal or personal use, beyond the free copying permitted under the 1978 U.S. Copyright Law (see statement below), is granted by the American Institute of Physics for users registered with the Copyright Clearance Center (CCC) Transactional Reporting Service, provided that the base fee of $23.00 per copy is paid directly to CCC, 222 Rosewood Drive, Danvers, MA 01923, USA. For those organizations that have been granted a photocopy license by CCC, a separate system of payment has been arranged. The fee code for users of the Transactional Reporting Services is: 0-7354-0302-3/06/$23.00

© 2006 American Institute of Physics

Permission is granted to quote from the AIP Conference Proceedings with the customary acknowledgment of the source. Republication of an article or portions thereof (e.g., extensive excerpts, figures, tables, etc.) in original form or in translation, as well as other types of reuse (e.g., in course packs) require formal permission from AIP and may be subject to fees. As a courtesy, the author of the original proceedings article should be informed of any request for republication/reuse. Permission may be obtained online using Rightslink. Locate the article online at http://proceedings.aip.org, then simply click on the Rightslink icon/"Permission for Reuse" link found in the article abstract. You may also address requests to: AIP Office of Rights and Permissions, Suite 1NO1, 2 Huntington Quadrangle, Melville, NY 11747-4502, USA; Fax: 516-576-2450; Tel.: 516-576-2268; E-mail: rights@aip.org.

L.C. Catalog Card No. 2005937215
ISBN 0-7354-0302-3
ISSN 0094-243X
Printed in the United States of America

CONTENTS

Preface ix
Committees xi
Photograph xiii

Progress in Chiral Perturbation Theory 1
    G. Ecker

Hadronic Decays of the Tau Lepton: $\tau^- \to (\pi\pi\pi)^- \nu_\tau$ within Resonance Chiral Theory 11
    D. Gómez Dumm, A. Pich, and J. Portolés

Latest Results on Kaon Physics 18
    A. Antonelli

Dynamical Consequences of Strong CP Breaking 26
    P. Faccioli

Properties of the Quark Correlator 33
    E. Di Salvo

Unitarity Structure of the QCD Sum Rules and KYN and $KY\Xi$ Couplings 40
    T. Aliev, A. Ozpineci, S. B. Yakovlev, and V. S. Zamiralov

Pentaquarks: The Latest Experimental Results 48
    M. Battaglieri, R. De Vita, V. Kubarovsky, and the CLAS Collaboration

Spin Physics Overview 57
    P. Di Nezza

Higgs Boson Production via Gluon Fusion at Hadron Colliders 67
    S. Catani

NNLO QCD Corrections for the Differential Higgs Boson Production Cross-Section in Gluon Fusion 75
    C. Anastasiou

Perturbative QCD Corrections to $b \to s\gamma$ 80
    L. Trentadue

CSW Diagrams and Electroweak Vector Bosons 89
    P. Mastrolia

QCD Effective Couplings in Minkowskian and Euclidean Domains 97
    D. V. Shirkov

Deconstructed Higgsless Models 104
    R. Casalbuoni

Could Spin-Charge Separation be the Source of Confinement? 114
    A. J. Niemi

Approximated Faddeev-Niemi Knotted Solitons 124
    A. Wereszczyński

Order, Disorder and Confinement 130
    M. D'Elia, A. Di Giacomo, and C. Pica

Topological Susceptibility in the SU(3) Gauge Theory 137
    L. Del Debbio

Finite Temperature Lattice Gauge Theories in External Fields 144
    L. Cosmai

CP Violation in B-Physics 151
    F. Ferroni

New Physics Search in *B* and Kaon Physics.............................. 164
    T. Hurth

Sum Rules for Leading and Subleading Form Factors in Heavy
Quark Effective Theory Using the Non-Forward Amplitude 173
    F. Jugeau, A. Le Yaouanc, L. Oliver, and J.-C. Raynal

Non-Factorizable Contributions to $\overline{B}_d^0 \to D_S^{(*)}\overline{D}_S^{(*)}$ from Chiral Loops
and Tree-Level $1/N_c$ Terms 183
    J. O. Eeg, S. Fajfer, and A. Prapotnik Brdnik

Relations for Direct CP Asymmetries in $B \to PP$ and $B \to PV$ Decays 190
    T. N. Pham

Final State Interactions in the *B* Meson Decay into Two Pions 197
    A. Deandrea, M. Ladisa, V. Laporta, G. Nardulli, and P. Santorelli

Charm Meson Resonances in *D* Semileptonic Decays 203
    S. Fajfer and J. Kamenik

How Charm Can Still be Charming: Some Recent Results from
FOCUS 210
    S. Malvezzi

Radiative Decays of Excited Charm Mesons: A Light-Cone QCD Sum
Rule Analysis 217
    P. Colangelo, F. De Fazio, and A. Ozpineci

Heavy Quarkonium Decays and Transitions in the Language of
Effective Field Theories 224
    A. Vairo

Model-Independent Study on Magnetic Dipole Transition in Heavy
Quarkonium 231
    Y. Jia

Quark-Antiquark Bound State Equation in the Wilson Loop
Approach with Minimal Surfaces 238
    F. Jugeau and H. Sazdjian

QCD at High Temperature: Results from Lattice Simulations with an
Imaginary μ 245
    M. D'Elia, F. Di Renzo, and M. P. Lombardo

Jet Quenching: RHIC Results and Phenomenology 252
    D. d'Enterria

Hadronic Modes in the Quark-Gluon Plasma 259
    M. Mannarelli and R. Rapp

Production of Multiply Heavy Flavoured Baryons from Quark Gluon
Plasma 266
    F. Becattini

Latest Results from the NA57 Experiment 272
    G. E. Bruno *(On behalf of the NA57 Collaboration)*

J/ψ Production in Indium-Indium Collisions at SPS Energies 279
    P. Pillot, R. Arnaldi, R. Averbeck, K. Banicz, J. Castor, B. Chaurand,
    C. Cicalo, A. Colla, P. Cortese, S. Damjanovic, A. David, A. de Falco,
    A. Devaux, A. Drees, L. Ducroux, H. En'yo, A. Ferretti, M. Floris,
    P. Force, N. Guettet, A. Guichard, H. Gulkanian, J. Heuser, M. Keil,

L. Kluberg, J. Lozano, C. Lourenço, F. Manso, A. Masoni, P. Martins,
A. Neves, H. Ohnishi, C. Oppedisano, P. Parracho, G. Puddu,
E. Radermacher, P. Ramalhete, P. Rosinsky, E. Scomparin, J. Seixas,
S. Serci, R. Shahoyan, P. Sonderegger, H. J. Specht, R. Tieulent, G. Usai,
R. Veenhof, and H. K. Wöhri

Indications of a Pseudogap in the Nambu Jona-Lasinio Model 286
    P. Castorina, G. Nardulli, and D. Zappalà

Color Superconductivity and the Strange Quark 293
    M. Alford

Smeared Gap Equations in Crystalline Color Superconductivity 303
    M. Ruggieri

Asymmetric Neutrino Emission from Spin-1 Color Superconductor 310
    A. Schmitt, I. A. Shovkovy, and Q. Wang

List of Participants 317
Author Index 327

Preface

On June 16-20, 2005, about 70 physicists from 15 countries met in the ancient and artistic town of Conversano, near Bari, Italy, to attend the workshop QCD@Work 2005, International Workshop on Quantum Chromodynamics: Theory and Experiment. This was the third edition of a workshop started in Martina Franca, Italy, on June 2001, and continued in Conversano, Italy, on June 2003.

The topics of all the editions were: Low Energy and Nonperturbative QCD; Deep inelastic scattering with polarized and unpolarized probes; Jet Physics and perturbative QCD; Heavy quark Physics; Physics of relativistic heavy ion collisions; QCD phase diagram at finite temperature and density. The list suggests the aims of the workshop; the results of the lively discussions during the Conversano 2005 edition are collected in this volume.

As organizers of the event, we would like to thank the Institutions and individuals who have contributed to its realization and success. First, MIUR, the Italian Ministry of Education, University and Research (Program "Frontier problems in the Theory of Fundamental Interactions"), and Istituto Nazionale di Fisica Nucleare (INFN). We acknowledge financial support from the Administrative Council and the Physics Department of Bari University. We thank the Municipality of Conversano and Mayor Francesco Iudice for having allowed us to organize the scientific sessions in the artistic environment of Pinacoteca Comunale in Castello Aragonese. We are grateful to Dr. Carlo Mansueto and all joung members of Cooperativa Armida, Conversano, for their support in the organization. We warmly thank Dr. Malinda Sassu for her enthusiastic collaboration. Finally, we are grateful to our colleagues Nicola Cufaro Petroni (violino) and Nicola Ippolito (pianoforte) for the concert "Two Physicists at Work".

Pietro Colangelo, Fulvia De Fazio, Eugenio Nappi, Giuseppe Nardulli

International Advisory Committee

G. Altarelli (CERN)
R. Gatto (University of Geneva, Chairman)
J. Harris (Yale University)
U. Heinz (Ohio State University)
N. Paver (University of Trieste)
R. Petronzio (University of Rome – Tor Vergata)
M. Shifman (Minnesota University)

Local Organizing Committee

P. Colangelo (INFN, Bari)
F. De Fazio (INFN, Bari)
E. Nappi (INFN, Bari)
G. Nardulli (University of Bari, Chairman)

Progress in Chiral Perturbation Theory

Gerhard Ecker

Inst. Theor. Physik, Univ. Wien, Boltzmanng. 5, A-1090 Wien, Austria

Abstract. After a short status report on chiral perturbation theory, I review recent progress in determining some of the low-energy couplings by matching the effective theory to QCD. Consequences for K_{l3} decays and for the extraction of the CKM matrix element V_{us} are reported. Hadronic vacuum polarization at low energies and its impact on the anomalous magnetic moment of the muon are discussed.

Keywords: Effective field theory, chiral perturbation theory, V_{us}, $(g-2)_\mu$
PACS: 11.30.Rd,12.39.Fe,12.15.Hh,14.60.Ef

INTRODUCTION

At a time when the LHC is getting ready to open a new era in particle physics, it is legitimate to ask why one should still be interested in QCD (more generally in the Standard Model) at low energies. There are at least two good reasons to pursue the study of QCD in the confinement regime.

- It is a challenge for theoretical particle physics to derive reliable results in the nonperturbative domain. An impressive example is pion pion scattering, one of the few examples in hadron phenomenology at low energies where theory is ahead of experiment [1]. Important information on the mechanism of spontaneous chiral symmetry breaking can be extracted from pion pion scattering, especially from S-wave scattering lengths. The study of QCD in the nonperturbative regime may turn out to be relevant even for LHC physics if the simple Higgs mechanism of the Standard Model turns out to be insufficient to describe electroweak symmetry breaking.
- The assessment of physics beyond the Standard Model will remain an important research topic even at much lower than LHC energies. The reduction in energy must be compensated by an increase in precision, both in experiment and in theory. In the long run, lattice gauge theories and low-energy effective field theories will survive as the most comprehensive and reliable approaches in this field.

In this talk, I present a short progress report on chiral perturbation theory (CHPT), which is precisely the effective field theory of the Standard Model at low energies. Green functions and amplitudes are dominated at low energies by the exchange of pseudoscalar mesons, the pseudo-Goldstone bosons of spontaneously broken chiral symmetry, allowing for a systematic expansion in momenta and quark masses. I discuss recent progress in determining some of the a priori unknown coupling constants of CHPT, a recent CHPT analysis of K_{l3} decays to extract the CKM matrix element V_{us} and, finally, very recent developments concerning the determination of hadronic vacuum polarization, a topic of great importance for comparing the Standard Model prediction of the muon magnetic moment with experiment. The extension of CHPT to the intermediate-energy region dominated by meson resonances is covered by J. Portolés [2].

STATUS OF CHIRAL PERTURBATION THEORY

The spontaneously and explicitly broken chiral symmetry of QCD is the key feature of CHPT. The corresponding Lagrangian is organized in an expansion in derivatives (vestige of spontaneous symmetry breaking) and in quark masses (explicit breaking). CHPT is a nonrenormalizable quantum field theory that must nevertheless be renormalized like any respectable quantum field theory. The main difference to renormalizable theories is the rapidly increasing number of low-energy constants (LECs) in higher orders of CHPT. As a low-energy effective field theory, CHPT can be applied to processes with momenta $\ll 1$ GeV.

In the mesonic sector, the original effective chiral Lagrangian of next-to-leading order [3, 4] has been extended to next-to-next-to-leading order [5] or $O(p^6)$ in the standard chiral counting. At this order, diagrams with up to two loops have to be taken into account for a consistent low-energy expansion (see Ref. [6] for a recent review).

Still in the meson sector, the formalism of CHPT has been extended to incorporate the nonleptonic weak interactions and to implement radiative corrections for strong processes as well as for semileptonic and nonleptonic weak decays. The corresponding Lagrangians and the associated number of LECs are displayed in Table 1. As the Table indicates, the state of the art for these extensions is next-to-leading order with at most one-loop amplitudes.

TABLE 1. The effective chiral Lagrangian of the SM in the meson sector. The numbers in brackets refer to the number of independent couplings for $N_f = 3$. The parameter-free Wess-Zumino-Witten action that cannot be written as the four-dimensional integral of an invariant Lagrangian must be added.

$\mathscr{L}_{\text{chiral order}}$ (# of LECs)	loop order
$\mathscr{L}_{p^2}(2) + \mathscr{L}^{\Delta S=1}_{G_F p^2}(2) + \mathscr{L}^{\text{em}}_{e^2 p^0}(1) + \mathscr{L}^{\text{emweak}}_{G_8 e^2 p^0}(1)$	$L = 0$
$+ \mathscr{L}_{p^4}(10) + \mathscr{L}^{\text{odd}}_{p^6}(32) + \mathscr{L}^{\Delta S=1}_{G_8 p^4}(22) + \mathscr{L}^{\Delta S=1}_{G_{27} p^4}(28)$ $+ \mathscr{L}^{\text{em}}_{e^2 p^2}(14) + \mathscr{L}^{\text{emweak}}_{G_8 e^2 p^2}(14) + \mathscr{L}^{\text{leptons}}_{e^2 p}(5)$	$L = 1$
$+ \mathscr{L}_{p^6}(90)$	$L = 2$

Effective chiral Lagrangians have also been employed for baryonic processes [7] and for light nuclei [8].

LOW-ENERGY CONSTANTS

As Table 1 shows, a major problem of CHPT is the abundance of LECs in higher orders of the chiral expansion. For a phenomenological determination of those constants, two types of LECs can be distinguished.

i. The associated contributions survive in the chiral limit. Such LECs govern the momentum dependence of amplitudes and are at least in principle accessible experimentally.

ii. The couplings are associated with explicit chiral symmetry breaking. Such LECs specify the quark mass dependence of amplitudes. They are difficult if not impossible to extract from experiment but they are accessible in lattice QCD.

However, at the present level of sophistication it is unrealistic to expect a phenomenological determination of all LECs even of type i only. Instead, some progress has been made recently in matching CHPT to QCD by investigating specific Green functions in the limit of large N_C. As in every effective field theory, the LECs are sensitive to the "heavy" degrees of freedom not represented by explicit fields in the Lagrangian. Experience shows that truncation of the infinitely many intermediate states (for $N_C \to \infty$) to the lowest-lying resonances is usually sufficient.

Instead of reviewing the matching procedure in general, I discuss two specific examples recently considered that have some impact on topics of current interest.

Radiative semileptonic decays

In the discussion of radiative corrections for semileptonic kaon decays the Lagrangian $\mathcal{L}^{\text{leptons}}_{e^2 p}$ [9] in Table 1 enters. In a two-step procedure, the Fermi theory of semileptonic decays was matched to both the Standard Model and CHPT [10] resulting in spectral representations for all five LECs in $\mathcal{L}^{\text{leptons}}_{e^2 p}$.

Let me concentrate here on one of those LECs (X_1) that will be relevant later on. The authors of Ref. [10] obtain the following representation for X_1,

$$X_1 = \frac{3i}{8} \int \frac{d^4 k}{(2\pi)^4} \left(\Gamma_{VV}(k^2) - \Gamma_{AA}(k^2) \right) / k^2, \qquad (1)$$

in terms of vertex functions (V^a_μ is an $SU(3)$ vector current and ϕ^c is a member of the pseudoscalar octet)

$$\Gamma_{VV}(k^2) \sim \lim_{p \to 0} \int d^4 x e^{ikx} \langle 0 | T V^a_\mu(x) V^b_\nu(0) | \phi^c(p) \rangle \qquad (2)$$

and similarly for $\Gamma_{AA}(k^2)$. The integral converges well and, when saturated with the lowest-lying V, A meson resonances, produces a value $X_1 = -0.0037$ [10] to be used for the analysis of K_{l3} decays.

Strong LECs of O(p^6)

The second example concerns LECs that appear in the K_{l3} amplitudes at $O(p^6)$. The Green function of interest is the three-point function of scalar and pseudoscalar densities:

$$i^2 \int dx dy\, e^{ipx+iqy+irz} \langle 0|TS^a(x)P^b(y)P^c(z)|0\rangle = d^{abc}\, \Pi_{SPP}(p^2,q^2,r^2) \,. \tag{3}$$

At low energies, Π_{SPP} is given in terms of LECs of $O(p^4)$ and $O(p^6)$ since loop contributions are subdominant for large N_C. At high momenta, the operator product expansion (OPE) fixes the behaviour of Π_{SPP} that vanishes in QCD perturbation theory as an order parameter of spontaneous chiral symmetry breaking. Additional constraints apply for (transition) form factors at large momentum transfer, with two external momenta on shell.

To interpolate between CHPT and QCD, a large-N_C motivated ansatz can be employed [11]:

$$\Pi^{\mathscr{SP}}_{SPP}(s,t,u) = \frac{P_0 + P_1 + P_2 + P_3 + P_4}{[M_S^2-s][-t][-u][M_P^2-t][M_P^2-u]} \,, \tag{4}$$

with polynomials P_n of degree n in s,t,u (altogether 21 parameters). The OPE limits $n \leq 4$ and lowest-order CHPT fixes the constant P_0. The high-energy conditions constrain the polynomials P_1, P_2 of direct relevance for the LECs. The final relations for the $O(p^6)$ LECs of interest are [11]

$$C^{\mathscr{SP}}_{12} = -\frac{F^2}{8M_S^4}\,, \quad C^{\mathscr{SP}}_{34} = \frac{3F^2}{16M_S^4} + \frac{d_m^2}{2}\left(\frac{1}{M_S^2} - \frac{1}{M_P^2}\right)^2 \tag{5}$$

in terms of the masses M_S, M_P of the lowest-lying (pseudo-)scalar nonets, the pion decay constant F and a resonance coupling $d_m \sim F/(2\sqrt{2})$. All parameters refer to the chiral limit.

The first interpretation of these results is not too encouraging. There are big uncertainties related to the value of M_S in particular and to the rather strong scale dependence of C_{12} and C_{34}, which is however inaccessible at leading order in $1/N_C$.

K_{l3} AND V_{us}

The analysis of K_{l3} decays allows for the presently most accurate determination of the CKM matrix element V_{us}. In general, two form factors characterize the decay matrix element:

$$\langle \pi^-(p_\pi)|\bar{s}\gamma_\mu u|K^0(p_K)\rangle = f_+^{K^0\pi^-}(t)(p_K+p_\pi)_\mu + f_-^{K^0\pi^-}(t)(p_K-p_\pi)_\mu \,. \tag{6}$$

Of special interest for the determination of V_{us} is the quantity $f_+^{K^0\pi^-}(0)$ with the following chiral expansion:

$$f_+^{K^0\pi^-}(0) = 1 + f_{p^4} + f_{e^2 p^2} + f_{p^6} + O[(m_u-m_d)p^4, e^2 p^4] \,. \tag{7}$$

The present status is as follows:

f_{p^4}	-0.0227 (no uncertainty)	[12]
$f_{e^2 p^2}$	radiative corrections (X_i)	[13]
f_{p^6}	loop contributions	[14, 15]
	tree contributions	$L_5^2, C_{12}+C_{34}$

For a first comparison with experiment, consider the ratio [13]

$$r_{+0} := \left(\frac{2\Gamma(K^+_{e3(\gamma)}) M_{K^0}^5 I_{K^0}}{\Gamma(K^0_{e3(\gamma)}) M_{K^+}^5 I_{K^+}}\right)^{1/2} = \frac{|f_+^{K^+\pi^0}(0)|}{|f_+^{K^0\pi^-}(0)|} . \tag{8}$$

The theoretical prediction for r_{+0} is independent of f_{p^6}. The only previously unknown LEC in r_{+0} is X_1. With the newly determined value for X_1 [10] and using quadratic fits for the form factors to extract $f_+(0)$ from the data, one finds [16, 17]

$$\begin{aligned} r_{+0}^{\text{th}} &= 1.023 \pm 0.003 \\ r_{+0}^{\text{exp}} &= 1.036 \pm 0.008 . \end{aligned} \tag{9}$$

A possible discrepancy between theory and experiment for r_{+0} could be due to several reasons: radiative corrections applied by experimentalists are not always state of the art, the lifetimes of K^+, K_L may still undergo revisions and the error in r_{+0}^{th} due to neglected effects of $O[(m_u - m_d)p^4, e^2 p^4]$ could be underestimated.

Turning now to $f_+^{K^0\pi^-}(0)$, the uncertainty in the $O(p^6)$ contribution f_{p^6} is mainly due to the LECs. Loop and local contributions are separately scale dependent. The loop contributions at the scale $\mu = M_\rho$ amount to [15]

$$f_{p^6}^{L=1,2}(M_\rho) = 0.0093 \pm 0.0005 . \tag{10}$$

The local contribution is given by

$$f_{p^6}^{\text{tree}}(M_\rho) = 8 \frac{(M_K^2 - M_\pi^2)^2}{F_\pi^2} \left[\frac{(L_5^r(M_\rho))^2}{F_\pi^2} - C_{12}^r(M_\rho) - C_{34}^r(M_\rho) \right] . \tag{11}$$

The results of large-N_c matching discussed in the previous section can be read off from Fig. 1. The separate contributions L_5^2 and $C_{12}+C_{34}$ depend strongly both on the uncertain scalar resonance mass M_S and on the renormalization scale. However, as shown in Fig. 1 for the M_S dependence, both uncertainties are substantially reduced for the relevant combination entering $f_{p^6}^{\text{tree}}(M_\rho)$.

A strong destructive interference between the two local contributions is observed. The final result (allowing for a second pseudoscalar multiplet P') is [11]

$$\begin{aligned} f_{p^6}^{\text{tree}}(M_\rho) &= -0.002 \pm 0.008_{1/N_C} \pm 0.002_{M_S}{}^{+0.000}_{-0.002}{}_{P'} \\ f_{p^6} &= 0.007 \pm 0.012 \\ f_+^{K^0\pi^-}(0) &= 0.984 \pm 0.012 . \end{aligned} \tag{12}$$

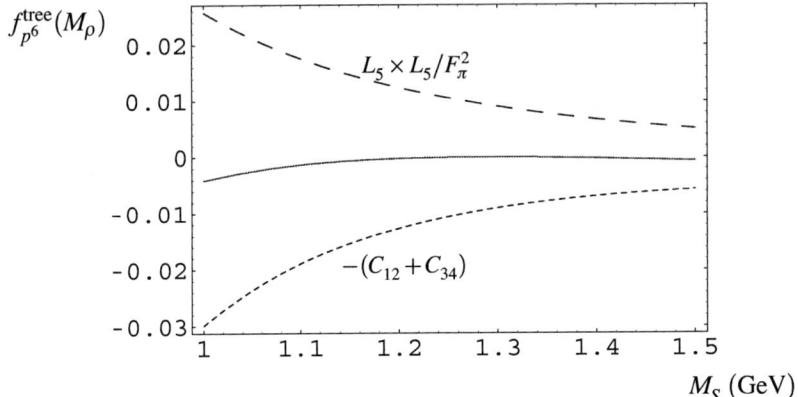

FIGURE 1. $f_{p^6}^{\text{tree}}(M_\rho)$ is displayed as a function of M_S for $M_P = 1.3$ GeV (solid line). The dashed line represents the term proportional to L_5^2, while the dotted line represents the term proportional to $-(C_{12}+C_{34})$.

We find less $SU(3)$ breaking in $f_+^{K^0\pi^-}(0)$ compared to Leutwyler and Roos [18], with f_{p^6} being dominated by the loop contribution. From the experimental result [16] $f_+^{K^0\pi^-}(0) \cdot |V_{us}| = 0.2160(10)$ one obtains

$$|V_{us}| = 0.2195 \pm 0.0027_{f_+(0)} \pm 0.0010_{\text{exp}} \,. \tag{13}$$

Before observing a possible conflict with CKM unitarity (the PDG value [19] for V_{ud} gives rise to $|V_{us}|^{\text{unitarity}} = 0.2265 \pm 0.0022$), the following remarks are in order.

i. A new result for the neutron lifetime [20] would prefer a value for V_{ud} in perfect agreement with $|V_{us}| = 0.2195$ and unitarity.
ii. A recent analysis of semileptonic hyperon decays [21] yields $|V_{us}| = 0.2199 \pm 0.0026$. After the Workshop, the uncertainties of extracting V_{us} from semileptonic hyperon decays have been reassessed in Ref. [22].
iii. To achieve an accuracy of better than 1% for V_{us}, the differences between K^+ and K^0 results must be straightened out.

An independent check of the theoretical estimate for the LECs of $O(p^6)$ is provided by the slope λ_0 of the scalar form factor (accessible in $K_{\mu 3}$ decays) that depends on the same LECs C_{12}, C_{34} as $f_+(0)$.

Ref.	Cirigliano et al. [11]	KTeV [23]
$\lambda_0 \cdot 10^3$	13 ± 3	13.72 ± 1.31

HADRONIC VACUUM POLARIZATION AND $(g-2)_\mu$

At present, the biggest uncertainty in the evaluation of the anomalous magnetic moment of the muon a_μ in the Standard Model is due to hadronic vacuum polarization at lowest order in α (shown in Fig. 2) that is directly related to the cross section $\sigma(e^+e^- \to$ hadrons). About 73 % of $a_\mu^{\text{vac.pol.}}$ comes from the $\pi^+\pi^-$ final state, the low-energy part being especially important.

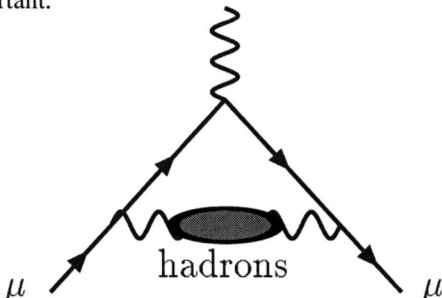

FIGURE 2. Contribution of lowest-order hadronic vacuum polarization to the muon magnetic moment.

In the isospin limit, the two-pion contribution to hadronic vacuum polarization can also be obtained from the decay $\tau^- \to \pi^-\pi^0 \nu_\tau$ [24]. At the level of accuracy needed for a comparison with the measured value of a_μ [25], isospin violating and electromagnetic corrections must be included [26, 27].

However, until recently the two-pion spectral functions from e^+e^- annihilation and from τ decays seemed to differ significantly especially above the ρ region, even after accounting for isospin violating effects. The value of $a_\mu^{\pi\pi}$ on the basis of the most precise e^+e^- data from the CMD-2 Collaboration [28] was then confirmed by KLOE [29] although the actually measured $\pi\pi$ cross sections are not in very good agreement. The consensus among many experts in the field was spelled out by Höcker at last year's High Energy Conference in Beijing [30]: until the origin of the discrepancy between e^+e^- and τ data is understood the τ data should be ignored for the evaluation of a_μ.

A recent analysis of Maltman [31] suggests a new perspective on this issue. He investigates so-called pinched FESR of the type

$$\int_0^{s_0} w(s)\rho(s)ds = -\frac{1}{2\pi i}\oint_{|s|=s_0} w(s)\Pi(s)ds \tag{14}$$

for current correlators $\Pi(s)$ with associated spectral functions $\rho(s)$. The spectral functions of interest here are the electromagnetic spectral function ρ_{em} measured in e^+e^- annihilation and the charged $I=1$ vector current spectral function $\rho_V^{I=1}$ accessible in τ decays. The weight function $w(s)$ is a positive definite analytic function in the complex s-plane for $|s| \leq s_0$, but otherwise arbitrary except for the constraint $w(s_0) = 0$ to minimize duality violations (pinching).

The left-hand side of the FESR (14) is evaluated with experimental input (CMD-2 [28] and ALEPH [32]) whereas the right-hand side is calculated from QCD with the help of the OPE. The freedom of choosing the weight function $w(s)$ can be employed to eliminate the dimension $D = 6$ OPE contributions altogether. The right-hand side is then mainly sensitive to the $D = 0$ perturbative part known up to $O(\alpha_s^3)$, with weaker dependences on m_s (in the $D = 2$ piece) and on $D = 4$ quark and gluon condensates. Effects with $D \geq 8$ can be kept under control by varying s_0. Discarding all low-energy input for the determination of $\alpha_s(M_Z)$ (such as the τ data that are to be tested with FESR), Maltman obtains a value

$$\alpha_s(M_Z) = 0.1200 \pm 0.0020 \tag{15}$$

to be used for the right-hand side of (14).

A first test performed in Ref. [31] consists in fitting $\alpha_s(M_Z)$ from the experimentally determined spectral integrals (left-hand side), leaving all other input for the right-hand side unchanged. The results for two typical weight functions w_1, w_6 are shown in Table 2, to be compared with the best value from high-energy data in Eq. (15). Taking into account that the weights are positive definite, the results in Table 2 indicate that the electromagnetic spectral density is too low whereas the τ spectral data are in perfect agreement with the canonical value of α_s.

TABLE 2. Fitted values of $\alpha_s(M_Z)$ from experimentally determined spectral integrals for two different weight functions [31].

weight	type	$\alpha_s(M_Z)$
w_1	em	$0.1138 \; ^{+0.0030}_{-0.0035}$
w_6	em	$0.1150 \; ^{+0.0022}_{-0.0026}$
w_1	τ	$0.1218 \; ^{+0.0027}_{-0.0032}$
w_6	τ	$0.1201 \; ^{+0.0020}_{-0.0022}$

A second independent consistency check of the data comes from a comparison of the two sides in Eq. (14) for different values of s_0 [31]. When plotting the spectral integrals as functions of s_0 one arrives at a similar conclusion as before: the slopes in the electromagnetic case differ by about 2.5 σ between data and QCD. On the other hand, the τ data show perfect consistency both for the slope and in absolute normalization (depending on α_s).

The conclusions of Ref. [31] are very convincing even if the statistical weight is not overwhelming: the sum rule tests clearly favour the τ over the e^+e^- data. The status of a_μ at the time of the Workshop can be summarized as follows [33]:

$$\left(a_\mu^{\text{exp}} - a_\mu^{\text{SM}}\right) \cdot 10^{10} = \begin{cases} 23.9 \pm 9.9 & (2.4\;\sigma) \; [e^+e^-] \\ 7.6 \pm 8.9 & (0.9\;\sigma) \; [\tau, e^+e^-] \end{cases}. \tag{16}$$

Using the isospin corrected τ data for the 2π and 4π final states thus leads to agreement between theory and experiment to better than 1 σ.

Two weeks after the Workshop, new $e^+e^- \to \pi^+\pi^-$ data were released [34] that appear to lie between the CMD-2 and the (isospin corrected) ALEPH data.

CONCLUSIONS

In the meson sector, chiral perturbation theory has been pushed to next-to-next-to-leading order. At this order, the main limitation for further progress is the abundance of coupling constants, an unavoidable feature of a nonrenormalizable effective field theory. Some progress has been made recently in estimating those constants by using large-N_C methods to interpolate between CHPT and QCD.

CHPT is the only reliable approach for calculating electromagnetic and isospin violating corrections for hadronic processes at low energies. This is in particular important for the analysis of K_{l3} decays in order to extract the CKM matrix element V_{us} to better than 1 % accuracy.

Recent sum rule tests [31] favour τ over e^+e^- data for evaluating the hadronic vacuum polarization at low and intermediate energies. As a consequence, there is at present no conflict between the Standard Model and experiment for the anomalous magnetic moment of the muon.

ACKNOWLEDGMENTS

This work has been supported in part by HPRN-CT2002-00311 (EURIDICE). I thank Helmut Neufeld for discussions and for numerical help with the ratio r_{+0}. Special thanks and congratulations to Pietro Colangelo, Fulvia De Fazio, Giuseppe Nardulli and their team for a very successful Workshop.

REFERENCES

1. G. Colangelo, J. Gasser and H. Leutwyler, Nucl. Phys. B **603** (2001) 125 [arXiv:hep-ph/0103088].
2. J. Portolés, these Proceedings.
3. J. Gasser and H. Leutwyler, Ann. Phys. **158** (1984) 142.
4. J. Gasser and H. Leutwyler, Nucl. Phys. B **250** (1985) 465.
5. J. Bijnens, G. Colangelo and G. Ecker, JHEP **9902** (1999) 020 [arXiv:hep-ph/9902437].
6. For a recent review see J. Bijnens, arXiv:hep-ph/0409068.
7. V. Bernard, N. Kaiser and U. G. Meißner, Int. J. Mod. Phys. E **4** (1995) 193 [arXiv:hep-ph/9501384].
8. P. F. Bedaque and U. van Kolck, Ann. Rev. Nucl. Part. Sci. **52** (2002) 339 [arXiv:nucl-th/0203055].
9. M. Knecht, H. Neufeld, H. Rupertsberger and P. Talavera, Eur. Phys. J. C **12** (2000) 469 [arXiv:hep-ph/9909284].
10. S. Descotes-Genon and B. Moussallam, Eur. Phys. J. C **42** (2005) 403 [arXiv:hep-ph/0505077].
11. V. Cirigliano, G. Ecker, M. Eidemüller, R. Kaiser, A. Pich and J. Portoles, JHEP **0504**, 006 (2005) [arXiv:hep-ph/0503108].
12. J. Gasser and H. Leutwyler, Nucl. Phys. B **250** (1985) 517.
13. V. Cirigliano, H. Neufeld and H. Pichl, Eur. Phys. J. C **35** (2004) 53 [arXiv:hep-ph/0401173].
14. P. Post and K. Schilcher, Eur. Phys. J. C **25** (2002) 427 [arXiv:hep-ph/0112352].

15. J. Bijnens and P. Talavera, Nucl. Phys. B **669** (2003) 341 [arXiv:hep-ph/0303103]; see also http://www.thep.lu.se/~bijnens/chpt.html.
16. F. Mescia, arXiv:hep-ph/0411097.
17. H. Neufeld, private communication.
18. H. Leutwyler and M. Roos, Z. Phys. C **25** (1984) 91.
19. S. Eidelman et al. [Particle Data Group Collaboration], Phys. Lett. B **592** (2004) 1.
20. A. Serebrov et al., Phys. Lett. B **605** (2005) 72 [arXiv:nucl-ex/0408009].
21. R. Acosta and R. Flores-Mendieta, arXiv:hep-ph/0501231.
22. V. Mateu and A. Pich, arXiv:hep-ph/0509045.
23. T. Alexopoulos et al. [KTeV Collaboration], Phys. Rev. D **70** (2004) 092007 [arXiv:hep-ex/0406003].
24. R. Alemany, M. Davier and A. Höcker, Eur. Phys. J. C **2** (1998) 123 [arXiv:hep-ph/9703220].
25. G. W. Bennett et al. [Muon g-2 Collaboration], Phys. Rev. Lett. **92** (2004) 161802 [arXiv:hep-ex/0401008].
26. V. Cirigliano, G. Ecker and H. Neufeld, JHEP **0208** (2002) 002 [arXiv:hep-ph/0207310]; Phys. Lett. B **513** (2001) 361 [arXiv:hep-ph/0104267].
27. M. Davier, S. Eidelman, A. Höcker and Z. Zhang, Eur. Phys. J. C **31** (2003) 503 [arXiv:hep-ph/0308213].
28. R. R. Akhmetshin et al. [CMD-2 Collaboration], Phys. Lett. B **578** (2004) 285 [arXiv:hep-ex/0308008].
29. A. G. Denig [KLOE Collaboration], Int. J. Mod. Phys. A **20** (2005) 1935.
30. A. Höcker, arXiv:hep-ph/0410081.
31. K. Maltman, arXiv:hep-ph/0504201.
32. R. Barate et al. [ALEPH Collaboration], Z. Phys. C **76** (1997) 15.
33. M. Davier and W. J. Marciano, Ann. Rev. Nucl. Part. Sci. **54** (2004) 115.
34. M. N. Achasov et al., arXiv:hep-ex/0506076.

Hadronic decays of the tau lepton : $\tau^- \to (\pi\pi\pi)^- \nu_\tau$ within Resonance Chiral Theory

D. Gómez Dumm*, A. Pich† and J. Portolés†

*IFLP, CONICET - Depto. de Física, Univ. Nac. de la Plata, C.C. 67, 1900 La Plata, Argentina
†Instituto de Física Corpuscular, IFIC, CSIC-Universitat de València, Edifici d'Instituts de Paterna, Apt. Correus 22085, E-46071 València, Spain

Abstract. τ decays into hadrons foresee the study of the hadronization of vector and axial-vector QCD currents, yielding relevant information on the dynamics of the resonances entering into the processes. We analyse $\tau \to \pi\pi\pi\nu_\tau$ decays within the framework of the Resonance Chiral Theory, comparing this theoretical scheme with the experimental data, namely ALEPH spectral function and branching ratio. Hence we get values for the mass and on-shell width of the $a_1(1260)$ resonance, and provide the structure functions that have been measured by OPAL and CLEO-II.

Keywords: Non-perturbative QCD, Chiral Symmetry, Tau decays
PACS: 12.38.Aw, 12.38.Lg, 13.35.Dx

INTRODUCTION

The implementation of Quantum Chromodynamics (QCD) in the energy region populated by light-flavoured resonances ($M_\rho \lesssim E \lesssim 2\,\text{GeV}$, being M_ρ the mass of the $\rho(770)$) is a demanding task that involves poorly known aspects such as bound and resonant states, duality and hadronization mechanisms. Though *ad hoc* Breit-Wigner parameterisations have been widely employed in the literature [1, 2] they are not necessarily consistent with the underlying theory, as they seem to violate the chiral symmetry of massless QCD [3, 4].

τ decays into hadrons allow to study the hadronization properties of vector and axial-vector QCD currents and, accordingly, to determine intrinsic properties of the dynamics generated by resonances [5]. At very low energies, typically $E \ll M_\rho$, Chiral Perturbation Theory (χPT) [6] is the Effective Field Theory of QCD. Still the $\tau \to \pi\pi\pi\nu_\tau$ decays, through their full energy spectrum, are driven by the $\rho(770)$ and $a_1(1260)$ resonances mainly, in an energy region where the invariant hadron momentum approaches the masses of the resonances. Hence χPT is no longer applicable to the study of the whole spectrum but only to the very low energy domain [7]. The standard procedure that has been followed [1] to deal with these decays has been to modulate the amplitudes with a Breit-Wigner parameterization, fixing the normalization in order to match the leading $\mathcal{O}(p^2)$ χPT. Nevertheless, its deviation of the chiral behaviour at higher orders [3, 4] could spoil any outcome provided by the analysis of data.

Lately several experiments have collected good quality data on $\tau \to \pi\pi\pi\nu_\tau$, such as branching ratios and spectra [8] or structure functions [9]. Their analysis within a model-independent framework is highly desirable if one wishes to collect information on the hadronization of the relevant QCD currents.

THE RESONANCE CHIRAL THEORY OF QCD

At energies $E \sim M_\rho$ the resonance mesons are active degrees of freedom that have to be properly included into the pertinent Lagrangian. The procedure, put forward in Refs. [10, 11] and known as Resonance Chiral Theory (RχT), is ruled by the approximate chiral symmetry of QCD, that drives the interaction of Goldstone bosons (the lightest octet of pseudoscalar mesons), and the $SU(3)_V$ assignments of the resonance multiplets. This construction is embedded within a comprehensive framework guided by the large number of colours (N_C) limit of QCD [12]. The $1/N_C$ expansion tells us that, at leading order, we should only consider the tree level diagrams given by a local Lagrangian with infinite zero-width states in the spectrum. This is precisely the role of RχT. However in most processes, like hadron tau decays, we need to include finite widths that only appear at next-to-leading order in the large-N_C expansion and, moreover, we will only include one multiplet of vector and axial-vector resonances in our theory. Thus, in practice, we have to model this large-N_C expansion to some extent.

The final hadron system in the $\tau \to \pi\pi\pi\nu_\tau$ decays spans a wide energy region, namely $3m_\pi \lesssim E \lesssim M_\tau$ that is populated by many resonances. RχT is the appropriate framework to work with and we consider the Lagrangian [4, 10] :

$$\mathscr{L}_{R\chi T} = \frac{F^2}{4}\langle u_\mu u^\mu + \chi_+ \rangle + \frac{F_V}{2\sqrt{2}}\langle V_{\mu\nu}f_+^{\mu\nu}\rangle + i\frac{G_V}{\sqrt{2}}\langle V_{\mu\nu}u^\mu u^\nu\rangle + \frac{F_A}{2\sqrt{2}}\langle A_{\mu\nu}f_-^{\mu\nu}\rangle$$
$$+ \mathscr{L}_{kin}^V + \mathscr{L}_{kin}^A + \sum_{i=1}^{5}\lambda_i \mathscr{O}_{VAP}^i, \qquad (1)$$

where all the couplings are real, being F the decay constant of the pion in the chiral limit, and the operators $\mathscr{O}_{VAP}^i$ are given by :

$$\begin{aligned}
\mathscr{O}_{VAP}^1 &= \langle [V^{\mu\nu}, A_{\mu\nu}]\chi_-\rangle, \\
\mathscr{O}_{VAP}^2 &= i\langle [V^{\mu\nu}, A_{\nu\alpha}]h_\mu^\alpha\rangle, \\
\mathscr{O}_{VAP}^3 &= i\langle [\nabla^\mu V_{\mu\nu}, A^{\nu\alpha}]u_\alpha\rangle, \\
\mathscr{O}_{VAP}^4 &= i\langle [\nabla^\alpha V_{\mu\nu}, A_\alpha^\nu]u^\mu\rangle, \\
\mathscr{O}_{VAP}^5 &= i\langle [\nabla^\alpha V_{\mu\nu}, A^{\mu\nu}]u_\alpha\rangle.
\end{aligned} \qquad (2)$$

The notation is that of Ref. [10]. Notice that we are using the antisymmetric tensor formulation to describe the spin 1 resonances and, consequently, we do not consider the $\mathscr{O}(p^4)$ χPT Lagrangian of Goldstone bosons [11].

Our Lagrangian theory has eight a priori unknown coupling constants, namely F_V, F_A, G_V and λ_i, $i = 1,...5$. Phenomenology could provide direct information on them; for instance F_V could be extracted from the measured $\Gamma(\rho^0 \to e^+e^-)$, G_V from $\Gamma(\rho^0 \to \pi^+\pi^-)$, F_A from $\Gamma(a_1 \to \pi\gamma)$ and the λ_i appear in $\Gamma(a_1 \to \rho\pi)$ or the $\tau \to \pi\pi\pi\nu_\tau$ processes themselves. It is conspicuous, though, that $\mathscr{L}_{R\chi T}$ is not QCD for arbitrary values of the couplings. Hence if we want to comprehend more about QCD in this non-perturbative regime we should try to learn about the determination of the couplings from the underlying theory [13]. On this account we will implement several known features of the strong interaction theory in the following.

The QCD ruled short–distance behaviour of the vector and axial-vector form factors in the large–N_C limit (approximated with only one octet of vector resonances) constrains the couplings of $\mathscr{L}_{R\chi T}$ in Eq. (1), which must satisfy [11]:

$$
\begin{aligned}
1 - \frac{F_V G_V}{F^2} &= 0, \\
2 F_V G_V - F_V^2 &= 0.
\end{aligned}
\tag{3}
$$

In addition, the first Weinberg sum rule, in the limit where only the lowest narrow resonances contribute to the vector and axial–vector spectral functions, leads to

$$F_V^2 - F_A^2 = F^2. \tag{4}$$

In this way all three couplings F_V, G_V and F_A can be written in terms of the pion decay constant: $F_V = \sqrt{2} F$, $G_V = F/\sqrt{2}$ and $F_A = F$. These results are well satisfied phenomenologically and we have adopted them. In the next section we will comment on an analogous study of the λ_i couplings.

THE AXIAL-VECTOR FORM FACTORS IN $\tau^- \to (\pi\pi\pi)^- \nu_\tau$

The decay amplitudes for the $\tau^- \to \pi^+ \pi^- \pi^- \nu_\tau$ and $\tau^- \to \pi^- \pi^0 \pi^0 \nu_\tau$ processes can be written as

$$\mathscr{M}_\pm = -\frac{G_F}{\sqrt{2}} V_{ud} \bar{u}_{\nu_\tau} \gamma^\mu (1-\gamma_5) u_\tau T_{\pm\mu}, \tag{5}$$

$$T_{\pm\mu}(p_1, p_2, p_3) = \langle \pi_1(p_1) \pi_2(p_2) \pi^\pm(p_3) | \mathscr{A}_\mu | 0 \rangle, \tag{6}$$

as in the isospin limit there is no contribution of the vector current to these processes. In $T_{\pm\mu}(p_1,p_2,p_3)$ the π^+ is the one in $\tau^- \to \pi^+ \pi^- \pi^- \nu_\tau$ and π^- that in $\tau^- \to \pi^- \pi^0 \pi^0 \nu_\tau$. The hadronic tensor can be written in terms of three form factors, F_1, F_2 and F_P, as [14]:

$$T^\mu = V_1^\mu F_1 + V_2^\mu F_2 + Q^\mu F_P, \tag{7}$$

where

$$
\begin{aligned}
V_1^\mu &= \left(g^{\mu\nu} - \frac{Q^\mu Q^\nu}{Q^2}\right)(p_1 - p_3)_\nu, \\
V_2^\mu &= \left(g^{\mu\nu} - \frac{Q^\mu Q^\nu}{Q^2}\right)(p_2 - p_3)_\nu, \\
Q^\mu &= p_1^\mu + p_2^\mu + p_3^\mu.
\end{aligned}
\tag{8}
$$

The form factors F_1 and F_2 have a transverse structure in the total hadron momenta Q_μ and drive a $J^P = 1^+$ transition. Bose symmetry under interchange of the two identical pions in the final state demands that $F_1(Q^2, s, t) = F_2(Q^2, t, s)$ where $s = (p_1 + p_3)^2$ and

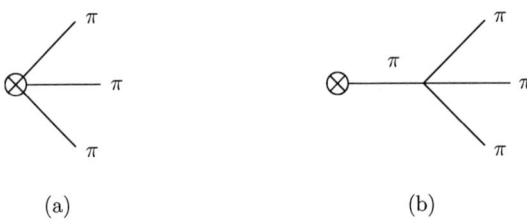

FIGURE 1. Diagrams contributing to the hadronic amplitude $T_{\pm\mu}$ at $\mathcal{O}(p^2)$ χPT.

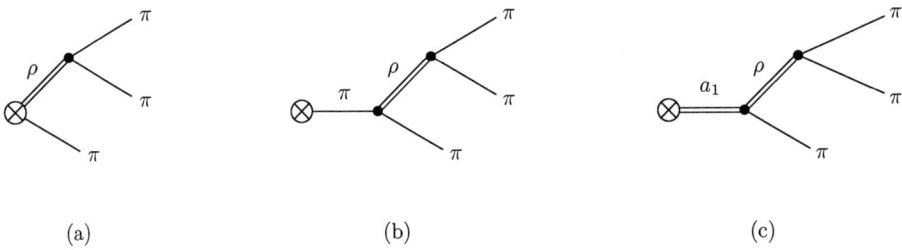

FIGURE 2. Resonance–mediated diagrams contributing to $T_{\pm\mu}$.

$t = (p_2 + p_3)^2$. Meanwhile F_P accounts for a $J^P = 0^-$ transition that carries pseudoscalar degrees of freedom and vanishes with the square of the pion mass. Its contribution to the spectral function of $\tau \to \pi\pi\pi\nu_\tau$ goes like m_π^4/Q^4 and, accordingly, it is very much suppressed with respect to that coming from F_1 and F_2. We will not consider it in the following.

In the low Q^2 region, the matrix element in Eq. (6) can be calculated using χPT. At $\mathcal{O}(p^2)$ one has two contributions, arising from the diagrams in Fig. 1. The sum of both graphs yields

$$T^\chi_{\pm\mu} = \mp \frac{2\sqrt{2}}{3F} \left\{ V_{1\mu} + V_{2\mu} \right\} . \tag{9}$$

We now include the resonance–mediated contributions to the amplitude, to be evaluated through the interacting terms in $\mathcal{L}_{R\chi T}$ Eq. (1). The relevant diagrams to be taken into account are those shown in Fig. 2. We get

$$\begin{aligned} T^R_{\pm\mu} =\ & \mp \frac{\sqrt{2} F_V G_V}{3F^3} \left[\alpha(Q^2,s,t) V_{1\mu} + \alpha(Q^2,t,s) V_{2\mu} \right] \\ & \pm \frac{4 F_A G_V}{3F^3} \frac{Q^2}{Q^2 - M_A^2} \left[\beta(Q^2,s,t) V_{1\mu} + \beta(Q^2,t,s) V_{2\mu} \right] . \end{aligned} \tag{10}$$

The functions $\alpha(Q^2,s,t)$ and $\beta(Q^2,s,t)$ are:

$$\alpha(Q^2,s,t) = -3 \frac{s}{s - M_V^2} + \left(\frac{2 G_V}{F_V} - 1 \right) \left\{ \frac{2Q^2 - 2s - u}{s - M_V^2} + \frac{u - s}{t - M_V^2} \right\}, \tag{11}$$

$$\beta(Q^2,s,t) = -3(\lambda'+\lambda'')\frac{s}{s-M_V^2} + F(Q^2,s)\frac{2Q^2+s-u}{s-M_V^2} + F(Q^2,t)\frac{u-s}{t-M_V^2}, \quad (12)$$

with

$$F(Q^2,s) = -\lambda_0 \frac{m_\pi^2}{Q^2} + \lambda' \frac{s}{Q^2} + \lambda'', \quad (13)$$

and depend on three combinations of the λ_i couplings in $\mathscr{L}_{R\chi T}$, that we call λ_0, λ' and λ''. Following the ideas outlined above we can get information on these combinations by implementing known aspects of asymptotic QCD. In particular we expect that the form factor of the axial-vector current into three pions should vanish at infinite transfer of momentum ($Q^2 \to \infty$). This is a consequence of the fact that its contribution to the spectral function of the axial-vector current correlator, being positive, has to add to other infinite hadronic positive contributions to reach the constant value evaluated within QCD [16]. Accordingly the proper behaviour of the $T_{\pm\mu}$ form factor imposes the constraints:

$$\begin{aligned} 2\lambda' - 1 &= 0, \\ \lambda'' &= 0. \end{aligned} \quad (14)$$

Hence there is only one combination of couplings left unknown, namely λ_0.

Finally an additional comment on the result for $T_{\pm\mu}^R$ is required. The form factors in Eq. (10) include zero-width $\rho(770)$ and $a_1(1260)$ propagator poles, leading to divergent phase-space integrals in the calculation of the $\tau \to \pi\pi\pi\nu_\tau$ decay width as the kinematical variables go along the full energy spectrum. The result can be regularized through the inclusion of resonance widths, which means to go beyond the leading order in the $1/N_C$ expansion, and implies the introduction of some additional theoretical inputs. This issue has been analysed in detail within the resonance chiral effective theory in Ref. [15] and, accordingly, we include off-shell widths for both resonances [4].

THEORY VERSUS EXPERIMENT

To analyse the experimental data we will only consider the dominating $J^P = 1^+$ driven axial-vector form factors, that satisfy $T_{+\mu} = -T_{-\mu}$ hence providing the same predictions for both $\tau^- \to \pi^+\pi^-\pi^-\nu_\tau$ and $\tau^- \to \pi^-\pi^0\pi^0\nu_\tau$ processes in the isospin limit.

We have fitted the experimental values for the $\tau^- \to \pi^+\pi^-\pi^-\nu_\tau$ branching ratio and normalized spectral function obtained by ALEPH [8] and we get a reasonable $\chi^2/d.o.f. = 64.5/52$. This is shown in Fig. 3. Hence we get the axial-vector $a_1(1260)$ parameters $M_A = (1.203 \pm 0.003)\,\text{GeV}$ and $\Gamma_{a_1}(M_A^2) = (0.48 \pm 0.02)\,\text{GeV}$, where the errors are only statistical. We also obtain a value for the still unknown combination of λ_i couplings: $\lambda_0 = 11.9 \pm 0.4$. However, as pointed out in Ref. [17], this value seems too large when additional QCD constraints are imposed. The origin of the discrepancy could be the small sensitivity of the tau decay amplitude to this parameter, as it only appears multiplied by the mass of the pion (13), together with an improvable implementation of the off-shell width of the $a_1(1260)$ resonance.

Ultimately we predict the integrated structure functions w_A, w_C, w_D and w_E [14], that we compare with the experimental results for $\tau^- \to \pi^-\pi^0\pi^0\nu_\tau$ in Fig. 4. In spite of the

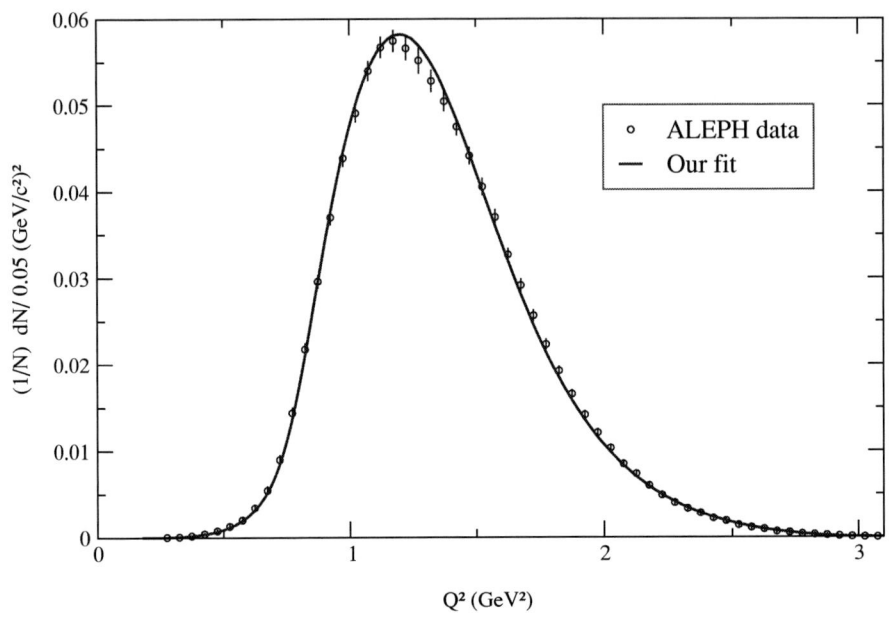

FIGURE 3. Fit to the ALEPH data [8] for the normalized $\tau^- \to \pi^+\pi^-\pi^-\nu_\tau$.

large errors the predictions follow notably the depicted trend. As analysed in Ref. [4] a variance between the ALEPH data on one side and the OPAL and CLEO-II on the other are at the origin of the seemingly inconsistent result for w_A in the high Q^2 region.

In conclusion it can be inferred that, within the present experimental errors and for the studied observables, there is no evidence of relevant contributions in $\tau \to \pi\pi\pi\nu_\tau$ decays beyond those of the $\rho(770)$ and $a_1(1260)$ resonances.

ACKNOWLEDGMENTS

J.P. wishes to thank Pietro Colangelo and Fulvia De Fazio for the remarkable organization of the QCD@Work 2005 meeting in Conversano (Italy). This work has been supported in part by MEC (Spain) under grant FPA2004-00996, by Generalitat Valenciana (Grants GRUPOS03/013, GV04B-594 and GV05/015) and by HPRN-CT2002-00311 (EURIDICE).

REFERENCES

1. A. Pich, "QCD Tests From Tau Decay Data," *Talk given at Tau Charm Factory Workshop, Stanford, Calif., May 23-27, 1989* ; J. H. Kuhn and A. Santamaria, Z. Phys. C **48** (1990) 445.
2. N. N. Achasov and A. A. Kozhevnikov, Phys. Rev. D **55** (1997) 2663 [arXiv:hep-ph/9609216].
3. J. Portolés, Nucl. Phys. Proc. Suppl. **98** (2001) 210 [arXiv:hep-ph/0011303];
4. D. Gómez Dumm, A. Pich and J. Portolés, Phys. Rev. D **69** (2004) 073002 [arXiv:hep-ph/0312183].

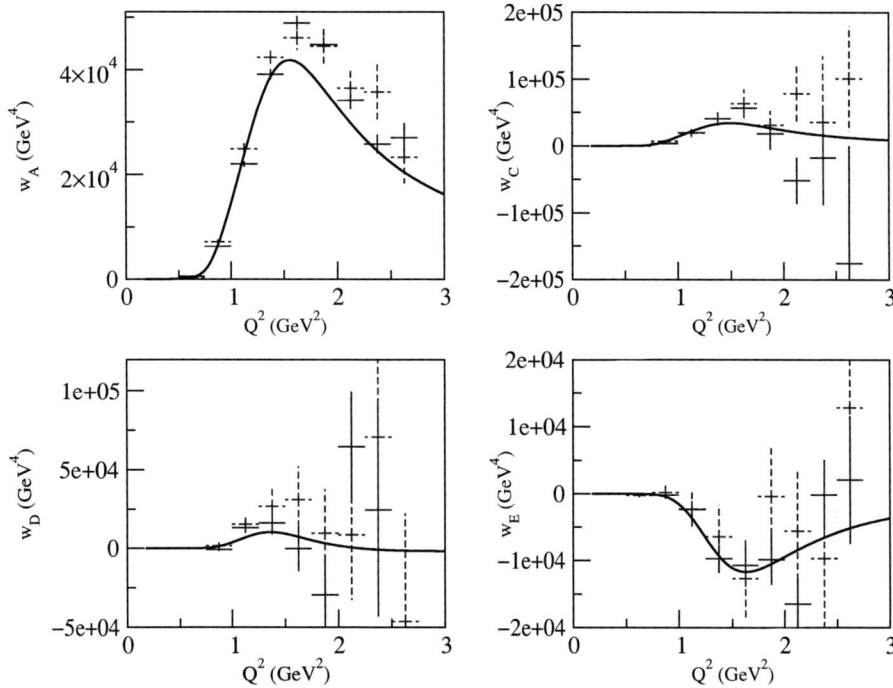

FIGURE 4. Theoretical values for the w_A, w_C, w_D and w_E integrated structure functions in comparison with the experimental data from CLEO-II (solid) and OPAL (dashed) [9].

5. J. Portolés, Nucl. Phys. Proc. Suppl. **144** (2005) 3 [arXiv:hep-ph/0411333].
6. S. Weinberg, PhysicaA **96** (1979) 327; J. Gasser and H. Leutwyler, Annals Phys. **158** (1984) 142.
7. G. Colangelo, M. Finkemeier and R. Urech, Phys. Rev. D **54** (1996) 4403 [arXiv:hep-ph/9604279].
8. R. Barate et al. [ALEPH Collaboration], Eur. Phys. J. C **4** (1998) 409.
9. K. Ackerstaff et al. [OPAL Collaboration], Z. Phys. C **75** (1997) 593; T. E. Browder et al. [CLEO Collaboration], Phys. Rev. D **61** (2000) 052004 [arXiv:hep-ex/9908030].
10. G. Ecker, J. Gasser, A. Pich and E. de Rafael, Nucl. Phys. B **321** (1989) 311.
11. G. Ecker, J. Gasser, H. Leutwyler, A. Pich and E. de Rafael, Phys. Lett. B **223** (1989) 425.
12. G. 't Hooft, Nucl. Phys. B **75** (1974) 461; E. Witten, Nucl. Phys. B **160** (1979) 57.
13. A. Pich, in Proceedings of the Phenomenology of Large N_C QCD, edited by R. Lebed (World Scientific, Singapore, 2002), p. 239, arXiv:hep-ph/0205030.
14. J. H. Kuhn and E. Mirkes, Z. Phys. C **56** (1992) 661 [Erratum-ibid. C **67** (1995) 364].
15. D. Gomez Dumm, A. Pich and J. Portolés, Phys. Rev. D **62** (2000) 054014 [arXiv:hep-ph/0003320].
16. E. G. Floratos, S. Narison and E. de Rafael, Nucl. Phys. B **155** (1979) 115.
17. V. Cirigliano, G. Ecker, M. Eidemuller, A. Pich and J. Portolés, Phys. Lett. B **596** (2004) 96 [arXiv:hep-ph/0404004].

Latest results on kaon physics

A. Antonelli

Laboratori Nazionali di Frascati- INFN

Abstract. In spite of a long history Kaon system is still a laboratory for interesting physics, from flavor physics to CP violation. Kaon rare decays are also a powerfull tool, complementary to B decays, to probe the physics beyond the Standard Model. In this paper I will review most of the recent results in kaon physics coming from KLOE, KTEV, NA48 experiment.

Keywords: kaons
PACS: 12.15Hh

SEMILEPTONIC KAON DECAYS, V_{us} DETERMINATION

The measurement of the semileptonic kaon decay widths allows us to test many fundamental aspects of the Standard Model.

The most precise test of unitarity of the CKM matrix comes from its first row: $1 - \Delta \simeq |V_{ud}|^2 + |V_{us}|^2$. Using $|V_{ud}|$ as extracted from nuclear beta decays, and $|V_{us}|$ as extracted from the semileptonic decay width of the kaon, a test on Δ with a precision of few parts per mil can be performed. Using the PDG 2004 values for V_{us} and V_{ud} one finds a 2 σ deviation from unitarity.

The value of V_{us} can be extracted from the measurement of the semileptonic kaon decay width using [1]:

$$V_{us} \cdot f_+^{K^0\pi^-}(0) = \left[\frac{\Gamma}{\mathcal{N} S_{ew} I_i(\lambda_+, \lambda_0, 0)}\right]^{1/2} \frac{1}{1 + \delta^i_{e^2p^2} + \frac{1}{2}\Delta I_i(\lambda_+, \lambda_0)}$$

with

$$\mathcal{N} = \frac{G_\mu^2 M_{K_i}^5}{192\pi^3}$$

where $f_+^{K^0\pi^-}$ is the vector form factor at zero momentum transfer, $I_i(\lambda_+, \lambda_0, 0)$ is the result of the phase space integration after factoring out $f_+^{K^0\pi^-}$, and both are evaluated in absence of radiative corrections. The radiative corrections for the form factor and the phase space integral are included via $\delta^i_{e^2p^2}$ and $\Delta I_i(\lambda_+, \lambda_0)$ respectively. λ_+ and λ_0 are the parameters that describe the momentum transfer dependence of the vector and scalar form factor.

The first hint that something is going on comes from the measurement of E865 experiment[2]. They measure a BR($K^+ \to \pi^0 e^+ \nu$) that is 2.7 σ above the previous measurements but which gives a value of V_{us} consistent with unitarity. The key issue in this measurement is the systematic control of the branching ratio. The detector was not optimised for photons, it was designed to search for $K \to \pi e \mu$ decays, and π^0

Dalitz decay is required in both signal and in normalisation ($K^+ \to \pi^0\pi^+$ channels). There was a lot of interest in seeing this result confirmed. KLOE, NA48 and KTEV experiment have recently measured semileptonic kaon branching ratios and form factors using different techniques.

RESULTS ON SEMILEPTONIC KAON DECAYS: KLOE, NA48, KTEV

Measurements of the absolute kaon branching ratios are a unique possibility of the ϕ factory. In ϕ decays pairs of monochromatic K_S K_L are produced and the K_L (K_S can be identified looking at the decay of the companion on the other side (tagging). The K_L is selected by identification of K_S decays while a K_S beam is tagged using the fraction of events in which the K_L interacts in the calorimeter (K_L crash). The absolute branching ratios can be determined by counting the fraction of K_L's that decay into each channel and correcting for acceptances, reconstruction efficiencies, and background. KLOE has measured the dominant K_L branching ratios using the K_L beam tagged by $K_S \to \pi^+\pi^-$ decays [3]. In the 2001/2002 data sample, $\sim 13 \times 10^6$ tagged K_L decays have been used for the measurement, and $\sim 4 \times 10^6$ to evaluate efficiencies. To measure the BR's for decays to charged particles, a K_L decay vertex is required in the DC fiducial volume.

The number of events of each type is obtained by fitting the distribution of missing momentum minus missing energy, in the $\pi\mu$ mass assignment, with the sum of the MC distributions for each of the decay channels. To select $K_L \to 3\pi^0$ events, at least three photons are required from the K_L decay vertex. The reconstruction efficiency and purity of the selected sample are both about 99%. The resulting BR's are:

$BR(K_L \to \pi e \nu(\gamma)) = 0.4049 \pm 0.0010_{stat} \pm 0.0031_{syst}$,
$BR(K_L \to \pi \mu \nu(\gamma)) = 0.2726 \pm 0.0008_{stat} \pm 0.0022_{syst}$,
$BR(K_L \to 3\pi^0) = 0.2018 \pm 0.0004_{stat} \pm 0.0026_{syst}$ and,
$BR(K_L \to \pi^+\pi^-\pi^0(\gamma)) = 0.1276 \pm 0.0006_{stat} \pm 0.0016_{syst}$,
where the errors are dominated by the error on τ_{K_L} through the geometrical acceptance. Taking the BR's for rare K_L decays to $\pi^+\pi^-$, $\pi^0\pi^0$, and $\gamma\gamma$ from the PDG [4] and imposing the constraint $\sum BR(K_L) = 1$, τ_{K_L} can be measured: $\tau_{K_L} = (50.72 \pm 0.17 \pm 0.33)$ ns. Imposition of this constraint also results in more precise measurements of the dominant BR's:

$BR(K_L \to \pi e \nu(\gamma)) = 0.4007 \pm 0.0006 \pm 0.0014$,
$BR(K_L \to \pi \mu \nu(\gamma)) = 0.2698 \pm 0.0006 \pm 0.0014$,
$BR(K_L \to 3\pi^0) = 0.1997 \pm 0.0005 \pm 0.0019$,
$BR(K_L \to \pi^+\pi^-\pi^0(\gamma)) = 0.1263 \pm 0.0005 \pm 0.0011$.

The K_L lifetime has been also measured directly [5], employing 10^7 $K_L \to 3\pi^0$ events selected from the 2001/2002 data sample. The result is $\tau_{K_L} = (50.92 \pm 0.17 \pm 0.25)$ ns, which together with that from the K_L BR measurements gives the KLOE average: $\tau_{K_L} = (50.84 \pm 0.23)$ ns

To extract V_{us} from the different BR measurements we use the values $\lambda'_+ = 0.0221 \pm 0.0011$, $\lambda''_+ = 0.00023 \pm 0.0004$, and $\lambda_0 = 0.0154 \pm 0.0008$, obtained from a combined

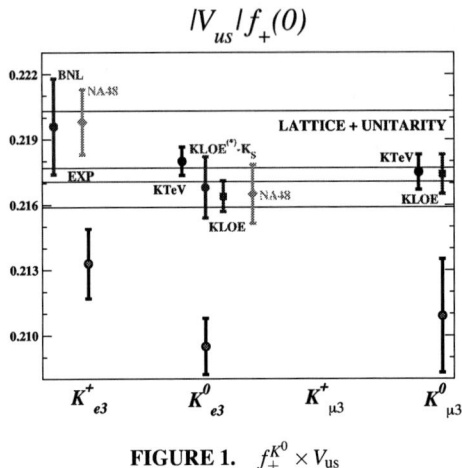

FIGURE 1. $f_+^{K^0} \times V_{us}$

fit to K_{Le3} and $K_{L\mu3}$ results from KTeV [6] and ISTRA+[8, 7]. We obtain:

$$f_+^{K^0} \times V_{us} = 0.21638 \pm 0.00067 \text{ from } K_{Le3},$$
$$f_+^{K^0} \times V_{us} = 0.21732 \pm 0.00087 \text{ from } K_{L\mu3},$$
$$f_+^{K^0} \times V_{us} = 0.21690 \pm 0.00170 \text{ from } K_{Se3},$$

KLOE has also measured the fully inclusive absolute $K^+ \to \mu^+ \nu_\mu$ BR. From about 9×10^5 $K^+ \to \mu^+ \nu_\mu$ decays obtained from a sample of about 250 pb^{-1} KLOE obtains $BR(K^+ \to \mu^+ \nu_\mu) = 0.6366 \pm 0.0009_{stat} \pm 0.0015_{syst}$ [9], corresponding to an overall fractional error of 0.27%. Using recent lattice results on the decay constants of pseudo-scalar mesons [10], this corresponds to the result $V_{us} = 0.2223 \pm 0.0026$. This result has a fractional error of about 1% that is dominated by the theoretical uncertainty.

NA48 and KTEV experiments have used a different approach: they have measured ratios of branching ratios since they do not have an absolute normalisation. NA48 has normalised $K_L \to \pi e \nu$ events to $K_L \to 2$ tracks. Using for BR($K_L \to 3\pi^0$) the average of the PDG value and the KTEV measurement they, obtain $BR(K_L \to \pi e \nu(\gamma)) = 0.401 \pm 0.004$ [11]. KTEV measures 5 K_L ratios $\Gamma_{e3}/\Gamma_{\mu3}$, Γ_{+-0}/Γ_{e3}, Γ_{000}/Γ_{e3}, Γ_{+-}/Γ_{e3} and Γ_{00}/Γ_{000}. The 6 decay modes in the above ratios account for 99.93% of K_L decays and the ratio can be combined to extract the semileptonic branching ratio [12]. The measurements from KLOE, KTeV and NA48 experiments agrees quite well and they disagree with the old PDG values for the $K_L \to \pi e \nu$ and $K_L \to 3\pi^0$ decays.

In Fig. 1 $f_+^{K^0} \times V_{us}$ is reported for the different experiments, the new measurements are well inside the unitarity band and they disagree with the old PDG values.

$K_S \rightarrow 3\pi^0$: TEST OF CP CPT

Observation of this decay signals CP violation in kaon mixing and/or in decay. In the Standard Model the transition width for $K_L \rightarrow 3\pi^0$ is related to that for $K_S \rightarrow 3\pi^0$, giving a predicted branching ratio of 1.9×10^{-9}. Untill recently the uncertainty on the amplitude for $K_S \rightarrow 3\pi^0$ limits the precision of a CPT invariance test via the unitarity relation for the kaon system: a limit on BR($K_S \rightarrow 3\pi^0$) at the level of 10^{-7} would translate in to a 2.5-fold improvement on the accuracy on $K^0 \bar{K}^0$ mass difference. In KLOE, the signature is an event with a K_L crash, six photon clusters, and no tracks from the interaction point. Background is mainly from $K_S \rightarrow \pi^0 \pi^0$ events with two spurious clusters from splittings or accidental activity. Based on an analysis of 450 pb^{-1} of 2001–2002 data, KLOE has obtained the 90% CL limit BR $\leq 1.2 \times 10^{-7}$ [13].

K_S SEMILEPTONIC DECAY

The knowledge of both the K_L and the K_S semileptonic decay branching ratios and lifetimes allows the validity of the $\Delta S = \Delta Q$ rule to be tested through the quantity

$$Re(x_+) = 1/2 \frac{\text{BR}(K_S \rightarrow \pi e\nu)/\tau_S - \text{BR}(K_L \rightarrow \pi e\nu)/\tau_L}{\text{BR}(K_S \rightarrow \pi e\nu)/\tau_S - \text{BR}(K_L \rightarrow \pi e\nu)/\tau_L} \quad (1)$$

In the Standard Model, $Re(x_+)$ is expected to be of the order of $G_F m_\pi^2 \sim 10^{-7}$, receiving contributions from second order weak transitions [14, 15]. At present, the most precise test of the $\Delta S = \Delta Q$ rule comes from an analysis of the time distribution of semileptonic strangeness-tagged kaon decays at CPLEAR [16]. The CPLEAR collaboration measures $Re(x_+)$ to be compatible with zero with an error of 6×10^{-3}.

Finally, discrete symmetries are tested through the measurement of the charge asymmetries for K_L and K_S decays, $A_{L,S}$, defined as

$$A_{L,S} = \frac{\Gamma\left(K_{L,S} \rightarrow \pi^- e^+ \nu\right) - \Gamma\left(K_{L,S} \rightarrow \pi^+ e^- \bar{\nu}\right)}{\Gamma\left(K_{L,S} \rightarrow \pi^- e^+ \nu\right) + \Gamma\left(K_{L,S} \rightarrow \pi^+ e^- \bar{\nu}\right)}$$

If *CP* symmetry is assumed, each of the two charge asymmetries are expected to be equal to $2 \times Re(\varepsilon) \simeq 3 \times 10^{-3}$, where ε is the parameter describing *CP* violation in the $K^0 - \bar{K}^0$ mass matrix. A difference between A_S and A_L signals *CPT* violation either in the mass matrix, or in the decay amplitudes. The value of A_L is known at present with a precision of 10^{-4} [17], while A_S has never yet been measured. KLOE has measured the K_S semileptonic branching ratio using the 2001-2002 data set (410 pb^{-1}). Identification of $K_S \rightarrow \pi e\nu$ events begins with the requirement of two tracks of opposite curvature. Electrons and pions are discriminated by time of flight (TOF). F inally the missing energy and momentum E_{miss}, p_{miss} are computed, using the K_S momentum measured from the K_L.

The value for BR($K_S \rightarrow \pi e \nu$) is obtained by normalising the number of signal events to the number of $K_S \rightarrow \pi^+ \pi^-$ events in the same data set Averaging the results obtained for each data set, we obtain the following branching ratios:

$$BR(K_S \to \pi^- e^+ \nu) = (3.54 \pm 0.06_{stat} \pm 0.04_{syst}) \times 10^{-4},$$
$$BR(K_S \to \pi^+ e^- \bar{\nu}) = (3.55 \pm 0.05_{stat} \pm 0.02_{syst}) \times 10^{-4},$$
$$BR(K_S \to \pi^{\pm} e^{\mp} \bar{\nu}(\nu)) = (7.09 \pm 0.08_{stat} \pm 0.05_{syst}) \times 10^{-4}.$$

The value of the combined BR for both charge states is in good agreement both with the previous KLOE value [18], and with the KLOE and KTeV [12] values for $K_L \to \pi^{\pm} e^{\mp} \bar{\nu}(\nu)$, assuming $\Delta S = \Delta Q$ and using the KLOE average value for τ_L. The value for the parameter $Re(x_+)$ obtained using KLOE results for K_L BR and lifetime is:
$Re(x_+) = (3.8 \pm 3.2_{stat} \pm 2.9_{syst}) \times 10^{-3}$

The charge asymmetry is found to be:

$$A_S = (-2 \pm 9_{stat} \pm 6_{syst}) \times 10^{-3}$$

This result is compatible with that for K_L semileptonic decays and with the expectation obtained assuming CPT symmetry, $A_S = 2Re(\varepsilon)$.

DIRECT CP VIOLATION IN $K^{\pm} \to 3\pi$

The study of CP violation (CPV) in the kaon system in different channels provides a powerful tool to overconstrain the CKM matrix and to search for new physics. A promising way to study direct CPV in the kaon system, complementary to $K_L \to 2\pi$, is the asymmetry in the K^+ and K^- decays into three pion. Due to absence of mixing, decays of charged kaons are sensitive to direct CP violation only.

The $K^{\pm} \to 3\pi$ matrix element is parameterised in terms of the slopes of two Dalitz variables u and v:

$$|M(u,v)|^2 \propto 1 + gu + hu^2 + kv^2 + O(u^3, v^3) \tag{2}$$

where $|h|, |k| \ll |g|$ and

$$u = \frac{s_3 - s_0}{m_\pi^2} \quad \text{and} \quad v = \frac{s_2 - s_1}{m_\pi^2} \tag{3}$$

with $s_i = (p_K - p_{\pi i})^2$ and $s_0 = \sum s_i / 3$ ($i = 1, 2, 3$ index $i=3$ for the odd pion). The measured quantity sensitive to direct CPV is the asymmetry:

$$A_g = \frac{g^+ - g^-}{g^+ + g^-} \tag{4}$$

The Standard Model estimates for A_g vary within an order of magnitude, form few 10^{-6} to 8×10^{-5} [19], and models beyond the SM predict substantial enhancement that is partially within the reach of NA48/2. The integrated decay rates asymmetry, A_Γ, is highly suppressed for symmetry reason and is expected to be more than one order of magnitude smaller.

The measurement is performed considering slopes of ratios of normalised u-distributions for K^+ and K^- decays. In case of the $\pi^\pm\pi^+\pi^-$ final state, given the current value [4] of $g = (-0.2154 \pm 0.0035)$, the ratio $N_{K^+}(u)/N_{K^-}(u)$ is proportional with sufficient precision to $(1 + \Delta g u)$. $A_g = \Delta g/2g$ is extracted from a linear fit to the ratio $N_{K^+}(u)/N_{K^-}(u)$. The above statement is valid insofar as the acceptance is the same for K^- and K^+, this has been obtained thanks to the simultaneous K^- and K^+ beams superimposed in space and with narrow momentum spectra.

The presence of magnets both in the beamline (achromats, focusing quadrupoles, etc.) and in the magnetic spectrometer introduces an unavoidable charge asymmetry of the apparatus. In order to equalise local effects on the K^+ and K^- beams the polarities of the relevant magnets were reversed frequently during data taking

Using the full sample of reconstructed $K^\pm \to \pi^\pm\pi^+\pi^-$ decays (1.6×10^9 events) collected by the experiment NA48/2 during 2003 data taking, the collaboration quotes a preliminary result for the asymmetry [20]:

$$A_g = (0.5 \pm 2.4_{\text{stat}} \pm 2.1_{\text{stat(trig)}} \pm 2.1_{\text{syst}}) \times 10^{-4}$$
$$= (0.5 \pm 3.8) \times 10^{-4}$$

This result is compatible with no CP violation and with Standard Model predictions but by more than an order of magnitude better precision than previous measurements [21]. Further improvements are expected in future after analysing the full data sample and refining the analysis.

CUSP EFFECT IN $K^+ \to \pi^+\pi^0\pi^0$ DECAY

The preliminary analysis of a sub-sample of $K^+ \to \pi^+\pi^0\pi^0$ decays collected by the NA48/2 experiment at CERN, has shown a change of the slope in the $\pi^0\pi^0$ invariant mass distribution located around $2m_\pi$. This effect was not seen before by lower precision experiments and is interpreted by N.Cabibbo [22] as the effect of the interference between the direct emission and the $\pi\pi$ re-scattering amplitudes that contribute to the total amplitude in $K^+ \to \pi^+\pi^0\pi^0$. Recent two-loops calculations allows accurate extraction of the difference between the I=0 and I=2 scattering lengths of the S-wave matrix, a0-a2, from the $\pi^0\pi^0$ invariant mass spectrum [23]. This can be compared to precise Chiral Perturbation Theory prediction.

NA48 performs the analysis of a sample of 2×10^8 $K^+ \to \pi^+\pi^0\pi^0$ decays, corresponding to 50 days of the 2003 data taking period. The best fit to the $\pi^0\pi^0$ invariant mass distribution is obtained with a 2-loop exchange model and a small contribution from pionium fixed to BR($K \to \pi$ pionium)/ BR($K \to 3\pi$)= 0.8×10^{-5}

The preliminary result is $(a0 - a2) \cdot m_\pi = 0.281 \pm 0.007_{\text{stat}} \pm 0.014_{\text{syst}} \pm 0.014_{\text{theor}}$. Further improvements are expected in reducing the theoretical and the systematic error.

FUTURE: RARE KAON DECAYS

Rare Kaon decays mediated by Flavor Changing Neutral Currents are the main experimental tool to probe the flavour structure of physics beyond the SM, independent from B mesons [24]. This is because there are no SM tree-level contributions, there is a strong suppression within the SM because of the CKM hierarchy, and their branching ratios are predicted with high precision within the SM if dominated by short-distance dynamics. The golden modes are $K_L \to \pi^0 \bar{\nu}\nu$ and $K^+ \to \pi^+ \bar{\nu}\nu$. The first is a direct measurement of η, the height of the unitarity triangle. Its detection and measurement would establish the second example of direct CP violation after the measurement of ε'/ε in the K^0 system, but with the advantage of a very clean theoretical analysis [25]. The rate of $K^+ \to \pi^+ \bar{\nu}\nu$ determines the absolute value of V_{td}. A measurement of this rate would offer a valid alternative to the measurement deduced from $B^0 \bar{B}^0$ oscillations, but with different, possibly smaller, theoretical uncertainties.

It is important to place these measurements in the world contest. The KTeV experiment has recently put an upper limit on $K_L \to \pi^0 \bar{\nu}\nu$ that is 4 order of magnitude from the SM prediction. We expect results from data collected by E391 at KEK (proposed SES$\sim 3 \times 10^{-10}$) and there is a proposed experiment at JPARC-Japan which aims at collecting 180 signal events.

So far the study of the decay $K^+ \to \pi^+ \bar{\nu}\nu$ has been performed with kaon decays at rest. BNL-AGS-E787 (E949) has collected data from 1995 until 1998 (2002) and has published a measurement of the branching ratio $BR(K^+ \to \pi^+ \nu \bar{\nu}) = 1.47^{+1.30}_{-0.89} \times 10^{-10}$ based on three events interpreted as signal. Plans to further pursue the decay-at-rest technique at the J-PARC have been expressed. As far as decay-in-flight is concerned there is a proposal to measure $K^+ \to \pi^+ \bar{\nu}\nu$ at CERN. The experiment NA48/3 aims at collecting O(100) Standard Model events in two years of data taking.

CONCLUSIONS

After more then 50 years of life kaon physics is still alive and is producing a lot of interesting results.The KLOE experiment is at present collecting more data with 2 fb^{-1} expected to be integrated for the end of 2005. This will allow refined measurement of kaon braching ratios and of K_S rare decays. NA48/2 is completing the analysis on the full data sample and this will improve the CPV measurement in the charged kaon system. New experiments are planned at CERN and JPARC in Japan to measure ultra-rare kaon decays.

ACKNOWLEDGMENTS

I would like to tanks NA48, KteV and KLOE collaborations for providing material for this presentation.

REFERENCES

1. M. Battaglia et al , arXiv:hep-ph/0304132.
2. A. Sher et al., *Phys. Rev. Lett.*,**91** 261802 (2003).
3. F. Ambrosino *et al.*, accepted by *Phys. Lett. B*, hep-ex/0508027
4. S. Eidelman *et al.* (PDG),*Phys. Lett. B***592**,1 (2004).
5. F. Ambrosino *et al.*, *Phys. Lett. B* **626**, 15-23 2005
6. KTEV Collaboration , *Phis. Rev. D*, **70**, 092007 (2004).
7. Yushchenko, O. P. *et al.*, *Phis. Lett. B*, **589**, 111 (2004).
8. Yushchenko, O. P. *et al.*, *Phis. Lett. B*, **581**, 31 (2004).
9. F. Ambrosino *et al.*, Submitted to *Phys. Lett.* **B**, hep-ex/0509045
10. C. Aubin *et al.*, MILC Collaboration, *Phys. Rev. D* **70**, 114501 (2004)
    W.J. Marciano, *Phys. Rev. Lett.* **93**, 231803 (2004)
11. NA48 Collaboration , *Phys. Lett. B*, **602**, 41 (2004).
12. KTEV Collaboration , *Phis. Rev. D*, **70**, 092006 (2004).
13. KLOE Collaboration, F. Ambrosino *et al.*, *Phys. Lett. B*, **619**, 61 (2005).
14. Dib, C. O. and Guberina, B *Phys. Lett. B*, **255**, 113 (1991).
15. Luke, M *Phys. Lett. B*, **256** ,256 (1991).
16. Angelopoulos, A. *et al.Phys. Lett. B*, **444**, 38 (1998).
17. Alavi-Harati, A. *et al.Phys. Rev. Lett.*, **88**, 181601 (2002).
18. KLOE Collaboration, A. Aloisio *et al.*, *Phys. Lett. B*, **535** , 37 (2002).
19. L. Maiani and N. Paver, The second DAΦNE Physics Handbook, INFN, LNF, Vol 1. , 51 (1995),
    E.P. Shabalin, hep-ph/0405229, *Phys. Atom. Nucl.*68882005,
    A.A. Belkov, A.V. Lanyov and G. Bohm, hep-ph/0311209,
    I. Scimemi, E. Gamiz and J. Prades, hep-ph/0405204,
    G. Fäldt and E. Shabalin, hep-ph/0503241.
20. Sozzi, M., Proceedings of the Rencontres de Physique de la Vallee d'Aoste (2005).
21. W. T. Ford *et al.*, *Phys. Rev. Lett.* **25**13701970.
    G. A. Akopdzhanov *et al.* (TNF-IHEP), hep-ex/0406008.
22. N. Cabibbo, hep-ph/0405001
23. N. Cabibbo ,G. Isidori, hep-ph/0502130
24. G. D'Ambrosio and G. Isidori, *Phys. Lett. B*. **530**,108 (2002)
25. L. Littenberg, *Phys. Rev. D* **39**, 3322 (1989).

Dynamical consequences of strong CP breaking

Pietro Faccioli

Dipartimento di Fisica, Universitá degli Studi di Trento, Via Sommarive 15, Povo (Trento); E.C.T.,
Strada delle Tabarelle 286, I-38050 Villazzano (Trento); I.N.F.N. Sezione Collegata di Trento*

Abstract. In this talk we review some recent results of an on-going investigation of the dynamical consequences induced by the θ-term in the QCD Lagrangian. By using an instanton model to account for the topological fluctuations in the vacuum, we show that the θ-term generates an effective flavor-dependent repulsion between matter and anti-matter. Such a non-perturbative interaction triggers an asymmetry in the baryonic charge distribution in hadronic systems, along the direction of quantization of the spin. This represents the microscopic mechanism giving rise to the neutron Electric Dipole Moment (EDM), at the semi-classical level.

We also present a numerical estimate the dependence of the neutron EDM on the value of the θ angle, based on Monte Carlo simulations in the Interacting Instanton Liquid Model. By comparing quenched and unquenched results we show that, for such an observable, the quenched approximation leads to an unphysical divergence in the chiral limit. This result poses the question of the reliability of the quenched lattice QCD calculations of the neutron EDM.

Keywords: Strong CP breaking, Instantons, Neutron Electric Dipole Moment
PACS: 12.38.Lg, 11.30.Er, 14.20.Dh

INTRODUCTION

In the Standard Model, the non-perturbative structure of the vacuum on the one hand, and the observation of CP-violation in weak interactions on the other hand, give rise to the so-called θ-term, which in the Euclidean formulation reads:

$$\mathscr{L}_{QCD} \to \mathscr{L}_{QCD} + \mathscr{L}_\theta, \qquad \mathscr{L}_\theta = i\bar{\theta}\frac{1}{32\pi^2}F_{\mu\nu}\tilde{F}_{\mu\nu}. \qquad (1)$$

$F_{\mu\nu}$ is the QCD stress tensor, and $\bar{\theta}$ is a free real parameter. At the moment, the most constraining bounds on $\bar{\theta}$ come from measurements of the neutron EDM [1], which indicate that $\bar{\theta} < 10^{-9}$ [2, 3].

In this talk, we shall not address the question why $\bar{\theta}$ is so small ("Strong CP Problem"), but rather explore the phenomenological consequences the θ-term in the QCD Lagrangian [4, 5]. There are several motivations for such a study: it can be shown that the contribution of the θ-term vanishes at any finite order in perturbation theory. Hence, such a term can be used as a theoretical probe to investigate the non-perturbative interactions associated to the topological structure of the QCD vacuum. On the other hand, understanding the quark and gluon CP-odd dynamics could be phenomenologically relevant even if $\bar{\theta} = 0$. In fact, large-N_c arguments suggest that near the deconfinement phase-transition QCD may develop CP-odd meta-stable "bubbles" [6].

In the present work, we focus on strong CP dynamics in the neutron and study the mechanism which gives rise to a finite EDM. In order to make calculation feasible, we work in the semi-classical limit, in which the mixing between degenerate classical QCD

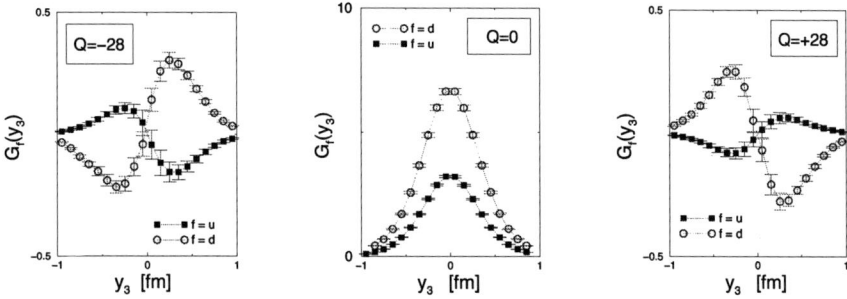

FIGURE 1. The baryon number density correlators for u and d quarks, Eq. (2), in units of $\bar{\theta} \times 10^{-5}$ fm^{-12}, calculated in canonical ensembles with topological charge $Q = -28$, $Q = 0$, $Q = +28$.

vacua is mediated by instantons. We shall show that the θ-term (1) generates an effective repulsion between matter and anti-matter in the neutron, triggered by the tunneling events. As a consequence, quarks and anti-quarks migrate in opposite directions giving rise to a finite EDM. Hence, at least on the semi-classical level, the EDM arises from the local separation of positive and negative baryonic charges.

BARYON AND ELECTRIC CHARGE DISTRIBUTIONS IN THE NEUTRON

In order to gain information about the structure of the non-perturbative CP-odd interactions at finite $\bar{\theta}$, we focus on density distributions of the baryonic and electric charges in the neutron. Our starting point is the neutron baryon number density correlator, defined as $G_B(\mathbf{y}, \tau) = 1/3\, G_d(\mathbf{y}, \tau) + 1/3\, G_u(\mathbf{y}, \tau)$, where

$$G_f(\mathbf{y}, \tau) = \langle 0 | \, \mathrm{Tr}[J_N(\mathbf{0}, 2\tau) J_4^f(\mathbf{y}, \tau) \bar{J}_N(\mathbf{0}, 0) \Gamma_3] | 0 \rangle. \qquad (2)$$

In (2) τ is the Euclidean time, $J_\mu^f(x) = \bar{q}_f \gamma_\mu q_f$, Γ_3 is a diagonal matrix of spinor indices $\Gamma_3 = \mathrm{diag}(1, -1, 1, -1)$ and $J_N(x) = \varepsilon_{abc}(u_a^T C \gamma_5 d_b) d_c$ is an interpolating field, which excites states with the quantum numbers of the neutron.

Similarly, we define the electric charge density correlator as $G_{e/m}(\mathbf{y}, \tau) = 2/3\, G_u(\mathbf{y}, \tau) - 1/3\, G_d(\mathbf{y}, \tau)$. The correlation function (2) measures the probability amplitude to find a quark of flavor f at the point $\mathbf{y}$, in a system with neutron quantum numbers.

The component of the EDM along the direction of the neutron spin can be extracted from the Green's function:

$$\Sigma(\tau) = \int d^3 y\, y_3\, G_{e/m}(\mathbf{y}, \tau), \qquad (3)$$

which, in the large time τ limit, reads:

$$\Sigma(\tau) \to D_z \times \int \frac{d^3\mathbf{p}}{(2\pi)^3} \frac{-2\Lambda_N^2 (p_1^2+p_2^2)}{e^{2\omega_\mathbf{p}\tau}(\mathbf{p}^2+M^2)}, \qquad (4)$$

where D_z is the component of the neutron EDM along the spin direction, M is the neutron mass and Λ_N is the coupling of the neutron to the interpolating operator, $\langle 0|J_N(0)|N\rangle = \Lambda_N u_\mathbf{p}$.

In QCD, a finite dipole can only arise from the CP-breaking interaction (1). To lowest order in $\bar{\theta}$, we can write (in an obvious notation):

$$G_f(\mathbf{y},\tau) \simeq \frac{-i\bar{\theta}}{32\pi^2} \langle 0|\text{Tr}\,[J_N J_4^f \bar{J}_N \Gamma_3] \int d^4 z F\widetilde{F} \,|0\rangle_{\bar{\theta}=0}. \qquad (5)$$

In the semi-classical limit, the topological charge is condensed around instantons and anti-instantons ($Q = N_I - N_A$), and the physics of the quantum mixing of the θ-vacuum can be formulated in terms of an intuitive pseudo-particle picture. The path-integral can then be computed by summing over the configurations of a statistical grand-canonical ensemble of pseudo-particles. In such an approach, a *quantitative* estimate of the matrix element (5) can only be performed in a model-dependent way, as we do not know from first principles the density and size distribution of pseudo-particles in the vacuum. This is the starting point of the Instanton Liquid Model (ILM) - for a review see [7] - in which one *assumes* an average instanton size of $\bar{\rho} \simeq 1/3$ fm and an average density $\bar{n} \simeq 1 fm^{-4}$. The ILM has been proved to be very successful in describing the phenomenology of the QCD vacuum and of light hadrons (for a review of recent results concerning the structure of light hadrons see [8], for recent lattice QCD results supporting the instanton picture see [9, 10, 11]).

Note however, that as long as we will be concern with *qualitative* effects generated by the dynamical interplay of the strong CP-breaking interaction (1) with the quantum structure of the θ-vacuum, our result will not depend on any model-dependent parameters, but only on ab-initio results concerning the semi-classical vacuum fluctuations.

In the instanton vacuum, matrix elements with one insertion of the topological charge operator can be written as a sum over the contributions of the different topological sectors [12]:

$$\left\langle \mathscr{O} \frac{1}{32\pi^2} \int d^4 z F_{\mu\nu}\widetilde{F}_{\mu\nu} \right\rangle = \sum_Q \mathscr{P}(|Q|) Q \langle \mathscr{O} \rangle_Q, \qquad (6)$$

where $\mathscr{O}$ is a generic operator, $\mathscr{P}(Q)$ denotes the relative occurrence of configurations with topological charge Q and $\langle \cdot \rangle_Q$ denotes the average performed in a canonical ensemble with total topological charge Q. This equation expresses the fact that CP-odd interactions are triggered only during tunneling between distinct topological vacua.

A detailed knowledge of $\mathscr{P}(|Q|)$ in (6) is not needed, as long as we are interested in qualitative phenomena. We shall therefore concentrate on the contribution from each topological sector (factor $Q\langle \mathscr{O} \rangle_Q$ in (6)). This can be done by evaluating averages in different canonical ensembles in which the number of instantons and anti-instantons is

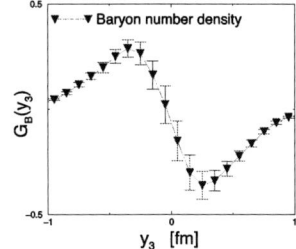 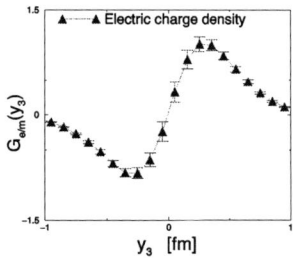

FIGURE 2. Contribution of the topological sectors $Q = -28$ and $Q = +28$ to the neutron baryon number density correlator (left panel) and to the neutron electric charge density correlator (right panel), in units of $\bar{\theta} \times 10^{-5}$ fm^{-12}. A negative baryon number density for $y_3 > 0$ denotes a local accumulation of anti-matter in the positive y_3 hemisphere.

fixed. In order to do so, we have used the Interacting Instanton Liquid Model (IILM), developed in [14]. We have averaged over 1000 configurations of an ensemble of 130 pseudo-particles in a periodic box of volume $V = 3.2 \times 4$ fm^4. For example, the contribution of the topological sector $Q = 10$ was obtained inserting 70 intantons and 60 anti-instantons in the box. The ensemble of collective coordinates of the pseudo-particles was then generated dynamically, i.e. including both bosonic and fermionic instanton-instanton interaction, through an usual accept/reject Metropolis algorithm. In topologically non-trivial sectors, the quark propagator receives contribution from $N_f|Q|$ *exact* zero-modes (semi-classical realization of the Index Theorem).

In Fig. 1 we report typical results of baryon density correlators (2) performed in canonical ensembles with negative, vanishing and positive topological charge (with **y** chosen along the $\hat{z}$ direction). Let us discuss the implications of these results. First of all, we note that the baryon density at zero net topology (center panel in Fig. 1) is even under parity transformation $y_3 \to -y_3$, as expected. On the other hand, contributions from the topologically non-trivial sectors are *P*-odd and lead to a non-vanishing EDM (left and right panels in Fig. 1).

Most notably, these results clearly display the separation of positive and negative baryonic charges in physical systems with neutron quantum numbers. The left and right panes of Fig. 1 imply that in sectors with non-vanishing topological charge, the baryonic charge density of u quarks is positive for $y_3 > 0$ and negative for $y_3 < 0$. Conversely, the baryonic charge density of d quarks is negative for $y_3 > 0$ and positive for $y_3 < 0$.

From these correlators we can construct the contribution to the baryonic number density (Fig. 2, left panel). We conclude that the tunneling produces a local separation of the baryonic charge, with an accumulation of matter in the $y_3 < 0$ hemisphere and of anti-matter in the $y_3 > 0$ hemisphere. Notice that the contribution to the total baryon charge of the neutron coming from configurations with $Q \neq 0$ vanishes, due to the odd symmetry of the correlators (2) under $y_3 \to -y_3$ transformations. In other words, the neutron *total* baryon number comes entirely from the topological sector with $Q = 0$. Hence, the baryon and electric charge asymmetries are associated with the sea quarks only.

In the right panel of Fig. 2 we show that the separation of the baryonic charge induced by the θ-term generates a disentanglement of positive and negative electric charge in the neutron. This is the microscopic dynamical mechanism underlying the EDM formation in QCD, at the semi-classical level.

Let us now discuss some phenomenological implications of the non-perturbative CP-breaking dynamics discussed above. One immediate consequence is that $\bar{\theta} > 0$ would imply an asymmetry in the photon-production of charged pions, with an excess of π^- produced in the $z < 0$ hemisphere. This effect is of the same type of the asymmetries discussed in [13], in the context of relativistic heavy ion collisions.

A second phenomenological implication of the present semi-classical description is that the neutron EDM must have characteristic frequency of oscillation. We have seen that the dissociation of the electric charge is realized through *periodic* currents, induced by *local* topological fluctuations in the vacuum. When the $Q = 0$ condition is restored, the electric charge distribution is relaxed to its symmetrical equilibrium state. It is possible to argue that the characteristic semi-classical period of oscillation of the neutron EDM is of the order $1/m_{\eta'}$ [4].

THE NEUTRON EDM IN THE ILM: COMPARING QUENCHED AND UNQUENCHED RESULTS.

In the previous section we have focused on qualitative results, which do not depend on the specific choice of the phenomenological paramenters of the ILM. In this section, we present the quantitative estimate for the neutron EDM and we investigate the reliability of the quenched approximation. Due to space constraints, here we only present the theoretical and numerical results of our study. For all the technical details of this calculation we refer the interested reader to the specific publication [5].

From an analysis of the semi-classical realization of the Index Theorem, we could show that the quenched and unquenched calculation of the neutron EDM in the ILM are expected to behave very differently in the chiral limit. When the fermionic determinant is included, the neutron EDM vanishes linearly in the quark mass, as it is also expected from other estimates based on chiral perturbation theory (see, e.g., Ref. [15]).

$$D^{(unq.)} \propto m . \qquad (7)$$

On the other hand, when the quenched approximation is adopted one expects a divergent behavior in the chiral limit of the form

$$D^{(q.)} \propto \frac{1}{m^{N_f}} . \qquad (8)$$

We stress the fact the mass dependencies given in Eqs. (7) and (8) do not depend on the particular values of the model parameters which define the ILM. These results rely only on the working assumption that the quantum mixing of the QCD vacuum can be described in terms of isolated tunneling events (instantons). In such a semi-classical limit, the Index Theorem is realized in a very specific way, with N_f exact zero-modes associated to each unit of topological charge.

On the other hand, the specific value of the neutron EDM at each m_q does depend on the key parameters of the model (average instanton size and density). The result of

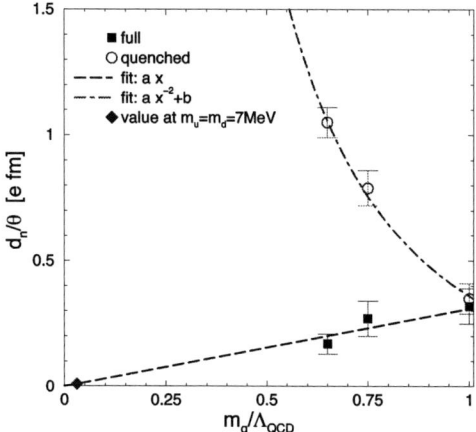

FIGURE 3. ILM results obtained at different values of the quark masses with quenched (circles) and unquenched (squares) simulations.

Monte Carlo simulations in the IILM are reported in Fig.3 and are consistent with the chiral behavior discussed above. Clearly, the existence of chiral divergence makes it impossible to perform any meaningful chiral extrapolation of the quenched results.

CONCLUSIONS

We have presented some recent results concerning the study of the dynamical mechanism for neutron EDM formation in QCD, in the presence of a θ-term. We have found that the baryon number carried by the sea quarks in the neutron periodically undergoes a *local* rearrangement. This fact can be interpreted as due to a flavor-dependent quark-antiquark repulsion, triggered by tunneling events in the θ-vacuum. This is the microscopic dynamical origin of the breaking the spherical symmetry of the electric charge distribution, at the semi-classical level. Notice that in our calculation, we have never used the fact that the neutron pole has been isolated in the Green's function we have computed. Hence, the same mechanism discussed here can be effective in excitations of the nucleon and at finite temperature.

In the second part of this talk, we have discussed our numerical estimate of the neutron EDM, in the IILM. By comparing quenched and unquenched results, we have shown that the quenched approximation gives rise to a well-defined power-law divergence in the chiral limit, as a consequence of the semi-classical realization of the index theorem. This fact suggests that quenched lattice QCD simulations of the neutron EDM (as well as other topologically driven observables) could be affected by similar pathologies.

An an outlook, we would like to stress that the CP-odd dynamical effects described here could have important cosmological implications, *if* CP-odd meta-stable bubbles where formed at the time of the QCD phase transition. A study to determine if such large CP-odd domains are realized in the instanton vacuum near T_c is currently being performed.

REFERENCES

1. P.G. Harris *et al.* Phys. Rev. Lett. **82**, 904(1999).
2. V. Baluni Phys. Rev. D **19**, 2227(1979).
3. R. Crewther, P. Di Vecchia, G. Veneziano and E. Witten, Phys. Lett. **B88**, 123 (1979). Errata Phys. Lett. **B91**, 487 (1980).
4. P. Faccioli, Phys. Rev. D **71**, 091502(2005).
5. P. Faccioli, D. Guadagnoli and S. Simula, Phys. Rev. D **70**, 074017(2004), [arXiv:hep-ph/0406336].
6. D. E. Kharzeev, R. D. Pisarski and M. H. G. Tytgat, [arXiv:hep-ph/0012012].
7. T. Schafer and E. V. Shuryak, Rev. Mod. Phys. **70**, 323 (1998).
8. P. Faccioli, [arXiv:hep-ph/0411088].
9. C. Gattringer, Phys. Rev. Lett. **88**, 221601 (2002). [arXiv:hep-lat/0202002].
10. C. Gattringer, M. Gockeler, P. E. L. Rakow, S. Schaefer and A. Schafer, Nucl. Phys. B **617**, 101 (2001) [arXiv:hep-lat/0107016].
11. P. Faccioli and T. A. DeGrand, Phys. Rev. Lett. **91**, 182001 (2003) [arXiv:hep-ph/0304219].
12. D. Diakonov, . V. Polyakov and C. Weiss, Nucl. Phys. B **461**, 539 (1996).
13. D. Kharzeev, [arXiv:hep-ph/0406125].
14. T. Schäfer and E.V. Shuryak, Phys. Rev. D **53**, 6522(1996).
15. R.J. Crewther *et al.*: Phys. Lett. **B88**, 123 (1979); *errata ibid.* **B91**, 487 (1980).

Properties of the Quark Correlator

E. Di Salvo

Dipartimento di Fisica and INFN - sez. Genova
Via Dodecaneso 33 - 16146 Genova - Italy

Abstract. I deduce some properties of the quark correlator, by exploiting the equations of motion. As a first result, at zero order in the coupling constant the correlator depends on just three independent functions; therefore approximate relations can be established among functions defined in the usual parametrizations of the correlator. Secondly, the azimuthal asymmetries which can be attributed to T-odd functions are suppressed by powers of the QCD hard scale Q. Lastly, higher twist terms are conveniently parametrized in a particular gauge.

Keywords: Quark Correlator
PACS: PACS 12.38.-t

INTRODUCTION

The quark correlator is an important theoretical tool in high energy reactions. It was introduced by Ralston and Soper in 1979[1]. Since then, various papers have been dedicated to the subject[2-5]. However, as I shall show, further properties could be deduced, especially by using the equations of motion. The aim of my talk is just to exhibit some new properties of the correlator and to illustrate their consequences. First of all, I define this quantity and recall its well-known properties. Then I show that the correlator may be conveniently split into a T-even and a T-odd term in a particular gauge. Thirdly, I apply the Politzer theorem[6] on equations of motion and find a power series solution to such equations. I show some consequences of this solution. Lastly, I draw a short conclusion.

THE CORRELATOR

The cross sections for reactions at high energies and high momentum transfers may be written as convolutive and/or matrix products of "hard" factors (which may be calculated perturbatively) times "soft" factors, *i.e.*,

$$d\sigma \propto (Hard) \otimes (Soft). \qquad (1)$$

This approximate formula relies on factorization theorems[7], which must be proved separately for any single reaction. The "soft" factors may be encoded in the correlator, defined as[2]

$$\Phi_{ij}(p;P,S) = \int \frac{d^4x}{(2\pi)^4} e^{ipx} \langle P,S|\bar{\psi}_j(0)\mathscr{L}(x)\psi_i(x)|P,S\rangle. \qquad (2)$$

Here ψ is the quark field, p the quark four-momentum and $|P,S\rangle$ is a state of some hadron with a given four-momentum P and Pauli-Lubanski four-vector S. Moreover

$$\mathscr{L}(x) = \text{P}exp\left[-ig\Lambda_{\mathscr{P}}(x)\right] \tag{3}$$

is the gauge link operator[2]. Here

$$\Lambda_{\mathscr{P}}(x) = \int_0^x \lambda_a A_\mu^a(z) dz^\mu, \tag{4}$$

"P" denotes the path-ordered product along the integration contour $\mathscr{P}$ and g, λ_a and A_μ^a are respectively the strong coupling constant, the Gell-Mann matrices and the gluon fields. The link operator depends on the choice of $\mathscr{P}$, which has to be fixed so as to make a physical sense[8-10]. According to previous treatments[2,11], I define $\mathscr{P}_\pm$ as a set of three pieces of straight line, from the origin to $x_{1\infty} \equiv (\pm\infty, 0, \mathbf{0}_\perp)$, from $x_{1\infty}$ to $x_{2\infty} \equiv (\pm\infty, x^+, \mathbf{x}_\perp)$ and from $x_{2\infty}$ to $x \equiv (x^-, x^+, \mathbf{x}_\perp)$, where the $\pm$ sign has to be chosen, according as to whether final or initial state interactions[8,11-13] are involved in the reaction. Here I have adopted a frame - to be used throughout this talk - whose z-axis is taken along the nucleon momentum, with $x^\pm = 1/\sqrt{2}(t \pm z)$. The correlator (2) may be viewed as a Green function with 2 nucleon legs, 2 quark legs and any number of gluon legs.

The correlator enjoys two important properties, due to the hermiticity condition and to parity conservation:

$$\Phi^\dagger = \gamma_0 \Phi \gamma_0, \qquad \Phi(p,P,S) = \gamma_0 \Phi(\bar{p}, \bar{P}, -\bar{S}) \gamma_0. \tag{5}$$

Here $\bar{p} \equiv (p_0, -\mathbf{p})$, having set $p \equiv (p_0, \mathbf{p})$; $\bar{P}$ and $\bar{S}$ are defined analogously. On the contrary, time reversal invariance does not give rise to any condition on Φ. Indeed, we may have T-even and T-odd functions, the latter being generated by interference between two amplitudes which behave differently under time reversal. It has been shown[14] that such terms are produced by two amplitudes with different quark helicities (generated by spontaneous chiral symmetry breaking[8]) and different components ($\Delta L_z = 1$) of the orbital angular momentum, with a phase shift, caused, for example, by one gluon exchange between the spectator partons and the active quark, either before or after the hard scattering. T-odd functions are not strictly universal, in that they change their sign according as to whether they arise in final or initial state interactions, as a consequence of the choice of the contour $\mathscr{P}_\pm$ defined above[8].

The correlator may be parametrized according to the 16 operators of the Dirac algebra; but, starting from the available vectors (p, P and S), for each operator one can define two independent functions, so that we have 32 functions in all, 12 of which are T-odd[5].

The "soft" functions can be extracted from the correlator by means of projections over Dirac components, defined as

$$\Phi^\Gamma = \frac{1}{2} \int dp^- tr(\Phi\Gamma), \tag{6}$$

where Γ is a Dirac operator. In particular I consider the three main densities of quarks in the nucleon, $i.e.$, the unpolarized density $f_1(x, \mathbf{p}_\perp^2)$, the longitudinally polarized density $g_{1L}(x, \mathbf{p}_\perp^2)$ and the transversity $h_{1T}(x, \mathbf{p}_\perp^2)$. These functions are related, respectively, to the projections Φ^{γ^+}, $\Phi^{\gamma_5\gamma^+}$ and $\Phi^{\gamma_5\gamma^+\gamma^i}$, $i = 1,2$; the first and the third projection include also T-odd contributions[3], i. e., respectively, the unpolarized quark density $f_{1T}^\perp(x,\mathbf{p}_\perp^2)$ in a transversely polarized nucleon (the Sivers[15] function) and the transversity $h_1^\perp(x,\mathbf{p}_\perp^2)$ in an unpolarized nucleon (analogous to the Collins[16] fragmentation function).

PARAMETRIZATION OF THE CORRELATOR

Here I parametrize the quantity

$$\Phi' = \int dp^- \Phi \tag{7}$$

according to the Dirac components. For the sake of simplicity, I consider a transversely polarized nucleon, limiting to twist 2 and twist 3 T-even functions and to twist 2 T-odd functions. I set

$$\Phi' = \Phi'_E + \Phi'_O, \tag{8}$$

where Φ'_E is even under time reversal and Φ'_O is odd under the same transformation. One has, in the above mentioned approximation[2],

$$\Phi'_E = \Phi'^\perp_E \simeq \Phi'_{E2} + \Phi'_{E3}, \tag{9}$$

where

$$\begin{aligned}
\Phi'_{E2} &= \frac{1}{2}p^+\left(f_1 \slashed{n}_+ + \lambda_\perp g_{1T}\gamma_5 \slashed{n}_+ + \frac{1}{2}h_{1T}\gamma_5[\slashed{S},\slashed{n}_+]\right) \\
&\quad + \frac{1}{4\sqrt{2}}\lambda_\perp h_{1T}^\perp \gamma_5[\slashed{p}_\perp,\slashed{n}_+],
\end{aligned} \tag{10}$$

$$\begin{aligned}
\Phi'_{E3} &= \frac{1}{2}M(e + g_T\gamma_5\slashed{S}) + \frac{1}{2}\left(f_1^\perp + \lambda_\perp g_T^\perp \gamma_5\right)\slashed{p}_\perp \\
&\quad + \frac{1}{4}\lambda_\perp \left(h_T^\perp \gamma_5[\slashed{S},\slashed{p}_\perp] + h_T \mu\gamma_5[\slashed{n}_-,\slashed{n}_+]\right).
\end{aligned} \tag{11}$$

On the other hand, the term Φ'_O reads (at leading twist)

$$\Phi'_O \simeq \Phi'_{O2} = \frac{1}{2\mu}\left[f_{1T}^\perp \varepsilon_{\mu\nu\rho\sigma}\gamma^\mu n_+^\nu p_\perp^\rho S^\sigma + ih_1^\perp \frac{1}{2}[\slashed{p}_\perp,\slashed{n}_+]\right]. \tag{12}$$

In formulae (10) to (12) I have used the notations of refs.[2,3] for the "soft" transverse momentum functions. Moreover M is the nucleon rest mass and $p \equiv (p^-,p^+,\mathbf{p}_\perp)$, taking the z-axis along the nucleon momentum. Furthermore I have set

$$\lambda_\perp = -S\cdot p_\perp/\mu, \qquad p_\perp \equiv (0,0,\mathbf{p}_\perp) \tag{13}$$

and $n_\pm$ are lightlike vectors, such that $n_+ \cdot n_- = 1$ and whose space components are directed along the nucleon momentum. Lastly, μ is a kinematical parameter with the dimensions of an energy, introduced for dimensional reasons, so that all "soft" functions have the dimensions of quark densities: as a consequence, some of these functions (like $f_{1T}^\perp$ and $h_1^\perp$) are multiplied by the dimensionless factor $|\mathbf{p}_\perp|/\mu$. The parameter μ is usually assumed equal to the rest mass of the hadron[1-3]. As we shall see below, this is not in agreement with the equations of motion.

SPLITTING OF THE CORRELATOR

I consider the splitting

$$\mathscr{L}(x) = \mathscr{L}_R(x) + \mathscr{L}_I(x), \tag{14}$$

where

$$\mathscr{L}_R(x) = P\cos\left[g\Lambda_{\mathscr{P}_+}(x)\right], \qquad \mathscr{L}_I(x) = -iP\sin\left[g\Lambda_{\mathscr{P}_+}(x)\right], \tag{15}$$

$\Lambda(x)_{\mathscr{P}_+}$ being defined by eq. (4). Inserting eq. (14) into eqs. (3) and (2), I get

$$\Phi(p) = \Phi_R(p) + \Phi_I(p), \tag{16}$$

where $\Phi_R(p)$ and $\Phi_I(p)$ are defined analogously to $\Phi(p)$, substituting $\mathscr{L}(x)$ respectively with $\mathscr{L}_R(x)$ and with $\mathscr{L}_I(x)$. In *an axial gauge with antisymmetric boundary conditions* it can be shown that $\Phi_R(p)$ generates only T-even functions, while $\Phi_I(p)$ gives rise just to T-odd functions. Therefore, in this gauge - to be referred to as $\mathscr{G}$-gauge in the following - the T-even (T-odd) part of the correlator consists of a series of even (odd) powers of the strong coupling constant g and contains an even (odd) number of gluon legs. An immediate advantage of this splitting in the $\mathscr{G}$-gauge is that we can extract separately T-even and T-odd functions, by projecting $\Phi_R(p)$ and $\Phi_I(p)$ over the various Dirac components.

EQUATIONS OF MOTION

Now I invoke the Politzer theorem on equations of motion[6], *i.e.*,

$$\langle P, S | \mathscr{F}(\psi)(i\slashed{D} - m_q)\psi(x) | P, S \rangle = 0. \tag{17}$$

Here m_q is the quark rest mass, $\mathscr{F}(\psi)$ a functional of the quark field and $D_\mu = \partial_\mu - ig\lambda_a A_\mu^a$ the covariant derivative. The result (17) survives renormalization.

Before applying this result to our case, it is worth observing that it is intuitive to expect that at zero order in g the correlator Φ' (eq. (7)) amounts to the density matrix of a free, on-shell quark, in agreement with the QCD parton model. This matrix reads

$$\rho = \frac{1}{2}(\slashed{p} + m_q)(f_1 + \gamma_5 \slashed{s}_\parallel^q g_{1L} + \gamma_5 \slashed{s}_\perp^q h_{1T}), \tag{18}$$

which involves the three main densities already introduced. I shall deduce this guess by means of eq. (17). I shall answer also other questions, like, e. g., in which way to take into account offshellness, T-odd terms and power corrections to the naive QCD parton model. To this end, I use eq. (17), with $\mathscr{F}(\psi) = \bar{\psi}(0)\mathscr{L}(x)$. Moreover I set, in the $\mathscr{G}$-gauge,

$$\Phi_R(p) = \Phi_E(p) = \Phi_E^{(0)}(p) + g^2\Phi_E^{(2)}(p) + O(g^4), \tag{19}$$
$$\Phi_I(p) = \Phi_O(p) = g\Phi_O^{(1)}(p) + O(g^3), \tag{20}$$

the suffixes E and O denoting respectively T-even and T-odd terms. I obtain, after some manipulations of eq. (17)[17],

$$(\not{p} - m_q)\Phi_E^{(0)}(p) = 0, \tag{21}$$
$$(\not{p} - m_q)\Phi_O^{(1)}(p) = -i\Psi_O(p), \tag{22}$$
$$(\not{p} - m_q)\Phi_E^{(2)}(p) = \Psi_E(p). \tag{23}$$

Here one has, in the $\mathscr{G}$-gauge,

$$(\Psi_O)_{ij} = \int \frac{d^4x}{(2\pi)^4} e^{ipx} \langle P,S | \bar{\psi}_j(0) [\not{A}_\infty(x)]_{ik} \psi_k(x) | P,S \rangle, \tag{24}$$

$$(\Psi_E)_{ij} = \mathscr{P}\frac{1}{(\not{p}-m_q)} \int \frac{d^4x}{(2\pi)^4} e^{ipx} \langle P,S | \bar{\psi}]_j(0) P[A_\infty^2(x)] \psi_i(x) | P,S \rangle. \tag{25}$$

I have set $A_\infty^\mu = A^\mu(x_{2\infty})$, while $\mathscr{P}$ denotes the Cauchy principal value. Incidentally, our result is in accord with the considerations of Ji and Yuan[11], who find that, using an axial gauge, the final (or initial) state interactions are confined at $x^- = \pm\infty$.

As a consequence of eqs. (19) to (25), the expansions of $\Phi_E(p)$ and $\Phi_O(p)$ in powers of g read[17]

$$\Phi_E(p) = \rho\delta(p^2 - m_q^2) + g^2\kappa_\perp^2 M_E(p) + O(g^4), \tag{26}$$
$$\Phi_O(p) = -ig\kappa_\perp M_O(p) + O(g^3), \tag{27}$$

where ρ is given by eq. (18),

$$M_E\kappa_\perp^2 = \mathscr{P}\frac{\Psi_E}{(\not{p}-m_q)}, \qquad M_O\kappa_\perp = \mathscr{P}\frac{\Psi_O}{(\not{p}-m_q)} \tag{28}$$

and

$$\kappa_\perp = \frac{|\mathbf{p}_\perp|}{p^+}. \tag{29}$$

Now I discuss the consequences implied by the equations just found. First of all, at zero order in g the correlator Φ' is gauge independent[17]. Furthermore it reduces to the density matrix of a free quark, as expected. Since this depends just on three functions (see eq. (18)), one deduces several approximate relations among the functions involved in the parametrization (10) - (11). In particular, I have, in the $\mathscr{G}$-gauge,

$$h_{1T}(x, \mathbf{p}_\perp^2) = g_{1T}(x, \mathbf{p}_\perp^2) + O(\alpha_s\kappa_\perp^2), \qquad h_{1L}(x, \mathbf{p}_\perp^2) = g_{1L}(x, \mathbf{p}_\perp^2) + O(\alpha_s\kappa_\perp^2). \tag{30}$$

Here g_{1T} is the helicity density of a quark in a transversely polarized nucleon, while h_{1L} is the transversity of a quark in a longitudinally polarized nucleon. These relations are especially important, because they relate chiral odd functions to chiral even ones, which are more easily determined experimentally. Such relations are not altered by perturbative QCD evolution, since the Politzer theorem survives renormalization. Moreover, in processes like Drell-Yan and semi-inclusive deep inelastic scattering (SIDIS), one has $\sqrt{2}p^+ \simeq Q$, where Q is the hard QCD scale. Then the first eq. (30) is particularly useful in determining experimentally the transversity, up to power corrections (in the $\mathscr{G}$-gauge) of order $g^2\pi_\perp^2/Q^2$, $\pi_\perp$ being the mean value of $|\mathbf{p}_\perp|$.

In the same gauge, important relations may be established also for other functions; in particular, we have

$$e = m_q f_1 + O(g^2\pi_\perp^2/Q^2), \qquad g_T = m_q h_1 + O(g^2\pi_\perp^2/Q^2), \qquad (31)$$
$$h_L = m_q g_1 + O(g^2\pi_\perp^2/Q^2). \qquad (32)$$

Here integration over transverse momentum has been performed. The second relation (31) is especially relevant. Indeed, it implies that, in the $\mathscr{G}$-gauge, the antisymmetric part of the operator product expansion in transversely polarized deep inelastic scattering derives contributions only from graphs with even numbers of gluon legs and that the main correction term of g_T decreases as Q^{-2}.

As regards T-odd functions, eq. (27) implies that they are of order $O(g\pi_\perp/Q)$. This conclusion turns out to be gauge independent[17]. Therefore I conclude, first of all, that such functions vanish in absence of a quark-gluon interaction, in agreement with the considerations by other authors[8,11,12,14-16]. Moreover they turn out to be inversely proportional to Q. Then those asymmetries in SIDIS and Drell-Yan which involve T-odd functions are suppressed at least as Q^{-n}, where n is the number of T-odd functions involved. This finds an experimental confirmation in the azimuthal asymmetry of unpolarized Drell-Yan[18]: indeed, this asymmetry is proportional to the convolutive product $h_1^\perp \otimes \bar{h}_1^\perp$, therefore it decreases as Q^{-2}, as confirmed by data[19,20].

The parameter μ, which appears in the parametrization of the correlator (see eqs. (8) to (13)), turns out to be equal to $p^+/\sqrt{2} \simeq Q/2$, as follows by comparing eqs. (26) and (27) with eqs. (10) - (12). This influences not only the T-odd functions, but also some T-even ones, like g_{1T}. In particular, the SIDIS double spin asymmetry considered by Kotzinian and Mulders[18] is proportional to the product $g_{1T} \otimes D_\pi$ and is predicted to decrease as Q^{-1}.

Two remarks are in order. Firstly, the zero order term refers to on-shell quarks, higher orders taking into account offshellness, analogously to the Qiu splitting[21], which is a refinement of the Ellis-Furmanski-Petronzio original treatment[22]. Secondly, the results exposed above can be extended to the fragmentation functions.

CONCLUSIONS

Here I summarize the main results just exposed.

1) At zero order, the correlator Φ' is T-even and gauge independent and is approximated by the density matrix, corresponding to the QCD parton model. Therefore, in this

approximation, one may establish relations among the "soft" functions involved in the parametrization of Φ', since the density matrix depends just on three functions. Especially relevant are the equalities between chiral-even and chiral-odd functions, which, in particular, may help in determining transversity.

2) First order power corrections include T-odd functions, which are proportional to $g(\pi_\perp/Q)$, independent of the gauge. This produces a Q^{-n} suppression for the asymmetries involving n such functions. This result is in accord with data of unpolarized Drell-Yan and, in principle, it could be compared with data of incoming HERMES, COMPASS and CLAS experiments.

3) Higher twist terms can be conveniently parametrized in an axial gauge with antisymmetric boundary conditions.

REFERENCES

1. J. Ralston and D.E. Soper, Nucl. Phys B **152**, 109 (1979).
2. P.J. Mulders and R.D. Tangerman, Nucl. Phys. B **461**, 197 (1996).
3. D. Boer, R. Jakob and P.J. Mulders, Nucl. Phys. B **564**, 471 (2000).
4. K. Goeke, A. Metz, P.V. Pobylitsa and M.V. Polyakov, Phys. Lett. B **567**, 27 (2003).
5. K. Goeke, A. Metz and M. Schlegel, Phys. Lett. B **618**, 90 (2005).
6. H. D. Politzer, Nucl. Phys. B **172**, 349 (1980).
7. See, *e. g.*, J.C. Collins: Phys. Rev. D **57**, 3051 (1998); "Perturbative QCD", A.H. Mueller ed., World Scientific, Singapore, 1989.
8. J.C. Collins, Phys. Lett. B **536**, 43 (2002).
9. J.C. Collins and D.E. Soper, Nucl. Phys. B **194**, 445 (1982).
10. J.C. Collins, Acta Phys. Polon. B **34**, 3103 (2003).
11. X. Ji and F. Yuan, Phys. Lett. B **543**, 66 (2002).
12. A.V. Belitsky, X. Ji and F. Yuan, Nucl. Phys. B **656**, 165 (2003).
13. D. Boer, P.J. Mulders and F. Pijlman, Nucl. Phys. B **667**, 201 (2003).
14. S.J. Brodsky, D.S. Huang and I. Schmidt, Phys. Lett. B **530**, 99 (2002); Nucl. Phys. B **642**, 344 (2002).
15. D.W. Sivers, Phys. Rev. D **41**, 83 (1990); Phys. Rev. D **43**, 261 (1991).
16. J.C. Collins, Nucl. Phys. B **396**, 161 (1993).
17. E. Di Salvo: in preparation
18. A. Kotzinian and P.J. Mulders, Phys. Rev. D **54**, 1229 (1996).
19. NA10 Coll., S. Falciano et al., Z. Phys. C - Particles and Fields **31**, 513 (1986); M. Guanziroli et al., Z. Phys. C - Particles and Fields **37**, 545 (1988).
20. E. Di Salvo, AIP Conf. Proc. **747**, 215 (2005).
21. J. W. Qiu, Phys. Rev. D **42**, 30 (1990).
22. R.K. Ellis, W. Furmanski and R. Petronzio, Nucl. Phys. B **212**, 29 (1983).

Unitary Structure of the QCD Sum Rules and KYN and $KY\Xi$ Couplings

T. Aliev*, A.Ozpineci†, S.B. Yakovlev** and V.S. Zamiralov**

*Physics Dept., Middle East Technical Univ., 06531, Ankara, Turkey
†INFN, Sezione di Bari, Bari, Italy [1]
**D.V. Skobeltsyn Institute of Nuclear Physics, Moscow State University, Moscow, Russia

Abstract. New relations between QCD Borel sum rules for strong coupling constants of K-mesons to baryons are derived. It is shown that starting from the sum rule for the coupling constants $g_{\pi\Sigma\Sigma}$ and $g_{\pi\Sigma\Lambda}$ it is straightforward to obtain corresponding sum rules for the g_{KYN}, $g_{KY\Xi}$ couplings, $Y = \Sigma, \Lambda$.

Keywords: QCD, Sum Rule, Baryon, Meson, Coupling Constant
PACS: 12.39.Th

INTRODUCTION

Meson-baryon couplings were studied for years thoroughly either for pion-baryon couplings or kaon-baryon baryon ones as these couplings are important parameters of strong interaction physics.

Since the advent of the $SU(3)$ symmetry all the meson-baryon coupling constants were usually expressed in terms of F and D constants which gave possibility to construct a reliable phenomenological aproach.

As soon as in [1] QCD sum rules (SR's) were proposed, they were used not only for baryon masses and magnetic moments starting from the works [2] but also for baryon-meson coupling constants. Naturally, a pion-nucleon coupling attracted the most attention (see, for example, [3], [4], [5]). Coupling constants of π^0- and η- mesons to baryons were studied recently in various QCD SR approaches [6], [7]. Also QCD sum rule for the η coupling to the Λ hyperon was written [8] which was usually absent in these approaches.

As for K-mesons they were also studied in the framework of the QCD sum rules (see, e.g., [9], [10], [11]). Usually these sum rules are not related straightforwardly to those treating π and η couplings to baryons.

We would like to propose here QCD sum rules for octet meson-baryon couplings through some universal $\mathscr{F}$ and $\mathscr{D}$ functions written in a unified manner. In order to be clear we choose for methodical reasons as a basis for our reasoning $SU(3)$ breaking QCD Borel sum rules proposed in [6].

---

[1] Present address: Middle East Technical University, Ankara,Turkey

RELATION BETWEEN $\pi^0\Sigma\Sigma$ AND $\pi\Sigma\Lambda$ CONSTANTS IN SU(3)

We begin as in [12] with a simple example. In the unitary model all the pion-baryon coupling constants can be expressed in terms of F and D coupling constants.

But coupling of the Σ-like baryons $B(qq,q'), q,q' = u,d,s$ to π^0 meson related in the quark model to the current $j^{\pi^0} = \frac{1}{\sqrt{2}}[\bar{u}\gamma_5 u - \bar{d}\gamma_5 d]$ can be put in the form

$$g(\pi^0 BB) == g_{\pi qq} 2F + g_{\pi q'q'}(F-D),$$

or, particle per particle:

$$g(\pi^0 pp) = g_{\pi uu} 2F + g_{\pi dd}(F-D) = \sqrt{\frac{1}{2}}(F+D);$$

$$g(\pi^0 \Sigma^+ \Sigma^+) = g_{\pi uu} 2F + g_{\pi ss}(F-D) = \sqrt{2}F,$$

and so on, where $g_{\pi uu} = +\sqrt{\frac{1}{2}}$, $g_{\pi dd} = -\sqrt{\frac{1}{2}}$ and $g_{\pi ss} = 0$ are just read off the quark current.

The only coupling which cannot be written immediately in this way is $\pi^0 \Sigma^0 \Lambda$. To overcome this difficulty let us write for $\pi^0 \Sigma^0 \Sigma^0$ coupling (which is equal to zero!):

$$g(\pi^0 \Sigma^0 \Sigma^0) = g_{\pi^0 uu} F + g_{\pi^0 dd} F + g_{\pi^0 ss}(F-D) = 0 \tag{1}$$

and change $(d \leftrightarrow s)$ and $(u \leftrightarrow s)$ to form two auxiliary quantities

$$g(\pi^0 \tilde{\Sigma}^{0,ds} \tilde{\Sigma}^{0,ds}) = g_{\pi^0 uu} F + g_{\pi^0 ss} F + g_{\pi^0 dd}(F-D) = \sqrt{\frac{1}{2}}D, \tag{2}$$

$$g(\pi^0 \tilde{\Sigma}^{0,us} \tilde{\Sigma}^{0,us}) = g_{\pi^0 dd} F + g_{\pi^0 ss} F + g_{\pi^0 uu}(F-D) = -\sqrt{\frac{1}{2}}D. \tag{3}$$

The following relation holds:

$$g(\pi^0 \tilde{\Sigma}^{0,ds} \tilde{\Sigma}^{0,ds}) - g(\pi^0 \tilde{\Sigma}^{0,us} \tilde{\Sigma}^{0,us}) = \sqrt{3} g(\pi^0 \Sigma^0 \Lambda). \tag{4}$$

The origin of this relation lies in the structure of $\Sigma^0(ud,s)$ and Λ wave functions in the NRQM. With the exchanges $d \leftrightarrow s$ and $u \leftrightarrow s$ one arrives at the corresponding U-spin and V-spin quantities, so

$$\begin{pmatrix} -\tilde{\Sigma}^0_{ds} \\ \tilde{\Lambda}_{ds} \end{pmatrix} = \begin{pmatrix} 1/2 & \sqrt{3}/2 \\ -\sqrt{3}/2 & 1/2 \end{pmatrix} \begin{pmatrix} \Sigma^0 \\ \Lambda \end{pmatrix}, \begin{pmatrix} -\tilde{\Sigma}^0_{us} \\ \tilde{\Lambda}_{us} \end{pmatrix} = \begin{pmatrix} 1/2 & -\sqrt{3}/2 \\ \sqrt{3}/2 & 1/2 \end{pmatrix} \begin{pmatrix} \Sigma^0 \\ \Lambda \end{pmatrix}. \tag{5}$$

It is easy now to show that the relation Eq.(4) follows and it shows us the way to proceed with the QCD sum rules.

KYN, *KY*Ξ AND $\pi\Sigma\Lambda$ COUPLINGS IN THE SU(3)

Now we consider kaon and charged pion couplings to baryons. They are given by $SU(3)$ symmetry formulae but we rewrite it in a way suitable for derivation of the corresponding Borel sum rules. Let us write coupling of pion to Σ^+ and Λ_{ds} given by the Eq.(5):

$$2[g(\pi^-\Sigma^+\bar{\Lambda}_{ds})] = -\sqrt{3}g(\pi^-\Sigma^+\bar{\Sigma}^0) + g(\pi^-\Sigma^+\bar{\Lambda}) = \quad (6)$$

$$-\sqrt{3}(-\sqrt{2}F) + \sqrt{\frac{2}{3}}D = \sqrt{\frac{2}{3}}(3F+D).$$

Now we perform $d \leftrightarrow s$ exchange. Our auxiliary baryon Λ_{ds} returns to real Λ while $\pi^-(\bar{d}u)$ changes to $K^-(\bar{s}u)$ and $\Sigma^+(uu,s)$ changes to $-p(uu,d)$, so that

$$2[g(\pi^-\Sigma^+\bar{\Lambda}_{ds})]_{ds} = -2[g(K^-p\Lambda)] = \sqrt{\frac{2}{3}}(3F+D). \quad (7)$$

This is the unitary symmetry result. In the same way we write the formal coupling of pion to Σ^+ and Λ_{us} given by the Eq.(5) and then perform $u \leftrightarrow s$ exchange to obtain

$$2[g(\pi^-\Sigma^+\bar{\Lambda}_{us})]_{us} = 2[g(K^0\Xi^0\bar{\Lambda})] = -\sqrt{\frac{2}{3}}(3F-D). \quad (8)$$

This is again the unitary symmetry result. Similarly one can show that

$$-2[g(\pi^-\Sigma^+\bar{\Sigma}^0_{ds})]_{ds} = 2[g(K^-p\bar{\Sigma}^0)] = \sqrt{2}(-F+D), \quad (9)$$

$$-2[g(\pi^-\Sigma^+\bar{\Sigma}^0_{us})]_{us} = 2[g(K^0\Xi^0\bar{\Sigma}^0)] = -\sqrt{2}(F+D).$$

Derivation of these coupling constants indicates us the way to proceed in the formalism of QCD sum rules.

QCD SUM RULES

We use as the example QCD sum rules based on the formalism developed in [6] where unitary symmetry is broken but formulae are rather transparent. The sum rule for the $\mathcal{M}\Sigma^0\Sigma^0$ coupling reads:

$$\frac{1}{\sqrt{2}}m_{\mathcal{M}}^2 \lambda_\Sigma^2 g(\mathcal{M}\Sigma^0\Sigma^0) e^{-(m_\Sigma^2/M^2)}[1+A_\Sigma M^2] =$$

$$g_{\mathcal{M}ss}m_{\mathcal{M}}^2 M^4 E_0(x)[\frac{\langle \bar{s}s \rangle}{12\pi^2 f_{\mathcal{M}}} + \frac{3f_{3\mathcal{M}}}{4\sqrt{2}\pi^2}]$$

$$-g_{\mathcal{M}ss}\frac{1}{f_{\mathcal{M}}}M^2(m_d\langle \bar{u}u \rangle + m_u\langle \bar{d}d \rangle)\langle \bar{s}s \rangle$$

$$-g_{\mathcal{M}ss}\frac{m_{\mathcal{M}}^2}{72 f_{\mathcal{M}}}\langle \bar{s}s \rangle \langle \frac{\alpha_s}{\pi}\mathcal{G}^2 \rangle$$

$$+\frac{1}{6f_{\mathcal{M}}}m_0^2[\langle\bar{s}s\rangle(m_d g_{\mathcal{M}uu}\langle\bar{u}u\rangle + m_u g_{\mathcal{M}dd}\langle\bar{d}d\rangle)$$
$$+m_s(g_{\mathcal{M}uu}+g_{\mathcal{M}dd})\langle\bar{u}u\rangle\langle\bar{d}d\rangle]. \tag{10}$$

where m_q, $q=u,d,s$ are current quark masses, $f_{\mathcal{M}}$ is a $\mathcal{M}$-meson decay constant, $\mathcal{M} = \pi^0, \eta$, quark condensates are $\langle\bar{u}u\rangle = \langle\bar{d}d\rangle = -(0.23)^3$ GeV^3, $\langle\bar{s}s\rangle/\langle\bar{d}d\rangle = 0.8$, while $m_0^2 = 0.8$ GeV^2, $\langle\bar{g}_c q\sigma\cdot Gq\rangle \equiv)m_0^2\langle\bar{q}q\rangle$. The factor $E_0(x) = (1-e^{-x})$ is used to subtract the continuum contribution, $x = W^2/M^2$ [2] (we take $W^2 = 2.0$ GeV^2). The overlap amplitude is taken as $\lambda_B^2 = C\cdot M_B^6$ GeV^6 [6], with $C = 5.48\times 10^{-4}$. We neglect in calculations $f_{3\mathcal{M}}$. Parameter A_B accounts for high-resonance contributions.

We define $\mathscr{D}^{(0)}(\mathcal{M};M^2;u,d;s)$ and $\mathscr{F}^{(0)}(\mathcal{M};M^2;u,d;s)$ (this shorthanded notation means that they depend on M^2, all quark masses and all condensates: $\mathscr{D}^{(0)}(\mathcal{M};M^2;u,d;s) \equiv \mathscr{D}^{(0)}(\mathcal{M};M^2;m_u,\langle\bar{u}u\rangle,...;m_d,\langle\bar{d}d\rangle,...;m_s,\langle\bar{s}s\rangle,...)$, similar for $\mathscr{F}$):

$$\mathscr{F}^{(0)}(\mathcal{M};M^2;u,d;s) = \frac{1}{6f_{\mathcal{M}}}m_0^2[\langle\bar{s}s\rangle(m_d\langle\bar{u}u\rangle + m_s\langle\bar{u}u\rangle\langle\bar{d}d\rangle],$$

$$\mathscr{D}^{(0)}(\mathcal{M};M^2;u,d;s) - \mathscr{F}^{(0)}(\mathcal{M};M^2;u,d;s) = -[m_{\mathcal{M}}^2 M^4 E_0(x)[\frac{\langle\bar{s}s\rangle}{12\pi^2 f_{\mathcal{M}}} + \frac{3f_{3\mathcal{M}}}{4\sqrt{2}\pi^2}]$$
$$-\frac{1}{f_{\mathcal{M}}}M^2(m_d\langle\bar{u}u\rangle + m_u\langle\bar{d}d\rangle)\langle\bar{s}s\rangle - \frac{m_{\mathcal{M}}^2}{72f_{\mathcal{M}}}\langle\bar{s}s\rangle\langle\frac{\alpha_s}{\pi}\mathscr{G}^2\rangle], \tag{11}$$

The righthand side (RHS) of the Eq.(10) can be written in a form

$$RHS(\mathcal{M}\Sigma^0\Sigma^0) = g_{\mathcal{M}uu}\mathscr{F}^0(\mathcal{M};M^2;u,d;s) + g_{\mathcal{M}dd}\mathscr{F}^0(\mathcal{M};M^2;d,u;s) +$$
$$\frac{1}{2}g_{\mathcal{M}ss}(\mathscr{F}^0(\mathcal{M};M^2;s,d;u) + \mathscr{F}^0(\mathcal{M};M^2;s,u;d)) -$$
$$\frac{1}{2}g_{\mathcal{M}ss}(\mathscr{D}^0(\mathcal{M};M^2;u,d;s) + \mathscr{D}^0(\mathcal{M};M^2;d,u;s)). \tag{12}$$

With isotopic invariance we construct Borel sum rule for the $\pi^-\Sigma^+\bar{\Sigma}^0$:

$$-m_\pi^2\lambda_\Sigma^2 g(\pi^-\Sigma^+\Sigma^0)e^{-(m_\Sigma^2/M^2)}[1+A_\Sigma M^2] =$$
$$\frac{m_0^2}{6f_\pi}[(m_u\langle\bar{s}s\rangle + m_s\langle\bar{u}u\rangle)(\langle\bar{u}u\rangle + \langle\bar{d}d\rangle)] \equiv \sqrt{2}\mathscr{F}^{(-)}(\pi^-;M^2;u,d;s) \tag{13}$$

and a similar sum rule for $\pi^+\Sigma^-\Sigma^0$ coupling (upon $u\leftrightarrow d$).

Using analogue of the Eq.(4)

$$\sqrt{3}RHS(\pi^0\Sigma^0\Lambda) = RHS(\pi^0\Sigma^0_{ds}\Sigma^0_{ds}) - RHS(\pi^0\Sigma^0_{us}\Sigma^0_{us})$$

we construct QCD Borel sum rule for $\pi^0\Sigma\Lambda$ coupling [8]

$$\sqrt{3}m_\pi^2\lambda_\Lambda\lambda_\Sigma g(\pi^0\Sigma^0\Lambda)\frac{M^2}{M_\Sigma^2 - M_\Lambda^2}(e^{-M_\Lambda^2/M^2} - e^{-M_\Sigma^2/M^2})[1+A_{\Sigma\Lambda}M^2] =$$

$$-m_\pi^2 M^4 E_0(x)[\frac{\langle\bar{d}d\rangle+\langle\bar{u}u\rangle}{12\pi^2 f_\pi}+\frac{3f_{3\pi}}{4\sqrt{2}\pi^2}]+\frac{m_\pi^2}{72f_\pi}[\langle\bar{d}d\rangle+\langle\bar{u}u\rangle]\langle\frac{\alpha_s}{\pi}\mathcal{G}^2\rangle$$

$$+\frac{1}{6f_\pi}(6M^2+m_0^2)[(m_s\langle\bar{u}u\rangle+m_u\langle\bar{s}s\rangle)\langle\bar{d}d\rangle+(m_d\langle\bar{s}s\rangle+m_s\langle\bar{d}d\rangle)\langle\bar{u}u\rangle] \quad (14)$$

The RHS of it with Eq.(12) can be put in the form

$$\sqrt{3}RHS(\pi^0\Sigma^0\Lambda)=\frac{1}{2\sqrt{2}}[\mathcal{D}^{(0)}(\pi^0;M^2;s,d;u)+\mathcal{D}^{(0)}(\pi^0;M^2;s,u;d)+$$

$$\mathcal{D}^{(0)}(\pi^0;M^2;u,s;d)+\mathcal{D}^{(0)}(\pi^0;M^2;d,s;u)]\to|_{exact \quad SU(3)} \quad \sqrt{2}D. \quad (15)$$

Isotopic invariance allows to deduce the corresponding expression for the $\pi^-\Sigma^+\bar{\Lambda}$ coupling:

$$\sqrt{3}m_\pi^2\lambda_\Lambda\lambda_\Sigma g(\pi^-\Sigma^+\bar{\Lambda})\frac{M^2}{M_\Sigma^2-M_\Lambda^2}(e^{-M_\Lambda^2/M^2}-e^{-M_\Sigma^2/M^2})[1+A_{\Sigma\Lambda}M^2]=$$

$$=[-m_\pi^2 M^4 E_0(x)[\frac{(\langle\bar{u}u\rangle+\langle\bar{d}d\rangle)}{12\pi^2 f_\pi}+\frac{3f_{3\pi}}{4\sqrt{2}\pi^2}]+$$

$$\frac{(m_0^2+6M^2)}{6f_\pi}[(m_u\langle\bar{s}s\rangle+m_s\langle\bar{u}u\rangle)(\langle\bar{u}u\rangle+\langle\bar{d}d\rangle)]+$$

$$+\frac{m_\pi^2}{72f_\pi}(\langle\bar{u}u\rangle+\langle\bar{d}d\rangle)\langle\frac{\alpha_s}{\pi}\mathcal{G}^2\rangle]\equiv\sqrt{2}\mathcal{D}^{(-)}(\pi^-;M^2;s,d;u)\to|_{exact \quad SU(3)} \quad \sqrt{2}D. \quad (16)$$

And now we are able to derive Borel sum rules for K-meson couplings to octet baryons starting from those for $\pi\Sigma\Lambda$ and $\pi\Sigma\Sigma$ couplings given by the Eqs.(13,16). We shall form auxiliary couplings upon using quantities $\Lambda_{ds},\Sigma_{ds}^0$ and $\Lambda_{us},\Sigma_{us}^0$ given by the Eq. (5), and then return to those usual ones performing transformations $d\leftrightarrow s$ and $u\leftrightarrow s$. First we construct a formal sum rule for the case where Λ is changed to Λ_{ds} just by using Eq.(5), and we retain for a moment only RHS of the corresponding sum rules:

$$RHS(\pi^-\Sigma^+\bar{\Lambda}_{ds})=-\frac{\sqrt{3}}{2}RHS(\pi^-\Sigma^+\bar{\Sigma}^0)+\frac{1}{2}RHS(\pi^-\Sigma^+\bar{\Lambda})=$$

$$\sqrt{\frac{1}{6}}(3\mathcal{F}^{(-)}(\pi^-;M^2;u,d;s)+\mathcal{D}^{(-)}(\pi^-;M^2;s,d;u)). \quad (17)$$

Performing transformation $(d\leftrightarrow s)$ we should change π^- to K^- and Σ^+ to $-p$ to obtain:

$$RHS((g(\pi^-\Sigma^+\bar{\Lambda}_{ds})_{ds})=-RHS(g(K^-p\bar{\Lambda})=$$

$$\sqrt{\frac{1}{6}}(3\mathcal{F}^{(-)}(K^-;M^2;u,s;d)+\mathcal{D}^{(-)}(K^-;M^2;d,s;u))$$

$$\to|_{exact \quad SU(3)} \quad \sqrt{\frac{1}{6}}(3F+D), \quad (18)$$

or in full notation

$$m_K^2 g_{K^-p\bar{\Lambda}} \frac{\lambda_\Lambda \lambda_N M^2}{(M_\Lambda^2 - M_N^2)}(e^{-M_N^2/M^2} - e^{-M_\Lambda^2/M^2})(1 + A_{\Lambda N}M^2)$$
$$= -\frac{1}{2\sqrt{3}}[-m_K^2 M^4 E_0(x)[\frac{(\langle\bar{u}u\rangle + \langle\bar{s}s\rangle)}{12\pi^2 f_K} + \frac{3f_{3K}}{4\sqrt{2}\pi^2}] +$$
$$\frac{(2m_0^2 + 3M^2)}{3f_K}[(m_u\langle\bar{d}d\rangle + m_d\langle\bar{u}u\rangle)(\langle\bar{u}u\rangle + \langle\bar{s}s\rangle)]$$
$$+ \frac{m_K^2}{72 f_K}(\langle\bar{u}u\rangle + \langle\bar{s}s\rangle)\langle\frac{\alpha_s}{\pi}\mathcal{G}^2\rangle]. \qquad (19)$$

Interchanging $(u \leftrightarrow d)$ one transforms it into the sum rule for $g_{K^0 n\bar{\Lambda}}$.
In a similar way constructing a formal sum rule with $\bar{\Lambda}_{us}$ we obtain:

$$m_K^2 g_{\bar{K}^0 \Xi^0 \bar{\Lambda}} \frac{\lambda_\Lambda \lambda_\Xi M^2}{(M_\Xi^2 - M_\Lambda^2)}(e^{-M_\Lambda^2/M^2} - e^{-M_\Xi^2/M^2})(1 + A_{\Lambda\Xi}M^2) =$$
$$-RHS((g(\pi^-\Sigma^+\bar{\Lambda}_{us})_{us}) = RHS(\bar{K}^0\Xi^0\bar{\Lambda}) =$$
$$\sqrt{\frac{1}{6}}(3\mathcal{F}^{(-)}(K^0; M^2; s, d; u) - \mathcal{D}^{(-)}(K^0; M^2; u, d; s))$$
$$\rightarrow |_{exact \; SU(3)} \quad \sqrt{\frac{1}{6}}(3F - D), \qquad (20)$$

Upon interchange $(u \leftrightarrow d)$ one get the sum rule for the coupling constant $\bar{K}^- \Xi^- \bar{\Lambda}$.
Analogous sum rules can be constructed for Σ^0 coupling with kaon. First using Eq.(5) and Eqs.(13,16) we construct RHS of the sum rule involving Σ^0_{ds}:

$$-2 \cdot RHS(\pi^-\Sigma^+\bar{\Sigma}^0_{ds}) = RHS(\pi^-\Sigma^+\bar{\Sigma}^0) + \sqrt{3}RHS(\pi^-\Sigma^+\bar{\Lambda}) =$$
$$-\sqrt{2}\mathcal{F}^{(-)}(\pi^-; M^2; u, d; s) + \sqrt{2}\mathcal{D}^{(-)}(\pi^-; M^2; s, d; u) \qquad (21)$$

and then return to real Σ^0 with the 2nd transformation $(d \leftrightarrow s)$ changing Σ^+ to $-p$ and π^- to K^-:

$$2m_K^2 g_{K^-p\bar{\Sigma}^0} \frac{\lambda_\Sigma \lambda_N M^2}{(M_\Sigma^2 - M_N^2)}(e^{-M_N^2/M^2} - e^{-M_\Sigma^2/M^2})(1 + A_{\Sigma N}M^2) =$$
$$-2 \cdot RHS((\pi^-\Sigma^+\bar{\Sigma}^0_{ds})_{ds}) = 2 \cdot RHS(K^-p\bar{\Sigma}^0) =$$
$$-\sqrt{2}\mathcal{F}^{(-)}(K^-; M^2; u, s; d) + \sqrt{2}\mathcal{D}^{(-)}(K^-; M^2; d, s; u)$$
$$\rightarrow |_{exact \; SU(3)} \quad -\sqrt{2}(F - D), \qquad (22)$$

As the last one we construct sum rule for the formal quantity involving Σ^0_{us} to obtain finally

$$2m_K^2 g_{\bar{K}^0 \Xi^0 \bar{\Sigma}^0} \frac{\lambda_\Sigma \lambda_\Xi M^2}{(M_\Xi^2 - M_\Sigma^2)}(e^{-M_\Sigma^2/M^2} - e^{-M_\Xi^2/M^2})(1 + A_{\Sigma\Xi}M^2) =$$

$$-\sqrt{2}\mathscr{F}^{(-)}(K^0;M^2;s,d;u) - \sqrt{2}\mathscr{D}^{(-)}(K^0;M^2;u,d;s)$$
$$\to |_{exact \ SU(3)} \ -\sqrt{2}(F+D), \qquad (23)$$

Sum rules for other $g_{\bar{K}N\Sigma}$ and $g_{\bar{K}\Xi\Sigma}$ couplings are obtained with isotopic tramsformations.

SUMMARY AND RESULTS

Thus we have constructed QCD sum rules with the Lorenz structure $i\gamma_5$ for K meson - baryon coupling constants g_{KNY} and $g_{K\Xi Y}$, $Y = \Sigma, \Lambda$ starting from those for $g_{\pi\Sigma\Sigma}$ and $g_{\pi\Sigma\Lambda}$ ones. We have calculated (absolute values of) coupling constants of K-mesons to octet baryons. The results are presented in the Tables 1,2. In order to control our results we recalculate sum rules for π couplings to baryons obtaining values close to those of [6]. As the Lorenz structure $i\gamma_5$ was chosen mostly for methodical reasons the results do not pretend to account for real quantities [6].

The sum rules confirm a known result that unitary picture in terms of the D and F constants is not suitable for meson-baryon couplings due to large symmetry breaking. At the same time these sum rules when expressed in terms of the generalized functions $\mathscr{F}$ and $\mathscr{D}$ reveal indeed a simple $SU(3)_f$ pattern, and this is one of the main results we present here. The relations obtained here indicate in what way one can change and use the concept of the unitary symmetry in the framework of QCD sum rules.

ACKNOWLEDGMENTS

We are grateful to F.Hussain, B.L.Ioffe, G.Thompson for useful discussions. One of us (V.S.Z.) is grateful to to Abdus Salam ICTP (Trieste, Italy) for financial support. This work was supported in part by a Presidential grant N 1619.2003.2 for support of leading scientific schools.

REFERENCES

1. M. A. Shifman, V. I. Vainshtein, V. I. Zakharov, *Nucl. Phys.*, **B147**,385 (1979).
2. V.M.Belyaev, B.L.Ioffe, *JETP*, **56**, 493 (1982); B.L.Ioffe, Smilga, *Nucl.Phys.*, **B232**, 109 (1984); I. I. Balitsky and A. V. Yung, *Phys. Lett.*, **B129**, 328 (1983).
3. H.Sciomi and T.Hatsuda, *Nucl.Phys.* **A594**, 294 (1995).
4. M.C.Birse and B.Krippa, *Phys.Lett.* **B 373**, 9 (1996) *Phys.Rev.C*, **54**,3240 (1996).
5. T.Doi, H.Kim and M.Oka, *Phys.Rev.C*, **62**, 055202 (2000).
6. H.Kim, T.Doi, M.Oka, S.H.Lee, *Nucl.Phys.*, **A662**, 371 (2000); ibid.A678, 295 (2000).
7. T.M.Aliev, A.Ozpineci, and M.Savci, *Phys.Rev. D*, **64**, 034001 (2001).
8. A.Ozpineci, S.B.Yakovlev, V.S.Zamiralov, NPI MSU Preprint 2004-19/758.
9. S.Choe, M.K.Cheoun, and Su H.Lee, *Phys.Rev. C* , **53**, 1363 (1996).
10. S.Choe , *Phys.Rev. C*, **57**, 2061 (1998).
11. M.E.Bracco, F.S.Navarra and M.Nielsen, *Phys.Lett.*, **B454**, 346 (1999).
12. A.Ozpineci, S.B.Yakovlev, V.S.Zamiralov, *Mod.Phys.Lett.*, **A 20**,1 (2005); *Phys.Atom.Nucl.*, **68**, 279 (2005).

TABLE 1. The best-fitted values of the coupling constants g_{KNY}, $g_{K\Xi Y}$ and corresponding values of A_{NY}, $A_{\Xi Y}$ are given together with the Borel windows for each sum rule, $Y = \Lambda, \Sigma$

Coupling	Borel Window M^2, GeV^2	g	Ag, GeV^{-2}	A, GeV^{-2}
$\pi^0 pp$	1.0-1.4	$13.4/\sqrt{2}$	5.75	0.62
$\bar{K}^0 \Xi^0 \Lambda$	1.3-2.3	-1.35	0.8	-0.59
$K^- p\Sigma^0$	1.1-2.1	1.18	2.53	2.14
$\bar{K}^0 \Xi^0 \Sigma^0$	1.5-2.5	-3.09	-0.94	0.30

TABLE 2. The values of the coupling constants g_{KNY}, $g_{K\Xi Y}$, $Y = \Lambda, \Sigma$, of this work as well as of several recent works are given

Coupling	$|g|$ [11]	g [9],[10]	g, this work
$\pi^0 pp$	-	-	$13.4/\sqrt{2}$(input)
$KN\Lambda$	2.37±0.09	-3.47	0.77
$\bar{K}\Xi\Lambda$	-	-	-1.35
$KN\Sigma$	0.025±0.015	1.17	1.18
$\bar{K}\Xi\Sigma$	-	7.02	-3.09

Pentaquarks: the latest experimental results

M.Battaglieri*, R. De Vita*, V.Kubarovsky[†,**] and the CLAS Collaboration

Istituto Nazionale di Fisica Nucleare - Via Dodecaneso 33 16139 Genova ITALY
[†]*Rensselaer Polytechnic Institute - Troy New York 12180-359*
[**]*Jefferson Laboratory - 12000 Jefferson Avenue Newport News 23606 Virginia*

Abstract. After the claim of the possible discovery of a pentaquark state, many experiments reported positive and negative results opening a discussion about the pentaquark existence. New experiments with high resolution and high statistics are needed in the reaction channels and for the kinematics of the positive results to solve the controversy. Jefferson Lab started a comprehensive program to search for pentaquark in photoproduction at threshold on proton and deuteron targets, collecting more than 10 times the existing statistics. The first experiment on the proton ($g11$) just finished to analyze the data, and the first results of the pentaquark search are reported here.

Keywords: Pentaquark, photoproduction, proton target.
PACS: 13.60 Rj; 25.20 Lj

INTRODUCTION

All known hadronic matter is composed of two kinds of quark configurations: baryons, such as protons and neutrons, which are combinations of 3 quarks (qqq), and mesons, such as pions, which are combinations of a quark and an anti-quark ($q\bar{q}$). However, states of matter consisting of four quarks and an anti-quark ($qqqq\bar{q}$), called pentaquarks, are expected by the theory of Quantum Chromodynamics. In particular, states where the anti-quark has a different flavor with respect to the others have quantum numbers that are not allowed for standard particles and clearly reveal their 'exotic' nature. In the past, experimental searches focused on the search for pentaquark states, did not provide any clear evidence, leading to the conclusion that discoveries in this field would not be possible in the short term. As a consequence, this particular experimental program was abandoned and no further data analysis was pursued for many years. However, the lack of experimental proof left an open question about the physical manifestation of low-energy QCD. In spite of that, theoretical interest in this field has continued and, recently, Diakonov and collaborators [1] made definite predictions about the masses and widths of a decuplet of pentaquark states (the so-called "antidecuplet"). The most intriguing aspect of such a multiplet is the presence of three states with exotic quantum numbers: the Θ^+, $S = +1$, and the Ξ^{--} and Ξ^+ with $S = -2$. In particular the Θ^+ strangeness, never observed in the baryon sector and not compatible with a qqq state, requires at least a pentaquark configuration of the type $uudd\bar{s}$. The mass of the Θ^+ was predicted to be quite low (1540 MeV) and its widths very narrow ($\sim$10-15 MeV) implying that if such state exists it should be directly visible in the measured invariant masses without need for more sophisticated Partial Wave Analysis.

THE POSITIVE RESULTS

The first evidence of the Θ^+ was obtained in the photoproduction measurement performed by the LEPS Collaboration [2]. The signal was found in the reaction $\gamma n \to K^-\Theta^+ \to K^-K^+n$. The target was a CH_2 scintillator located just downstream of the primary hydrogen (LH_2) target. The signal was seen in the missing mass spectrum of the K^-, after cutting out sources of background from ϕ and $\Lambda(1520)$ production. Corrections to compensate for the neutron's Fermi motion in the carbon nucleus were applied. The final peak was found at a mass of 1540 ± 10 MeV, with a width less than 25 MeV, and a gaussian significance of 4.6 σ. This result was shortly thereafter confirmed by the CLAS Collaboration [3] who analyzed existing data on deuteron target looking at the channel $\gamma d \to K^-K^+np$. Events with the K^-K^+p in the final state were selected and the missing mass technique was used to identify the neutron. A peak was observed in the nK^+ invariant mass at a value of 1542 ± 5 MeV, with a width less than 21 MeV, and a statistical significance of more than 5 σ. Evidence of a Θ^+ candidate was also found in the reaction $\gamma p \to K^-\pi^+\Theta^+ \to K^-\pi^+K^+n$ [4]. DIANA Collaboration re-analyzed the old K-Xenon bubble-chamber data [5] ($K^+n \to (\bar{K}^0)\Theta^+ \to (\bar{K}^0)\pi^+\pi^-p$) and reported a narrow peak (~ 10 MeV), located at 1.54 GeV with a statistical significance of 4σ. The SAPHIR Collaboration reported about the reaction $\gamma p \to \bar{K}^0\Theta^+ \to K_S K^+(n)$ where a 5σ peak was isolated in the K_S missing mass. The peak was found to be at 1540 ± 5 MeV and a large production cross section of 300 nb was quoted in the paper [6]. Evidence of a narrow structure in the $(K_S p)$ system was also reported in neutrino and anti-neutrino collisions with nuclei [7] at CERN. A peak of less than 20 MeV width resonance was found at a mass of 1533 ± 5 MeV with a statistical significance of 6.7σ. A narrow 4σ peak in the same invariant mass spectrum, located at 1526 ± 3 MeV, was also found by the Hermes Collaboration [8] in the analysis of quasi-real photo production on a deuteron target. Another positive evidence was also found in pp scattering at few GeV beam energy (COSY-TOF Collaboration [9]). Positive results were then reported analyzing collisions of high energy proton beams against electron beam (ZEUS Collaboration [10]) and fixed targets (SVD-2 Collaborations [11]).

Are all these evidences enough to prove that the pentaquark does exist? The possible discovery of a state beyond ordinary matter triggered the interest of the scientific community, resulting in the publication of more than 500 related papers in the past two years.

THE NEGATIVE RESULTS

First doubts on the pentaquark existence, arise from the fact that, the mass of the Θ^+ measured in the different experiments were not compatible within the quoted errors. In particular, there is a systematic difference for the mass value found in the two possible decay modes: reported values were lower for $\Theta^+ \to pK^0$ than for $\Theta^+ \to nK^+$. The difference in the background or interference effects that systematically affect the two decay modes in a different way could explain the difference. One can simply assume that the quoted errors on the measured masses were underestimated but a possible indication of a serious concern about the Θ^+ existence should not be ignored.

In the past year reanalysis of data collected in high energy experiments [12, 13, 14, 15, 16, 17, 18, 19, 20, 21, 22, 23, 24, 25] show no evidence for pentaquarks. Data were collected in different laboratories (SLAC, DESY, CERN, FERMILAB, BNL) using a variety of different probes (e^+e^- colliders, heavy-ion and high-energy proton-proton and proton-nuclei interactions) and, for the most part of the results, analyzing a huge statistics.

Was it enough to prove that the pentaquark did not exist? The answer is not so simple. As an example we report the argument to exclude the significance of the e^+e^- experiments. It is well known that hadroproduction abundance in e^+e^- collisions as a function of the produced hadron mass has a very steep slope that depends on the number of quark-pairs that have to be generated from the vacuum. Pseudo-scalar mesons (one $q\bar{q}$ pair) are produced with a slope of $\sim 10^{-2}$/GeV while baryons (two $q\bar{q}$ pairs more) have $\sim 10^{-4}$/GeV. Following this logic, one can predict that pentaquarks (four $q\bar{q}$ pairs more) should have a slope of $\sim 10^{-8}$/GeV that makes the huge statistics collected by e^+e^- colliders, useless to draw any conclusion.

SECOND GENERATION EXPERIMENTS

In summary, twelve different laboratories, using different probes (photons, electrons, protons, neutrinos) and targets (protons, neutrons, nuclei) reported positive results. Some of them, with a high statistical significance. Limiting aspects of the so-called 'low energy' experiments were: the most part of the observed structures contained only few counts, all peaks did not show at the same mass, the background shape, necessary to define a possible signal over it, was not theoretically understood and the limited statistics prevented to derive a smooth curve directly from the data. Moreover many analyses used strong kinematic cuts to enhance the signal-to-background ratio that may distort the spectra and some experiment did not tag the strangeness of the final state.

Ten 'high energy' experiments reported null results analyzing high statistics data sets deriving stringent upper limits on the Θ^+ production but the different kinematic conditions, which likely involved dissimilar production mechanisms, the unknown relative weight of background reactions that could blind the small pentaquark production at high energies, and the different acceptance of the different detectors make very difficult direct comparisons of the results of the different experiments, preventing a definitive conclusion about the pentaquark's existence.

A second generation of dedicated experiments, optimized for the pentaquark search, was undertaken at Jefferson Lab. These experiments cover the few GeV beam energy region where most of the positive evidence were reported, with each collecting at least an order of magnitude more statistics than any of the previous measurements. The mass resolutions are approximately a few MeV and the accuracy of the mass determination is approximately 1-2 MeV, allowing precise determination of any possible narrow peaks in the decay distributions. All used real photon in different energy ranges and different targets. Two experiments ($g10$ and $eg3$) were in the range 1.0-3.0 GeV and 4.0-5.4 GeV respectively and used a liquid deuteron target while one experiment ($g11$) used a proton target with a photon beam of 1.6-3.8 GeV. The three experiments collected data and now they are completing the physics analyses.

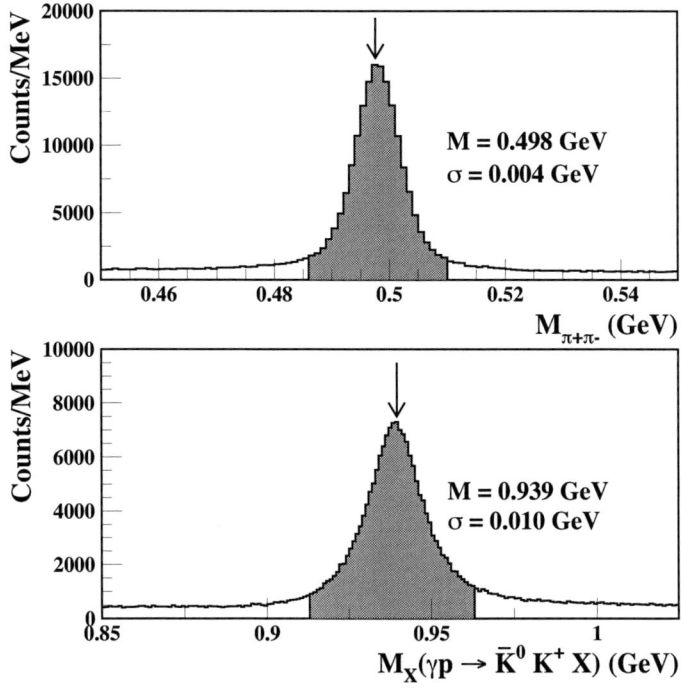

FIGURE 1. Bottom: missing mass for the reaction $\gamma p \to \bar{K}^0 K^+ X$ after $\bar{K}^0$ selection showing a peak at the neutron mass. The mass positions and widths of the measured peaks are given. For comparison, the arrows indicate the accepted value for the mass position. The shaded area corresponds to the events used in the analysis.

THE G11 EXPERIMENT AT JLAB

Here we present the first result of of the $g11$ experiment reporting the result of the exclusive measurement of the reaction $\gamma p \to \bar{K}^0 K^+ n$. The reconstruction of all participating particles (exclusive measurement) allows one to tag the strangeness of the reaction which clearly identifies the exotic nature of the baryon produced in association with the $\bar{K}_0$. This channel was previously investigated at ELSA by the SAPHIR collaboration in a similar photon energy range, finding positive evidence for a narrow Θ^+ state with M=1540 MeV and full width half maximum (FWHM) $\Gamma < 25$ MeV. They originally quoted a total production cross section of the order of 300 nb (later reduced to 50 nb [26]). For the first time, the new results put previous findings to a direct test.

The measurement was performed using the CLAS detector in Hall B with a bremsstrahlung photon beam produced by a continuous 60 nA electron beam of $E_0 = 4.0$ GeV impinging on a gold foil 8×10^{-5} radiation lengths thick. A bremsstrahlung tagging system with a photon energy resolution of 0.1% E_0 was used to tag photons in the energy range from $1.6 - 3.8$ GeV. A liquid hydrogen target was contained in a mylar cylinder cell 4 cm in diameter and 40 cm long. Outgoing hadrons were detected in the

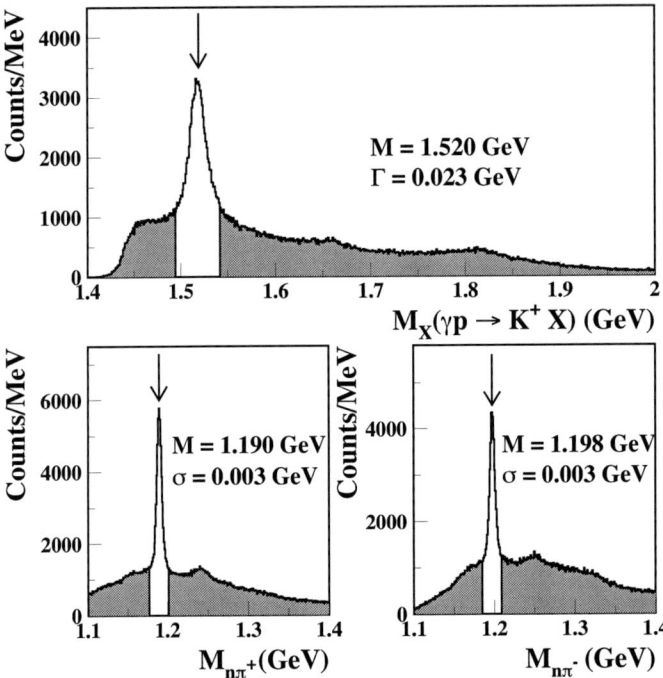

FIGURE 2. Top: K^+ missing mass distribution with the $\Lambda^*(1520)$ peak. Bottom: $n\pi^+$ (left) and $n\pi^-$ (right) invariant mass distributions with $\Sigma^+(1189)$ and $\Sigma^-(1197)$ peaks. The mass position and width of the measured peaks are indicated. For comparison, the arrows indicate the accepted value for the mass position. The shaded area corresponds to the events used in the analysis.

CLAS [27] spectrometer. Momentum information for charged particles was obtained via tracking through three regions of multi-wire drift chambers immersed in a toroidal magnetic field (~ 0.5 T), which was generated by six superconducting coils. The field was set to bend the positive particles away from the beam into the acceptance region of the detector. Time-of-flight scintillators (TOF) were used for hadron identification. The interaction time between the incoming photon and the target was measured by the Start Counter (ST), consisting of a set of 24, 2.2 mm thick plastic scintillators surrounding the hydrogen cell. The CLAS momentum resolution is of the order of 0.5-1% depending on the kinematics. The detector geometrical acceptance for each positive particle in the relevant kinematic region is about 40%. It is somewhat less for low energy negative hadrons, which can be lost at forward angles because they are bent out of the acceptance by the toroidal field. Coincidences between the photon tagger and the CLAS detector triggered the recording of the events. The trigger in CLAS was defined requiring the coincidence between the TOF system and the ST in at least two sectors. We took data for 50 days during June and July 2004 collecting more than 7G triggers corresponding to an integrated luminosity of 70 pb^{-1}. This is probably the highest statistics ever collected in experiments with tagged photons.

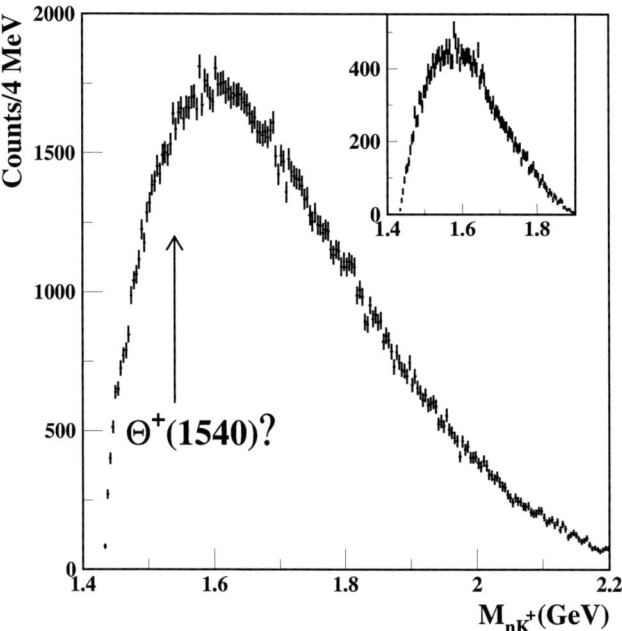

FIGURE 3. The nK^+ invariant mass distribution after all cuts. It is smooth and no narrow structures are evident. The arrow shows the position where evidence for the Θ^+ was found by previous experiments. The inset shows the nK^+ mass distribution with specific cuts to reproduce the SAPHIR analysis [6] as described in the text.

FIRST RESULTS

The reaction $\gamma p \rightarrow K^-\Theta^+ \rightarrow \bar{K}^0 K^+ n$ was isolated as follows. The K^+ was detected directly in the magnetic spectrometer, and the K^0 was reconstructed from its $\pi^+\pi^-$ decay. The momentum and energy of the neutron were reconstructed from the known incident photon energy and measurements of all other particles in the event. The quality of the channel identification is shown in Figure 1 where the K^0_S and the missing neutron peaks are seen above a small background. Reactions involving decay of hyperons, contribute to the same final state. The most sizable are: $\gamma p \rightarrow K^+ \Lambda^*(1520) \rightarrow K^+ \bar{K}^0 n$, $\gamma p \rightarrow \pi^- K^+ \Sigma^+$, and $\gamma p \rightarrow \pi^+ K^+ \Sigma^-$. Figure 2 shows the background hyperon peaks: $\Lambda^*(1520)$ in the K^+ missing mass spectrum and the Σ^+, Σ^- peaks in the $n\pi^+$ and $n\pi^-$ invariant mass spectra respectively. The mass region of each of these hyperon peaks was excluded from the final data set, as shown from the shaded regions in the same figure. While they represent a background to the pentaquark search, but easily removable in our analysis with a cut around their masses, they can be used as a check of the entire procedure, e.g. by extracting their production cross sections and comparing to the world data. After all cuts, the data sample contains approximately 0.17×10^6 events out of the

7×10^9 in the original data set. The resulting nK^+ invariant mass distribution is shown in Fig. 3. The distribution is smooth and structureless. In particular, no evidence for a peak or an enhancement is observed at masses near 1540 MeV, where signals associated with the Θ^+ were previously reported. To enhance a possible resonance signal not visible in the integrated distribution, we selected different center-of-mass angle intervals. Monte Carlo studies of the CLAS acceptance for this reaction showed that we could detect events over the entire angular range (0°-180°), with some reduction of efficiency at forward angles ($\theta_{\bar{K}^0}^{CM} < 30°$). No structures were found in the distribution when specific angular ranges were selected.

Since no signal was found, an upper limit for the Θ^+ production cross section in this reaction channel was extracted. The nK^+ mass distribution was fit using the maximum likelihood procedure to the sum of a narrow Gaussian function and a 5th-order polynomial that parameterizes, respectively, the Θ^+ contribution and a smooth background. The resonance position was varied from 1520 to 1600 MeV in 5 MeV steps while the width was fixed at 2.5 MeV, a value derived by Monte Carlo simulation for a long-lived resonance and following the same analysis chain used to process the data. The measured yields were then used to evaluate an upper limit on the Θ^+ yield using the Feldman and Cousins approach [28]. The upper limit on the number of Θ^+ events at the 95% confidence level was compared with the number of observed $\Lambda^*(1520)$'s shown in the upper panel of Fig. 2, which was estimated using a Breit-Wigner resonance shape fit. The ratio between the two yields is $\sim 220/100k = 0.22\%$. The upper limit on the yields was then transformed into an upper limit on the Θ^+ production cross section taking into account the luminosity of incident photons and target, the CLAS detection acceptance, the Θ^+ branching ratio to nK^+ of 50%, and several models for the production mechanism. The upper panel in Fig. 4 shows the upper limit on the total cross section as a function of the Θ^+ mass obtained in the most conservative scenario (forward peaked center-of-mass production). An upper limit of 0.8 nb was found for M=1540 MeV including a systematic uncertainty of 20% estimated by comparing different analysis procedures. The process to extract the yield described above was repeated for each angular bin to derive the 95% CL upper limit on the $\Theta^+(1540)$ differential cross section. The result is shown in the lower panel of Fig. 4. The cross section upper limit remains within about 1-2 nb for most of the angular range and rises at forward angles due to the reduced CLAS acceptance.

Our upper limit on the cross section is in clear disagreement with the findings of Ref.[6] which reported a Θ^+ signal of 55 events at a mass of 1540 MeV corresponding to the published total cross section of 300 nb. In order to better compare with that experiment, we repeated the analysis applying the same cuts reported in that paper: the photon energy was limited to 2.6 GeV, only events with a forward-emitted $\bar{K}^0$ ($\theta_{CM}^{\bar{K}^0} > 60°$) were used and the hyperons were not cut. The resulting mass distribution is shown in the inset of Fig. 2: it remains smooth and structureless.

CONCLUSIONS

Spurred by an initial report by the LEPS Collaboration in 2003, many experimental collaborations reported possible evidence for a pentaquark state. The observed signals

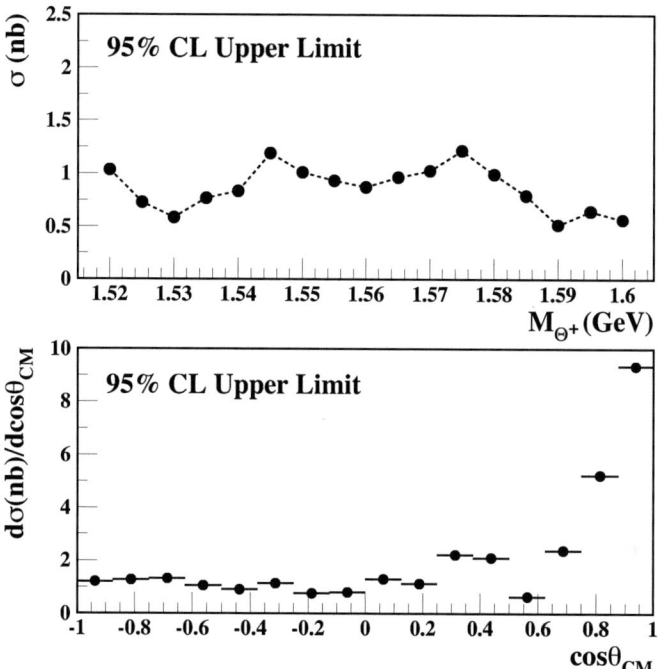

FIGURE 4. The 95% CL upper limit on the total cross section as a function of the Θ^+ mass (top) and on the differential cross section $d\sigma/d\cos\theta_{CM}^{\bar{K}^0}$ (bottom) for the reaction $\gamma p \to \bar{K}^0 \Theta^+$ for an assumed Θ^+ mass of 1540 MeV. The dotted line in the top plot is to guide the eye.

suffered from low statistics and were challenged by null results reported last year by many other experiments in different kinematic regions and reactions. Our analysis of a dedicated high-statistics and high-resolution measurement conducted under similar kinematic conditions and with more than ten times the statistics of a previously reported positive result found that there is no evidence for a Θ^+ pentaquark in $\bar{K}_0 K^+ n$ reaction channel setting an upper limit of less than 1nb on the production cross section. Analysis of other reaction channels on the new JLab data are still underway and expected to release results in a short time.

REFERENCES

1. D. Diakonov, V. Petrov and M. Polyakov, Z. Phys. A **359**, 305 (1997).
2. T. Nakano et al. (LEPS Collaboration), Phys. Rev. Lett. **91**, 012002 (2003).
3. S. Stepanyan et al. (CLAS Collaboration), Phys. Rev. Lett. **91**, 252001 (2003).
4. V. Kubarovsky et al. (CLAS Collaboration), Phys. Rev. Lett. **92**, 032001 (2004).
5. V. V. Barmin et al. (DIANA Collaboration), Phys. Atom. Nucl. **66**, 1715 (2003).
6. J. Barth et al. (SAPHIR Collaboration), Phys. Lett. B **572**, 127 (2003).
7. A.E. Asratyan, A.G. Dolgolenko, and M.A. Kubantsev, Phys. Atom. Nucl. **67**, 682 (2004).

8. A. Airapetian *et al.* (HERMES Collaboration), Phys. Lett. B **585**, 213 (2004).
9. M. Abdel-Bary *et al.* (COSY-TOF Collaboration), Phys. Lett. B **595**, 127 (2004).
10. S. Chekanov *et al.* (ZEUS Collaboration), Phys. Lett. B **591**, 7 (2004).
11. A. Aleev *et al.* (SVD Collaboration), Phys. At. Nucl. 68, 974 (2005) and hep-ex/0509033.
12. A. Aktas *et al.* (H1 Collaboration), Phys. Lett. B **588**, 17 (2004).
13. S. Schael *et al.* (ALEPH Collaboration), Phys. Lett. B **599**, 1 (2004).
14. B. Aubert *et al.* (BABAR Collaboration), arXiv:hep-ex/0408064.
15. K. Abe *et al.* (Belle Collaboration]), arXiv:hep-ex/0411005.
16. J. Z. Bai *et al.* (BES Collaboration), Phys. Rev. D **70**, 012004 (2004).
17. I. V. Gorelov (CDF Collaboration), arXiv:hep-ex/0408025; D. O. Litvintsev (CDF Collaboration), Nucl. Phys. Proc. Suppl. **142**, 374 (2005).
18. K. Stenson (FOCUS Collaboration), arXiv:hep-ex/0412021.
19. I. Abt *et al.* (HERA-B Collaboration), Phys. Rev. Lett. **93**, 212003 (2004); K. T. Knopfle, M. Zavertyaev and T. Zivko (HERA-B Collaboration), J. Phys. G **30**, S1363 (2004).
20. M. J. Longo *et al.* (HyperCP Collaboration), Phys. Rev. D **70**, 111101 (2004).
21. J. Napolitano, J. Cummings and M. Witkowski, arXiv:hep-ex/0412031.
22. S. R. Armstrong, Nucl. Phys. Proc. Suppl. **142**, 364 (2005).
23. C. Pinkenburg (PHENIX Collaboration), J. Phys. G **30**, S1201 (2004).
24. Y. M. Antipov *et al.* (SPHINX Collaboration), Eur. Phys. J. A **21**, 455 (2004).
25. M. I. Adamovich *et al.* (WA89 Collaboration), arXiv:hep-ex/0405042.
26. M. Ostrick, Prog. Part. Nucl. Phys. **55**, 337 (2005).
27. B. Mecking et al.,*Nucl. Instrum. and Meth.* **A503**, 513 (2003).
28. G. J. Feldman and R. D. Cousins, Phys. Rev. D **57** 3873 (1998).

Spin Physics Overview

Pasquale Di Nezza

INFN - Laboratori Nazionali di Frascati, via E.Fermi 40, I-00044 Frascati, Rome, Italy
E-mail: Pasquale.DiNezza@lnf.infn.it

Abstract. A selection of data obtained by the spin physics community is presented, which provides new insight into the QCD structure of the nucleon. New or more precise results and an improved theoretical understanding in several recent topics of interest in this field are now evident, especially in the emerging area of transversity distributions.

Keywords: Spin, Polarisation, Asymmetry, Transversity
PACS: 13.60.-r,13.88.+e,14.20.Dh,14.65.-q

Introduction

The particular issue how to perform measurements that disentangle the spin structure of the nucleon has received much attention in the last decades. Conceptually the spin of the nucleon can be decomposed into the spin of its constituents, in particular into the contribution coming from quarks, from gluons and from the total orbital angular momenta of the quarks and gluons, respectively. More recent treatments of data and theory provide a picture where terms like "transversity" must be included too for a more consistent picture of the present understanding of QCD.

Inclusive spin asymmetries

Precise results on the spin-dependent structure functions $g_1^p(x,Q^2)$ and $g_1^d(x,Q^2)$ were measured by HERMES while a new measurement of $g_1^d(x,Q^2)$ from COMPASS extends these results to lower values of $x-Bjorken$ and higher Q^2. HERMES data show the most complete dataset on g_1 including results from proton, deuterium and neutron targets, Fig.1. The latter comes both from the ^3He measurement and from the difference of the proton data from the deuterium data. These results adopt a multidimensional unfolding to take into account both radiative background and intra-bin migration. The technique implies the elimination of the systematic correlation between different kinematic bins [1]. COMPASS combined the 2002 and 2003 data taking for an integrated luminosity of $\sim 1.5\ nb^{-1}$ covering a Q^2 range up to 100 GeV^2. Data significantly improve statistical accuracy in the low x region (down to 10^{-3}) and does not show the tendency of the SMC data of negative g_1^d values [2], Fig.2.

The first measurement of the tensor asymmetry A_{zz} and the tensor structure function $b_1^d(x,Q^2)$ [3] by HERMES shows that, as a nuclear-polarized deuteron target always carries a large tensor polarization, it is a priori not justified to neglect the effect of the tensor asymmetry in g_1^d measurements. Here, b_1 represents the difference in the quark distribution between the helicity-0 and the averaged non-zero helicity states of the deuteron.

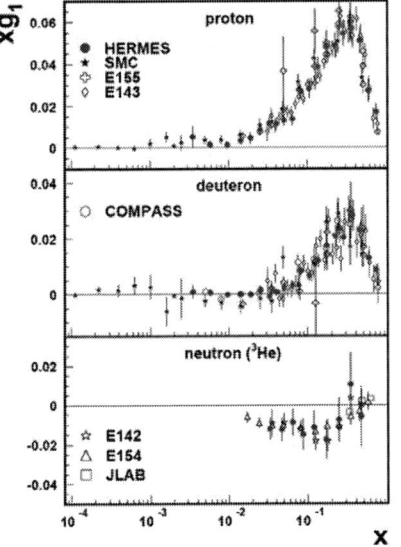

Our understanding of the high x region has improved with new precise data from experiments at Jefferson Laboratory. Results on A_1^n [4] show, for the first time, a significant positive value for high x, as would be expected with $SU(6)$ breaking, Fig.2. In summary, although the statistical accuracy is definitely good enough to allow a detailed analysis of the overall behavior of the helicity structure functions, there still remain regions at high and low x where our present knowledge is insufficient to distinguish among various models.

1: HERMES xg_1 data for the proton, deuteron and the neutron at the measured Q^2 and compared to world data.

Flavour decomposition

By means of the technique of flavour tagging, individual spin contributions can be determined directly from spin asymmetries of hadrons with the appropriate flavour content. The measured semi-inclusive photon-nucleon asymmetry, A_1^h, for the hadron of type h is related to the quark and antiquark polarisations $\Delta q_f(x)$.

HERMES has performed, for the first time, a global analysis of the inclusive spin asymmetries and semi-inclusive spin asymmetries for π^+, π^-, K^+ and K^-, measured for longitudinally polarised targets of hydrogen, and deuterium [5]. The results of the decomposition are presented in Fig. 3 and compared to LO QCD analyses [6, 7]. In

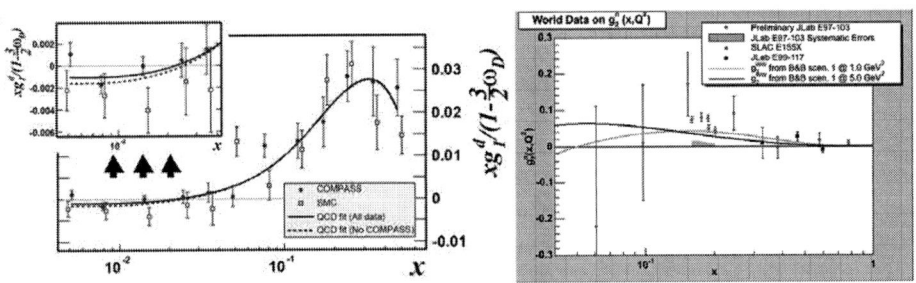

FIGURE 2. Right: COMPASS xg_1 data for deuterium including a QCD fit to world data. Left: Data for $g_2(x,Q^2)$ structure function on neutron.

the analysis $\Delta \bar{s} = 0$ is assumed. The results show that $x\Delta u(x)$ is positive and $x\Delta d(x)$ is negative, and that both reach their maximum in magnitude in the valence quark region. The polarised sea quark distributions are all consistent with zero, and show a hint for a small positive s-quark polarisation.

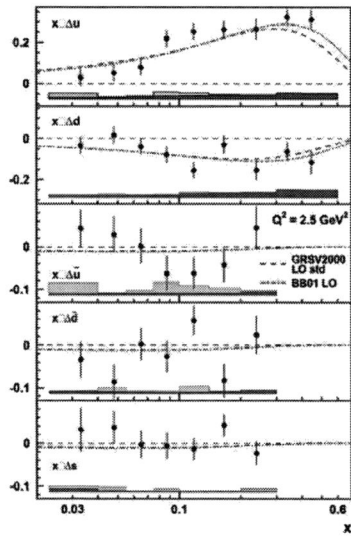

FIGURE 3. Quark helicity distributions $x\Delta q(x, Q_0^2)$ by HERMES evaluated at a common value of $Q_0^2 = 2.5$ GeV^2 as a function of x. The curves show LO QCD analyses of polarised inclusive data.

Transverse asymmetries

The trasversity distribution $h_1(x)$ completes the mapping of the spin structure of the nucleon at leading twist. Several features (i.e. the chiral-odd nature) makes transversity an elusive subject for detailed experimental investigations. As is well known, two main mechanisms affect azimuthal single-spin asymmetries on transversely polarized targets. The so-called Collins effect is probably the most representative of the different observables involving trasversity. Here a chiral-odd fragmentation function allows the observation of chiral-odd $h_1(x)$, which is otherwise impossible to access directly in a DIS process. On the other hand, a correlation between the intrinsic transverse momentum of an unpolarized quark and the direction of the transverse spin of its parent nucleon can exist and is described by the naïve t-odd Sivers distribution function $f_{1T}^\perp(x)$. Probably the most interesting features is that the latter requires a non-zero orbital angular momentum of the unpolarized quark.

HERMES, after the publication [8] of the first evidence for azimuthal single spin asymmetries in the semi-inclusive production of charged pions on transversely polarized target, has shown even more significant signals for both Collins and Sivers mechanism analyzing the data collected during the period 2002-2004 with a pure hydrogen target. In Fig. 4 asymmetries are shown as a function of x, the hadron momentum fraction

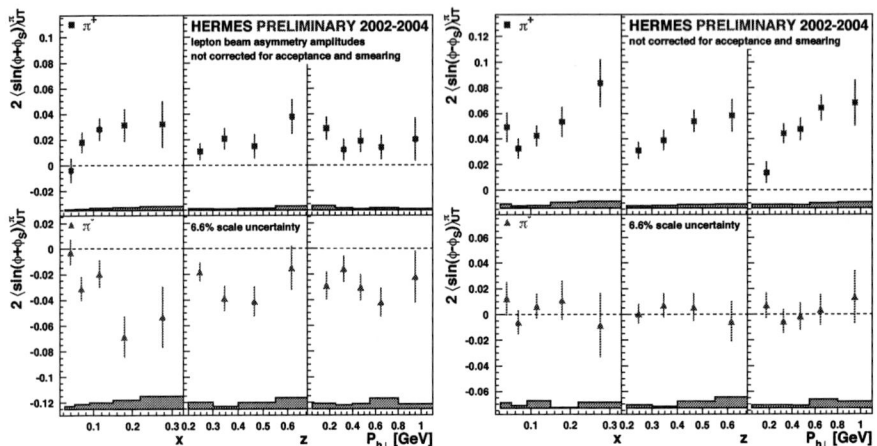

FIGURE 4. HERMES virtual-photon Collins (left) and Sivers (right) moments for charged pions as a function of x, z and $P_{h\perp}$.

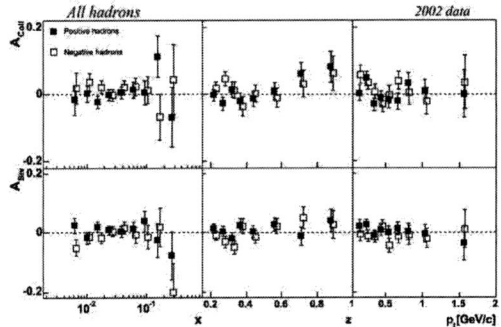

FIGURE 5. COMPASS Collins and Sivers asymmetries for positive (full points) and negative (open points) hadron as a function of x, z and $P_{h\perp}$.

z and the hadron transverse momentum $P_{h\perp}$. The average Collins moment is positive for π^+ and negative for π^-. Surprisingly, the magnitude of the π^- moment appears to be at least as large as that for π^+. The average Sivers moment is significantly positive for π^+. This shows the first measurement of naive T-odd function in DIS which also requires a non-zero orbital angular momentum of quarks inside the nucleon. For π^- the averaged Sivers moment is consistent with zero. COMPASS has devoted 20% of the 2002 running time to transversely polarized ^6LiD (deuteron) target. In Fig. 5 are shown results for positive and negative charged hadrons selecting all hadrons [9]. Within the accuracy of the measurement, both the Collins and Sivers asymmetries turned out to be compatible with zero. This result indicates a cancellation of the contributions of u and d quarks in case of a deuteron target [10]. In 2006 the COMPASS collaboration plans to

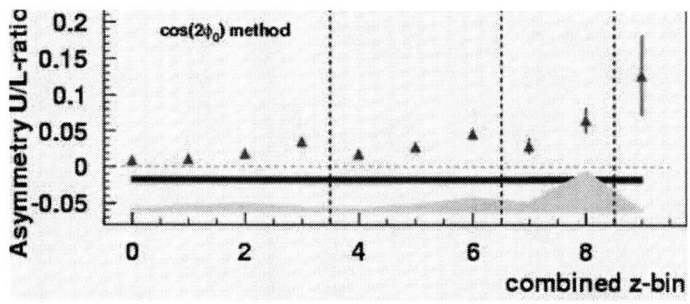

FIGURE 6. BELLE results for the asymmetry related to the Collins function.

take data on a transversely polarized proton target (NH_3).

A complementary way to approach transversity has been investigated by HERMES and COMPASS by measuring single spin asymmetries in two-hadron production in semi inclusive DIS. Although this method comes at the expense of a larger statistical uncertainty, it has the advantage to relate, at leading twist, directly to the product of h_1 and the fragmentation function, whereas the single-hadron method shows a convolution of that product with the transverse momentum of the hadron. Moreover the two-hadron measurement involves a completely different fragmentation function as compared to the single-hadron one providing an independent method of measuring transversity. Specifically the fragmentation functions involved describe the interference between different wave states of the pion pair. HERMES has measured this asymmetry on transversely polarized hydrogen target in the invariant mass region $0.51 < M_{\pi^+\pi^-} < 0.97$ GeV, showing a significant non zero results. The asymmetry is clearly positive over the mentioned entire invariant mass range and largest in the region of the ρ^0 mass. At the end of 2005 HERMES expects to collect a full data sample to lead to a decrease of the uncertainty on the asymmetry with approximately a factor $\sqrt{2}$ allowing a multi-dimensional analysis studying the x and z dependence. COMPASS has investigated the same asymmetry choosing hadron pairs selecting all combinations of positive and negative hadrons showing results compatible with zero over the whole kinematic range. COMPASS expects new results including hadron identification using RICH information resulting in a cleaner hadron sample and a complementary measurement on a proton target for the planned run in 2006.

As previously explained, due to the chiral-odd nature, both the transversity and the Collins fragmentation function cannot be measured directly in DIS processes. The results from HERMES and COMPASS are a convolution of these two objects. However, in order to extract transversity distributions one needs that particular fragmentation function (FF) to be precisely known. BELLE, using hadron production from e^+e^- interactions, has shown the first signature of the Collins FF using a combined measurement of a quark and an anti-quark distribution. In fact, the number density for producing two unpolarized hadrons from transversely polarized quarks contains two times the chiral-odd FF, resulting in a chiral-even object accessible in electromagnetic interaction directly. A clear signature is visible, with the data also showing increase with rising fractional energy z, Fig. 6.

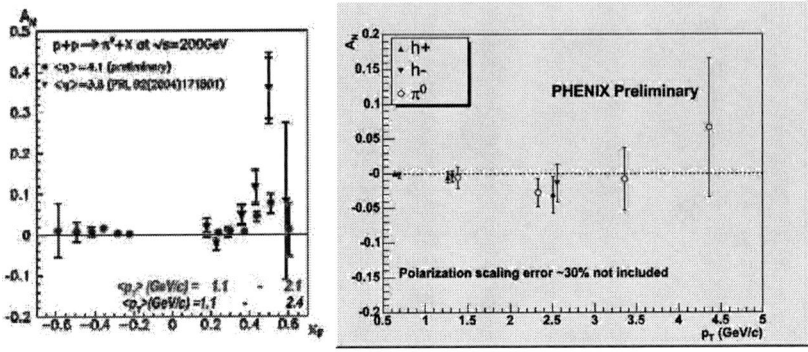

FIGURE 7. STAR (left) measurement of single spin asymmetry for π^0. PHENIX (right) inclusive asymmetry for π^0 and charged hadrons vs. $p_\perp$ at $x_F=0$.

In polarised hadron-hadron scattering, the first non zero single spin asymmetries were observed in pion production by E704 [11]. They have generated much interest since leading-twist collinear factorized pQCD predicts these asymmetries should be small. Such asymmetries have been shown to persist even at high energy of RHIC, Fig. 7. However the asymmetries receive contribution from Collins and Sivers mechanism, as well as contributions of the higher twist initial-state or final state interactions.

In general, the performance of the polarized RHIC collider has been steadily improving. Transverse polarization data sets from 2003 and 2004 were both obtained with about half Picobarn^{-1} of beam. The average proton beam polarization in 2003 was about 20%. In 2005, online polarization values in excess of 50% have been seen. Moreover a luminosity of about one order of magnitude greater than the one used to extract the presented results have been collected. A much higher statistics measurement of these transverse asymmetries can be expected in the near future. STAR has shown the first measurements at $\sqrt{s}=200$ GeV indicating that the π^0 production exhibits a large single spin asymmetry which increases rapidly for x_F above about 0.3 [12]. With improving forward calorimetry working in the rapidity range $-1<\eta<4$, used in conjunction with the existing more central STAR detector system, this experiment will be able to distinguish the signatures of the Collins and Sivers effects. PHENIX, with a limited dataset of 0.15 pb^{-1} had calculated the transverse single spin asymmetry for π^0 in the central arm of the spectrometer ($|\eta|<0.35$) and non-identified charged hadrons at $x_F=0$ and transverse hadron momentum up to $p_\perp=5$ GeV, Fig. 7. In the results there is an overall scale uncertainty of $\pm 35\%$ from uncertainty in the absolute polarization, and an absolute uncertainty of 0.2% on all points. The asymmetry at mid-rapidity is consistent with zero, to the few percent level, for all $p_\perp$ for both the π^0 and charged hadrons. In a near future PHENIX will access the measurement of back-to-back di-hadron azimuthal correlation to decouple contributions to SSA from the Sivers function. The BRAHMS experiment is well suited to study single spin asymmetries for identified hadrons at moderate x_F because of the PID coverage up to momenta of 40 GeV and the option to measure at $\eta \simeq 4$. The π^+ measured asymmetries are positive while the π^- are negative i.e. the same sign and magnitude as seen in the E704 data at lower energy. In addition the protons are found to have an asymmetry consistent with zero. BRAHMS has also compared the data with extrapolations of twist 3 (initial state) calculations [17]. The pQCD is apriori not valid at the lower values of $p_\perp$ covered in the presented measurement. Nevertheless it gives a good estimate how kinematic cuts may effect predictions as to give rise to a near constant asymmetry in a limited range of x_F. Both the magnitude and the x_F dependence is in reasonable agreement with the data. All the RHIC data show the clear potential to provide an independent means of determining the transversity distribution.

Besides the transversity distribution h_1 itself, there are other objects describing the transverse polarization of quarks but this time in longitudinally polarized targets. Hall B at Jefferson Lab has calculated single-spin asymmetries in semi-inclusive electroproduction of pions with a polarized NH_3 target using the CEBAF 6 GeV polarized electron beam [13]. In particular for such kind of process the only azimuthal asymmetry arising in leading order is the $sin2\phi$ moment involving the transverse momentum dependent Collins fragmentation function and the Mulders distribution function $h_{1L}^\perp$. The data for π^+ show a clear $sin\phi$ and $sin2\phi$ modulations and the x-dependence is consistent with the expectation.

Gluon polarization

Another important topic addressed in the talk was the measurement of the gluon polarization and the new results shown by PHENIX and COMPASS, Fig. 8. The PHENIX experiment has started making measurement of double spin asymmetries that relate to the polarised gluon distribution. The measurements were done for π^0 production in longitudinally polarised proton-proton collisions. For this data set PHENIX has reported the unpolarised cross section for π^0 production which is described extremely well by the next-to-leading-order pQCD calculations over eight orders of magnitude. The same calculations proved the estimation of the relative contribution of the gluon-gluon and gluon-quark subprocess to the double spin asymmetries necessary to extract the gluon polarisation in polarised gluon distribution in a near future. The data are consistent with $\Delta G/G=0$, albeit with large errors. Due to the uncertainties of theoretical nature, it is still not possible to rule out large values of gluon polarization. Significant results will be available soon, with improved statistics and the beam polarization.

A complementary way to measure the gluon polarization has been used by HERMES, SMC and COMPASS based on the helicity asymmetry of the photon-gluon fusion process (PGF). The theoretically cleanest signal of PGF is open charm production (i.e. by D^0 or D^* production). However, it is experimentally hard to access due to the small cross section and the experimental capabilities for charm detection. Experimentally cleaner is the selection of the PGF process by spin asymmetry of events for which a pair of large transverse momentum hadrons is produced. However this asymmetry has many competitive processed which are estimated using a Pythia Monte Carlo and are subtracted. This introduces large systematics which substantially decreases the significance of the results. New results from Compass are shown together with earlier measurements of SMC and HERMES in Fig. 8. Analysing both the kinematic regions for $Q^2<1$ and $Q^2>1$ GeV^2 data show no significant gluon polarization for an average x_g=0.095. An agree-

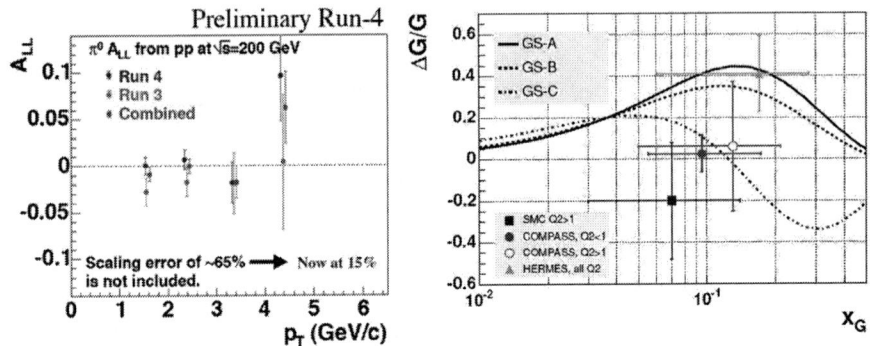

FIGURE 8. The PHENIX (left) measurement of double spin asymmetry in π^0 production in polarised pp collisions. The COMPASS extraction (right) of $\Delta G/G$ for different Q^2 range in comparison with the HERMES and SMC results.

ment in that specific x_g, at the level of 1.5 σ, has been shown with the models [14, 15]. It is worth mentioning that the large positive gluon polarization shown previously by HERMES [16] using the same method is not in contradiction with the COMPASS one because of the different kinematic region where this result was extracted and because it suffers of the same large uncertainties from theoretical modeling.

Hard exclusive processes

An exciting new field is opened with measurements of hard exclusive production of mesons and real photons (Deeply Virtual Compton Scattering). The parton correlation functions (known as Generalised Parton Distributions, GPDs) accessible in these processes are related to the total angular quark momentum contribution in the nucleon [18]. Thus, the orbital angular momentum of the quarks L^q may become accessible.

HERMES has measured the total cross section for exclusive π^+ production shown in Fig. 9 as function of Q^2 for three different x-ranges. These preliminary data have not yet been corrected for radiative effects which have been roughly estimated to be as large as 20% with little dependence on x or Q^2. The data is compared to calculations for the longitudinal part of the cross section computed from a GPD-model [19].

Hard exclusive electroproduction of a real photon appears to provide the theoretically cleanest access to GPDs. DVCS amplitudes can be determined through a measurement of the interference between the DVCS and Bethe-Heitler processes, in which the photon is radiated from a parton and from the lepton, respectively. Measuring the ϕ-dependence of a cross section asymmetry with respect to the charge (spin) of the lepton beam provides information about the real (imaginary) part of the DVCS amplitude.

HERMES has already measured azimuthal asymmetries with respect to the beam helicity [20]. Fig. 10 shows the first measurement of the beam-charge asymmetry A_C from hydrogen and deuterium targets. The left hand side presents A_C for the proton as function of the azimuthal angle ϕ. As the two beams have different average polarisations,

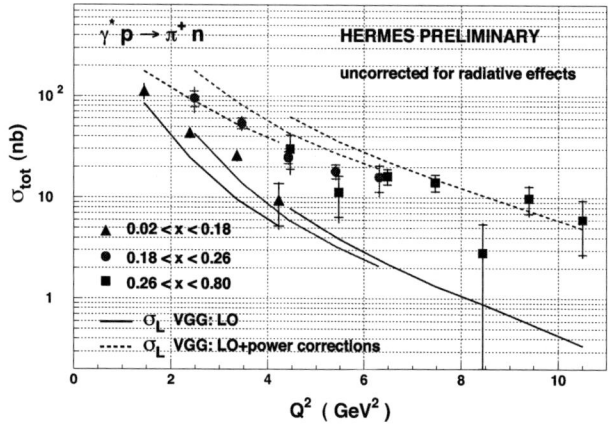

FIGURE 9. HERMES total cross section for exclusive π^+ production as function of Q^2 for three different x ranges, integrated over t. The curves represent calculations based on a GPD-model.

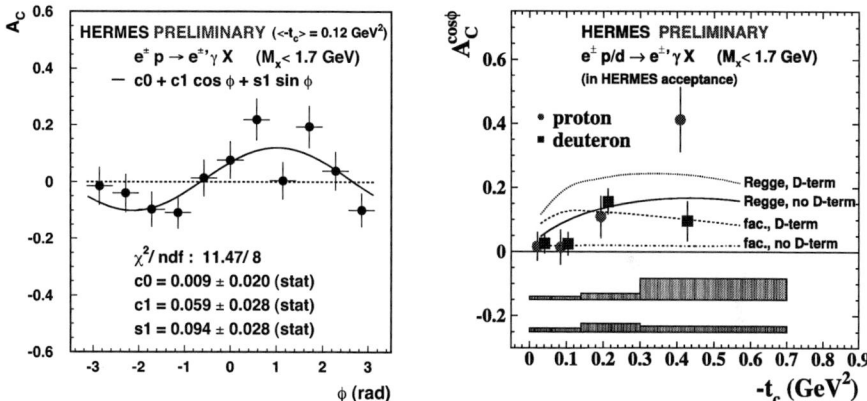

FIGURE 10. Left: HERMES Beam-Charge Asymmetry (BCA) for hard electroproduction of photons off the proton as function of the azimuthal angle ϕ. Right: HERMES $\cos\phi$ amplitude of the BCA for proton and deuterium as function of $-t$. The curves represent model calculations based on different GPD parameterisations.

also the $\sin\phi$-dependent part of the interference term contributes, and the asymmetry is fitted with the function $P_1 \cos\phi + P_2 \sin\phi$. The right hand side of Fig. 10 shows the $\cos\phi$ amplitudes derived from the two-parameter fits and corrected for background as function of the four-momentum transfer to the target $-t$ for the proton and the deuteron. The data are compared to GPD-based model calculations [19] assuming a factorised or Regge-inspired t-dependence with and without the D-term contribution [21]. It is apparent that measurements of the beam-charge asymmetry have high sensitivity to different GPD parameterizations. Data cannot yet distinguish between different model assumptions, however new data collected with electrons in 2005 will improve the statistics by at least a factor of 5.

Conclusions

The physics goals in the field of spin physics were presented. The evolution of new topics and the steady improvements to topics that have been studied previously show the ability of this field to explore missing elements and address new theoretical concepts. Our understanding of the spin structure of the hadronic matter continues to improve and we are looking forward to an exiting future.

REFERENCES

1. HERMES Coll., A. Airapetian et al., Phys. Rev. D71 (2005) 012003.
2. COMPASS Coll., Phys. Lett. B612 (2005) 154.
3. HERMES Coll., A. Airapetian et al, submitted to Phys. Rev. Lett., hep-ex/0506018.
4. X.Zeng et al., Phys. Rev. Lett. 92 (2004) 012004; Phys. Rev. C70 (2004) 065207.
5. HERMES Collab., A. Airapetian et al., Phys. Rev. Lett. 92, 012005 (2004); Phys. Rev D 71 (2005) 012003.
6. M. Glück, E. Reya, M. Stratmann, W. Vogelsang, Phys. Rev. D63, 094005 (2001).
7. J. Blümlein and H. Böttcher, Nucl. Phys. B636, 225 (2002).
8. HERMES Coll., A. Airapetian et al., Phys. Rev. Lett. 94 (2005) 012002; Phys. Lett. B622 (2005) 14-22.
9. COMPASS Coll, V.Yu.Alexakhin et al., Phys. Rev. Lett. 94 (2005) 202002.
10. M.Anselmino et al., Phys. Rev. D71 (2205) 074006.
11. D.L.Adams et al., Phys. Rev. D53 (1996) 4747.
12. STAR Coll., J.Adams et al., Phys. Rev. Lett. 92 (2004) 171801.
13. CLAS Coll., H.Avakian et al., Phys. Rev. D69 (2004) 112004.
14. M.Hirai et al., Phys. Rev D69 (2004) 054021.
15. E.Leader et al., Eur. Phys. J. C23 (2002) 479.
16. HERMES Coll.,A. Airapetian et al., Phys. Rev. Lett. 84 (2000) 2584.
17. J.Qiu and G.Sterman, Phys. Rev. D59 (1998) 014064.
18. X. Ji, Phys. Rev. Lett. 78, 610 (1997); Phys. Rev. D55, 7114 (1997).
19. M. Vanderhaegen, P.A.M. Guichon, M. Guidal, Phys. Rev. D60, 094017 (1999).
20. HERMES Collab., A. Airapetian et al., Phys. Rev. Lett. 87, 182001 (2001).
21. M. Polyakov, C. Weiss, Phys. Rev. D60, 114017 (1999).

Higgs boson production via gluon fusion at hadron colliders

Stefano Catani

INFN, Sezione di Firenze and Dipartimento di Fisica, Università di Firenze, I-50019 Sesto Fiorentino, Florence, Italy

Abstract. We consider Higgs boson production, through gluon fusion, in hadron collisions at the Tevatron and the LHC. We present a brief summary of the calculations of QCD radiative corrections to this process. We show available theoretical predictions for the total cross section and the transverse-momentum distribution of the Standard Model Higgs boson. The perturbative QCD uncertainty on these predictions is discussed.

1. INTRODUCTION

Within the Standard Model (SM) of electroweak interactions, the Higgs boson is responsible for the mechanism of the electroweak symmetry breaking. This particle has so far eluded experimental discovery. The next search for Higgs boson(s) will be carried out at hadron colliders, namely, the Fermilab Tevatron [1, 2] and the CERN LHC [3, 4].

The main production mechanism of the SM Higgs boson H at hadron colliders is the gluon fusion process $gg \to H$, through a heavy-quark (mainly, top-quark) loop. When combined with the decay channels $H \to \gamma\gamma$, $H \to WW$ and $H \to ZZ$, this production mechanism is one of the most important for Higgs boson searches and studies over the entire mass range, $100 \text{ GeV} \lesssim M_H \lesssim 1 \text{ TeV}$, to be investigated at the LHC [3]. The gluon fusion mechanism is also important for Higgs boson searches at the Tevatron: in the mass range $140 \text{ GeV} \lesssim M_H \lesssim 180 \text{ GeV}$, the process $gg \to H \to WW \to \ell^+\ell^-\nu\bar{\nu}$ can be exploited as main discovery channel [1], provided the background from $t\bar{t}$ production is suppressed by applying a veto cut on the transverse momenta of the jets accompanying the final-state leptons.

To fully exploit the physics potential of the gluon fusion process, it is relevant to provide reliable theoretical predictions for the corresponding total cross section and for the associated distributions, such as, for instance, the Higgs transverse-momentum (q_T) distribution. The dominant source of theoretical uncertainties on these quantities is the effect of QCD radiative corrections, which, therefore, have to be carefully investigated.

2. TOTAL CROSS SECTION

In QCD perturbation theory, the leading order (LO) contribution to the total cross section for Higgs boson production by gluon fusion is proportional to α_S^2, α_S being the QCD coupling. The next-to-leading order (NLO) QCD corrections to this process were found to be large [5, 6]: their effect increases the LO cross section by about 80–100%, thus leading to very uncertain predictions. Approximate evaluations [7] of higher-order terms suggested that they could still be sizeable.

Recent years have witnessed much theoretical progress [8]–[17] on the computation of QCD radiative corrections beyond the NLO. The total cross section has been computed at the next-to-next-to-leading order (NNLO) [12]. The Higgs boson rapidity distribution has been computed at the NLO [14] and at the NNLO [15, 16]. The effects of a jet veto on the NLO and NNLO cross sections have been studied [11, 13, 15]. Note that all these results [8]–[17] use the large-M_t approximation, M_t being the mass of the top quark (the validity of this approximation is discussed below).

A key point of this theoretical activity is that the QCD dynamics of the gluon fusion mechanism has been understood. The origin of the dominant perturbative contributions to the total cross section has been identified: the bulk of the radiative corrections is due to virtual and soft-gluon terms [7, 9, 10, 11]. On the contrary, as shown and discussed in Sects. 4.1 and 5.1 of Ref. [17] (see, in particular, Figs. 1 and 2 in [17]), the contribution from hard-gluon radiation to the total cross section is marginal: it is of $\mathcal{O}(10\%)$ and of $\mathcal{O}(1\%)$ at NLO and NNLO, respectively.

The dominance of virtual and soft-gluon terms has a twofold relevance. On one side, it explains the observation [7] of the validity of the large-M_t approximation (M_t is certainly much larger than the energy scale of the soft gluons that mainly contribute to the QCD radiative corrections) in the calculation at the NLO, and, therefore, it justifies the use of the same approximation at and beyond the NNLO. On the other side, it allows to estimate higher-order QCD contributions by supplementing the NNLO calculation with an all-order resummation of the logarithmically-enhanced terms due to multiple soft-gluon emission. Having these terms under control allows us to reliably predict the value of the cross section and, more importantly, to reduce the associated perturbative uncertainty.

The general method to systematically resum classes of logarithmically-enhanced soft-gluon contributions to hadroproduction cross sections is known [18, 19].

In Ref. [17], the total cross section at NNLO is supplemented with the resummation of the soft-gluon contributions up to next-to-next-to-leading logarithmic (NNLL) accuracy. The NNLO and NNLL cross sections at the LHC are reported in Fig. 1. The central curves are obtained by fixing the factorization (μ_F) and renormalization (μ_R) scales at the default value $\mu_F = \mu_R = M_H$. The bands are obtained by varying μ_F and μ_R simultaneously and independently in the range $0.5 M_H \leq \mu_F, \mu_R \leq 2 M_H$ with the constraint $0.5 \leq \mu_F/\mu_R \leq 2$. Tables with detailed numerical values of Higgs boson cross sections (using also different sets of NNLO parton densities) at the LHC and at the Tevatron can be found in Ref. [17]. The NNLL cross sections are larger than the NNLO ones; the increase is of about 6% at the LHC and varies from about 12% (when $M_H = 100$ GeV) to about 15% (when $M_H = 200$ GeV) at the Tevatron. For comparison, we recall that the NNLO results increase the NLO results by about 20% at the LHC and about 40% at the Tevatron. At the LHC, the NNLO scale dependence ranges from about $\pm 10\%$ when $M_H = 120$ GeV, to about $\pm 9\%$ when $M_H = 200$ GeV. At NNLL order, it is about $\pm 8\%$ when $M_H \lesssim 200$ GeV. At the Tevatron, when $M_H \lesssim 200$ GeV, the NNLO scale dependence is about $\pm 13\%$, whereas the NNLL scale dependence is about $\pm 8\%$. It is important to observe that the NNLO and NNLL bands overlap. Furthermore, the central value of the NNLL bands lies inside the corresponding NNLO bands. This gives us confidence in using scale variations at NNLO and at NNLL order to estimate the theoretical

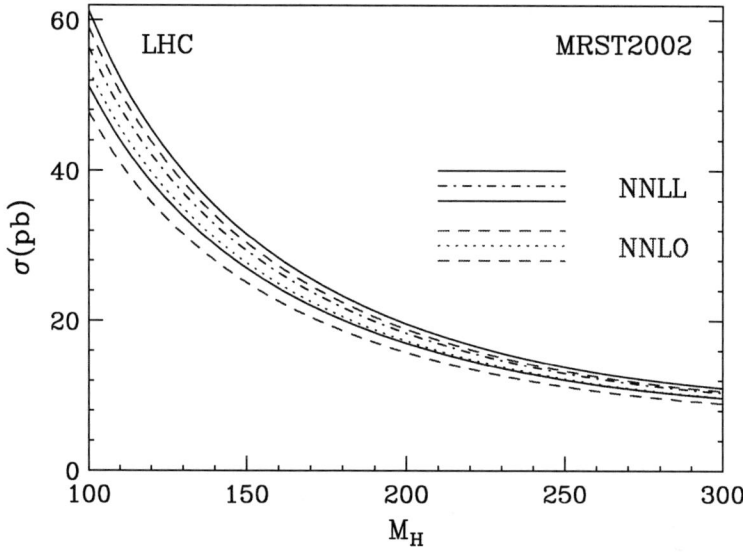

FIGURE 1. NNLL and NNLO cross sections at the LHC, using MRST2002 [20] parton densities.

uncertainty due to uncalculated perturbative terms at still higher orders. Another method to estimate the size of higher-order corrections is to compare the results at the highest order that is available with those at the previous order. Considering the differences between the NNLO and NNLL cross sections, we obtain results that are consistent with the uncertainty estimated from scale variations.

In summary, after inclusion of both NNLO corrections and soft-gluon resummation at the NNLL level, we can reliably predict the Higgs boson production cross section. In the low-mass range ($M_H \lesssim 200$ GeV), the estimated theoretical uncertainties of perturbative QCD origin are below 10% [17].

The NNLL+NNLO results are nicely confirmed by the recent computation [21] of part of the N^3LO and N^3LL perturbative contributions (see also Ref. [22, 23]).

3. TRANSVERSE-MOMENTUM DISTRIBUTION

When studying the q_T distribution of the Higgs boson in QCD perturbation theory, it is convenient to start by considering separately the large-q_T and small-q_T regions.

The large-q_T region is identified by the condition $q_T \sim M_H$. In this region, the perturbative series is controlled by a small expansion parameter, $\alpha_S(M_H^2)$, and calculations based on the truncation of the series at a fixed order in α_S are theoretically justified. SM Higgs boson production at large q_T via gluon fusion has to be accompanied by the radiation of at least one recoiling parton, so the LO term for this observable is proportional to α_S^3. The LO calculation was reported in Ref. [24]; it shows that the large-M_t approximation works well as long as $M_H \lesssim 2M_t$ and $q_T \lesssim M_t$. Similar results on the validity of the large-M_t approximation were obtained in the case of the associated production of a Higgs boson plus 2 jets (2 recoiling partons at large transverse momenta) [25]. In the framework of the large-M_t approximation, the NLO QCD corrections to the transverse-momentum distribution of the SM Higgs boson were computed in Refs. [26]–[29]. Corrections to the large-M_t approximation are considered in Ref. [30]. The numerical programs of Refs. [26, 29] can also be used to evaluate arbitrary infrared- and collinear-safe observables up to NLO in the large-q_T region and, in the case of Ref. [29], up to NNLO when $q_T = 0$.

At large values of q_T, the NLO calculation increases the LO result by about 60%. This implies that further higher-order terms (e.g. the NNLO term) in QCD perturbation theory have to be evaluated to improve the accuracy of the predictions in the large-q_T region.

In the small-q_T region ($q_T \ll M_H$), where the bulk of events is produced, the convergence of the fixed-order expansion is definitely spoiled, since the coefficients of the perturbative series in $\alpha_S(M_H^2)$ are enhanced by powers of large logarithmic terms, $\ln^m(M_H^2/q_T^2)$. These logarithmic terms are due to multiple emission of soft and collinear gluons (i.e. gluons with low transverse momentum). To obtain reliable perturbative predictions, these terms have to be resummed to all orders in α_S. The method to systematically perform all-order resummation of classes of logarithmically-enhanced terms at small q_T is known [31]–[37]. In the case of the SM Higgs boson, resummation has been explicitly worked out at leading logarithmic (LL), next-to-leading logarithmic (NLL) [38, 39] and next-to-next-to-leading logarithmic (NNLL) [40] level.

The fixed-order and resummed approaches at small and large values of q_T can then be matched at intermediate values of q_T, to obtain QCD predictions for the entire range of transverse momenta. Phenomenological studies of the SM Higgs boson q_T distribution have been performed in Refs. [39, 41]–[49], by combining resummed and fixed-order perturbation theory at different levels of theoretical accuracy. A comparison of theoretical calculations and of results from parton shower Monte Carlo generators is presented in Ref. [50].

In Refs. [46, 49], the Higgs boson q_T distribution at the LHC is computed by combining the most advanced perturbative information that is available at present: NNLL resummation at small q_T and NLO perturbation theory at large q_T. The formalism used in the calculation is illustrated in detail in Ref. [49]. The resummed and fixed-order contributions are matched by avoiding double-counting of perturbative terms. A constraint of perturbative unitarity is imposed on the resummed terms, so that the total cross section at the nominal NNLO (or NLO) accuracy is recovered upon integration over q_T of the transverse-momentum spectrum at NNLL+NLO (or NLL+LO) accuracy. This unitarity constraint has the purpose of reducing the effect of unjustified higher-order contributions (those introduced by the resummed terms) at large values of q_T and, especially, at intermediate values of q_T.

An example of the quantitative studies[1] presented in Ref. [49] is reported in Fig. 2. Considering Higgs boson ($M_H = 125$ GeV) production at the LHC, the q_T spectrum at

---

[1] The calculation of Ref. [49] is implemented in the numerical code HqT, which can be downloaded from the url http://arturo.fi.infn.it/grazzini/codes.html

NNLL+NLO accuracy is compared to that at the previous order, namely NLL+LO accuracy, in the resummed perturbative expansion. The computed q_T spectra have a physical behaviour when $q_T \to 0$ and a kinematical peak at $q_T \sim 12$ GeV. The NNLL+NLO and NLL+LO bands are obtained by performing variations of the factorizatione and renormalization scales as in Fig. 1 (see also Sect. 2). At NLL+LO accuracy, the scale dependence increases from about $\pm 15\%$ at the peak to about $\pm 20\%$ at $q_T = 100$ GeV. At NNLL+NLO accuracy, the scale dependence is about 8% at the peak and increases to about 20% at $q_T = 100$ GeV. The inset plot shows the NNLL+NLO band normalized to the NLL+LO result at central value of the scales. From the inset plot, we see that the q_T spectrum at NNLL+NLO accuracy is slightly harder than at NLL+LO accuracy; the ratio between the NNLL+NLO and NLL+LO cross section is stable, around the values 1.1–1.2, in the region 5 GeV $\lesssim q_T \lesssim 50$ GeV. Note that the NNLL+NLO band is smaller than the NLL+LO one, and the former is included in (overlaps with) the latter at $q_T \lesssim 50$ GeV ($q_T \lesssim 100$ GeV): this suggests a good convergence of the resummed perturbative expansion. The results in Fig. 2 are obtained in a purely perturbative framework. Non-perturbative effects on the q_T spectrum are estimated in Refs. [49, 52]: they are definitely below 10% when $q_T \gtrsim 10$ GeV and decrease very rapidly as q_T increases.

In summary, the comparison of the NLL+LO and NNLL+NLO results from small (around the peak region) to intermediate (say, roughly, $q_T \lesssim M_H/3$) values of transverse momenta shows a nice convergence of the resummed QCD predictions for the q_T spectrum of the Higgs boson at the LHC. From this comparison and from the effects of scale variations, we conclude [49] that the perturbative QCD uncertainty of the NNLL+NLO results is *uniformly* of about 10% *over* this range of transverse momenta.

4. CONCLUDING REMARKS

In recent years much theoretical progress has been made on the calculation of QCD radiative corrections to the gluon fusion mechanism for Higgs boson production at hadron colliders. Reliable and accurate theoretical predictions are presently available for the total cross section and the transverse-momentum distribution of the SM Higgs boson.

After inclusion of both NNLO corrections and soft-gluon resummation at the NNLL level, the theoretical uncertainties of perturbative origin in the calculation of the total cross section at the Tevatron and at the LHC are below 10% in the low-mass range ($M_H \lesssim 200$ GeV). The differences (and the uncertainties) obtained by using various sets of available parton densities can, however, reach values that, at present, are larger than 10% (see Ref. [17]).

The transverse-momentum distribution of the Higgs boson can be computed [49] by including the resummation (at the NNLL level) of the logarithmically-enhanced terms at small q_T and the fixed-order expansion (at the NLO level) of the perturbative terms at large q_T. The integral over q_T of the transverse-momentum distribution at NNLL+NLO level reproduces the total cross section at NNLO. At the LHC, the perturbative QCD uncertainty on the q_T spectrum at NNLL+NLO accuracy is uniformly of about 10% from small (around the peak region) to intermediate (say, roughly, $q_T \lesssim M_H/3$) values of q_T. The uncertainty due to perturbative and non-perturbative effects increases at smaller

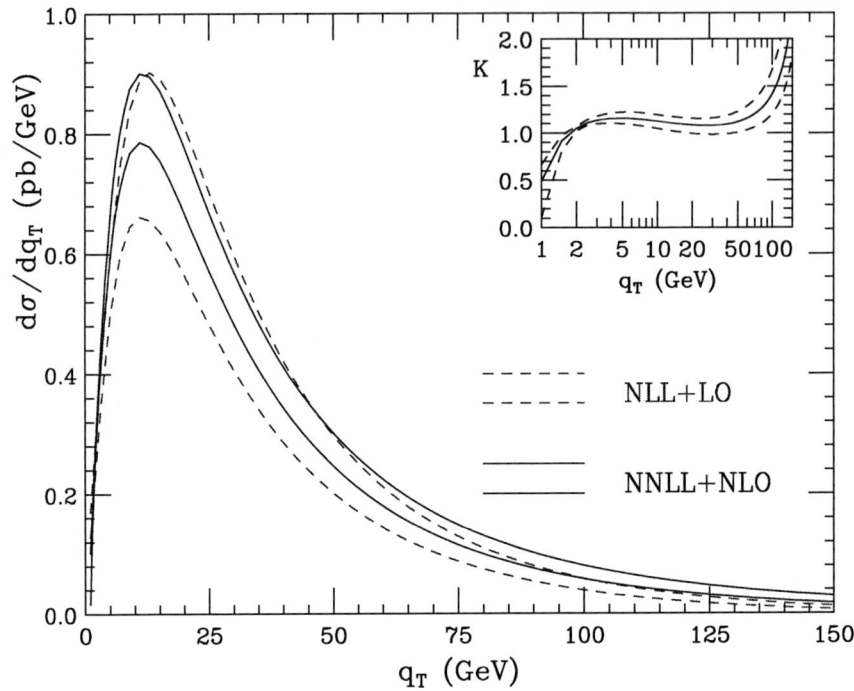

FIGURE 2. The q_T spectrum of the Higgs boson ($M_H = 125$ GeV) at the LHC. Comparison of the NNLL+NLO and NLL+LO results, using MRST2004 [51] parton densities. The inset plot shows the NNLL+NLO band normalized to the central value of the NLL+LO result.

values of q_T; the perturbative uncertainty increases also at larger values of q_T.

ACKNOWLEDGMENTS

I am very pleased to thank Pietro Colangelo, Fulvia De Fazio, Eugenio Nappi and Beppe Nardulli for their great organization of this nice Workshop. I wish to thank Daniel, Giuseppe, Massimiliano and Paolo for the scientific collaboration on the topics discussed in this talk.

REFERENCES

1. M. Carena et al., *Report of the Tevatron Higgs working group*, hep-ph/0010338; CDF and D0 Collaborations, *Results of the Tevatron Higgs Sensitivity Study*, report FERMILAB–PUB–03/320-E.
2. W. M. Yao [CDF and D0 Collaborations], report FERMILAB–CONF–04/307-E [hep-ex/0411053].
3. CMS Coll., *Technical Proposal*, report CERN/LHCC/94-38 (1994); ATLAS Coll., *ATLAS Detector and Physics Performance: Technical Design Report*, Vol. 2, report CERN/LHCC/99-15 (1999).
4. S. Abdullin et al., report CMS-NOTE-2003-033; S. Asai et al., Eur. Phys. J. C **32S2** (2004) 19.

5. S. Dawson, Nucl. Phys. B **359** (1991) 283; A. Djouadi, M. Spira and P. M. Zerwas, Phys. Lett. B **264** (1991) 440.
6. M. Spira, A. Djouadi, D. Graudenz and P. M. Zerwas, Nucl. Phys. B **453** (1995) 17.
7. M. Kramer, E. Laenen and M. Spira, Nucl. Phys. B **511** (1998) 523.
8. R. V. Harlander, Phys. Lett. B **492** (2000) 74; V. Ravindran, J. Smith and W. L. van Neerven, Nucl. Phys. B **704** (2005) 332.
9. S. Catani, D. de Florian and M. Grazzini, JHEP **0105** (2001) 025.
10. R. V. Harlander and W. B. Kilgore, Phys. Rev. D **64** (2001) 013015.
11. S. Catani, D. de Florian and M. Grazzini, JHEP **0201** (2002) 015.
12. R. V. Harlander and W. B. Kilgore, Phys. Rev. Lett. **88** (2002) 201801; C. Anastasiou and K. Melnikov, Nucl. Phys. B **646** (2002) 220; V. Ravindran, J. Smith and W. L. van Neerven, Nucl. Phys. B **665** (2003) 325.
13. S. Catani, D. de Florian and M. Grazzini, hep-ph/0206052, in *2002 QCD and hadronic interactions*, Proceedings of the 37th Rencontres de Moriond: QCD and Hadronic Interactions, 37th, Les Arcs, France, March 2002.
14. C. Anastasiou, L. J. Dixon and K. Melnikov, Nucl. Phys. Proc. Suppl. **116** (2003) 193.
15. C. Anastasiou, K. Melnikov and F. Petriello, Phys. Rev. Lett. **93** (2004) 262002.
16. C. Anastasiou, these Proceedings.
17. S. Catani, D. de Florian, M. Grazzini and P. Nason, JHEP **0307** (2003) 028.
18. G. Sterman, Nucl. Phys. B **281** (1987) 310; S. Catani and L. Trentadue, Nucl. Phys. B **327** (1989) 323, Nucl. Phys. B **353** (1991) 183.
19. S. Catani, M. L. Mangano, P. Nason and L. Trentadue, Nucl. Phys. B **478** (1996) 273.
20. A. D. Martin, R. G. Roberts, W. J. Stirling and R. S. Thorne, Phys. Lett. B **531** (2002) 216.
21. S. Moch and A. Vogt, report DESY-05-152 [hep-ph/0508265].
22. E. Laenen and L. Magnea, report DFTT-26-2005[hep-ph/0508284].
23. A. Idilbi, X. d. Ji, J. P. Ma and F. Yuan, report RBRC–562 [hep-ph/0509294].
24. R. K. Ellis, I. Hinchliffe, M. Soldate and J. J. van der Bij, Nucl. Phys. B **297** (1988) 221; U. Baur and E. W. Glover, Nucl. Phys. B **339** (1990) 38.
25. V. Del Duca, W. Kilgore, C. Oleari, C. Schmidt and D. Zeppenfeld, Nucl. Phys. B **616** (2001) 367, Phys. Rev. D **67** (2003) 073003.
26. D. de Florian, M. Grazzini and Z. Kunszt, Phys. Rev. Lett. **82** (1999) 5209.
27. V. Ravindran, J. Smith and W. L. Van Neerven, Nucl. Phys. B **634** (2002) 247.
28. C. J. Glosser and C. R. Schmidt, JHEP **0212** (2002) 016.
29. C. Anastasiou, K. Melnikov and F. Petriello, report UH-511-1066-05 [hep-ph/0501130].
30. J. Smith and W. L. van Neerven, Nucl. Phys. B **720** (2005) 182.
31. Y. L. Dokshitzer, D. Diakonov and S. I. Troian, Phys. Rep. **58** (1980) 269.
32. G. Parisi and R. Petronzio, Nucl. Phys. B **154** (1979) 427.
33. G. Curci, M. Greco and Y. Srivastava, Nucl. Phys. B **159** (1979) 451.
34. J. C. Collins and D. E. Soper, Nucl. Phys. B **193** (1981) 381 [Erratum-ibid. B **213** (1983) 545], Nucl. Phys. B **197** (1982) 446; J. C. Collins, D. E. Soper and G. Sterman, Nucl. Phys. B **250** (1985) 199.
35. J. Kodaira and L. Trentadue, Phys. Lett. B **112** (1982) 66, report SLAC-PUB-2934 (1982), Phys. Lett. B **123** (1983) 335.
36. S. Catani, D. de Florian and M. Grazzini, Nucl. Phys. B **596** (2001) 299.
37. E. Laenen, G. Sterman and W. Vogelsang, Phys. Rev. D **63** (2001) 114018.
38. S. Catani, E. D'Emilio and L. Trentadue, Phys. Lett. B **211** (1988) 335.
39. R. P. Kauffman, Phys. Rev. D **45** (1992) 1512.
40. D. de Florian and M. Grazzini, Phys. Rev. Lett. **85** (2000) 4678, Nucl. Phys. B **616** (2001) 247.
41. I. Hinchliffe and S. F. Novaes, Phys. Rev. D **38** (1988) 3475.
42. R. P. Kauffman, Phys. Rev. D **44** (1991) 1415.
43. C. P. Yuan, Phys. Lett. B **283** (1992) 395; C. Balazs and C. P. Yuan, Phys. Lett. B **478** (2000) 192.
44. C. Balazs, J. Huston and I. Puljak, Phys. Rev. D **63** (2001) 014021.
45. E. L. Berger and J. w. Qiu, Phys. Rev. D **67** (2003) 034026, Phys. Rev. Lett. **91** (2003) 222003.
46. G. Bozzi, S. Catani, D. de Florian and M. Grazzini, Phys. Lett. B **564** (2003) 65.
47. A. Kulesza and W. J. Stirling, JHEP **0312** (2003) 056.
48. A. Kulesza, G. Sterman and W. Vogelsang, Phys. Rev. D **69** (2004) 014012.
49. G. Bozzi, S. Catani, D. de Florian and M. Grazzini, report LPSC 05–63 [hep-ph/0508068].

50. C. Balazs, M. Grazzini, J. Huston, A. Kulesza and I. Puljak, hep-ph/0403052, published in M. Dobbs *et al.*, hep-ph/0403100, p. 51, Proceedings of the Les Houches 2003 Workshop on *Physics at TeV Colliders*.
51. A. D. Martin, R. G. Roberts, W. J. Stirling and R. S. Thorne, Phys. Lett. B **604** (2004) 61.
52. G. Bozzi, S. Catani, D. de Florian and M. Grazzini, in K. A. Assamagan *et al.* [Higgs Working Group Collaboration], hep-ph/0406152, p. 14, proceedings of the Les Houches 2003 Workshop on *Physics at TeV Colliders*.

NNLO QCD corrections for the differential Higgs boson production cross-section in gluon fusion

Charalampos Anastasiou

*Institute for Theoretical Physics,
ETH, 8093 Zürich, Switzerland*

Abstract.
I describe a recent computation of the NNLO QCD corrections for the fully differential cross-section for Higgs boson production in the gluon fusion channel. This result is an application of a new method for calculating perturbative corrections beyond the next-to-leading order.

Keywords: QCD, NNLO calculations, Higgs boson phenomenology
PACS: 12.20.-m,12.38.Bx,14.80.Bn

INTRODUCTION

The Large Hadron Collider (LHC) will probe directly the phenomena of TeV scales; this is an energy region where exciting new discoveries could be made for the breaking mechanism of the electroweak symmetry. Within the Standard Model, it is predicted that Higgs boson interactions will produce visible signals. Significant efforts have been made for the computation of the cross-sections for the corresponding signal and background processes.

The Higgs boson will be produced predominantly via gluon-fusion. Detailed analyses [1] for its signals have already been made with the use of leading order matrix-elements and parton-shower Monte-Carlo programs. However, these studies are insufficient for the gluon fusion process. Explicit calculations of the cross-section at next-to-leading-order in QCD have demonstrated that higher order contributions are very important [2, 3]. Depending on the choice of parton densities and the renormalization and the factorization scale, the corrections increase the cross-section by more than 70%

$$\sigma \sim \text{Leading Order} \times \left[1 + 0.7 + \mathcal{O}\left(\alpha_s^2\right)\right] \qquad (1)$$

This result is rather worrisome since the perturbative correction from one order to the next is much larger than the expansion parameter α_s. Clearly, the leading order result does not qualify as an approximation of the cross-section. It also remains uncertain if the perturbative expansion is likely to converge close to the NLO value, or whether it is meaningful to compute the cross-section perturbatively. To improve this situation it is important to compute the NNLO QCD corrections.

NNLO CONTRIBUTIONS FOR HIGGS PRODUCTION VIA GLUON-FUSION

The required NNLO corrections are very difficult to compute. At NLO already, two-loop diagrams are required; at NNLO very complicated tree-loop integrals emerge. However, in the heavy top-quark limit $M_h \ll 2M_{\text{top}}$, the coupling of the Higgs-boson to gluons at leading-order can be replaced by an effective tree-type vertex [4]. The exact NLO result [3] demonstrates that the approximation is excellent for a light Higgs boson. This approximation simplifies dramatically the calculation of the NNLO corrections; there are three different types of contributions:

- Interference of two-loop and tree diagrams:

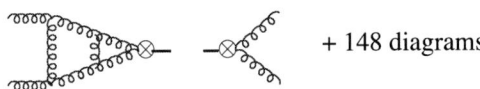 + 148 diagrams

These contributions were first computed by Harlander [5], when the calculation of two-loop diagrams was considered the stumbling block towards NNLO cross-sections. Two-loop diagrams still pose today serious technical challenges; however, it is well understood how to organize their evaluation successfully with the advent of new powerful methods [6].

- Interference of one-loop and tree diagrams with an additional radiated parton.

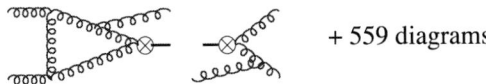 + 559 diagrams

These contributions require the integration of one-loop box diagrams, and the integration over the phase-space of the additional parton. Both integrations can be performed with straightforward methods which have been used in simpler NLO computations.

- Interference of tree diagrams with double real-radiation in the final state:

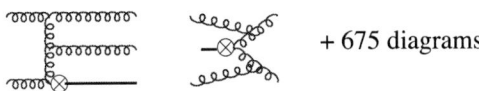 + 675 diagrams

These are the most challenging NNLO contributions, due to the complicated structure of the infrared singularities. In some regions of the phase-space, the double real radiation cannot be distinguished from final states with single or no radiation. An explicit integration of the matrix-elements in the unresolved regions yields universal poles in the dimension parameter $\varepsilon = 2 - d/2$ for every infrared-safe observable. The $1/\varepsilon$ poles should cancel against opposite in sign poles from the loop integrations, yielding a finite result. After infrared cancellations are made manifest, the integration over the remaining phase-space, where the double radiation is well separated (resolved), can be performed numerically.

At NNLO: (i) it is rather complicated to separate the resolved from the unresolved regions of the phase-space, and (ii) it is difficult to make the $1/\varepsilon$ poles manifest by integrating analytically the singular part of the phase-space.

NNLO CROSS-SECTION

We can obtain an estimate of the magnitude of NNLO effects in Higgs signal observables by computing the total cross-section. For this inclusive calculation, it is not necessary to find the boundaries of the unresolved double real radiation. In addition, the required phase-space integrations over the full phase-space are simpler than for generic observables, and they can be performed analytically. As it is suggested from the optical theorem, they are similar to loop integrals which can be computed systematically.

The NNLO inclusive cross-section for Higgs boson production in gluon fusion was computed by three different groups [7]. The corrections are significant:

$$\sigma \sim \text{Leading Order} \times \left[1 + 0.7 + 0.3 + \mathscr{O}\left(\alpha_s^3\right)\right] \quad (2)$$

However, they are comfortably smaller than the corrections at NLO indicating that the perturbative expansion (slowly) converges. Other resummed higher order contributions from threshold [8] do not change the cross-section significantly, consolidating the validity of the perturbative NNLO result. Recently, most of the leading threshold contributions were computed at NNNLO [9]; they add a modest $\sim 5\%$ correction to the above result.

For phenomenological purposes, the result for the inclusive cross-section is not sufficient. Let us consider, as an example, the search for the Higgs boson in the two-photon decay channel. A number of drastic cuts is applied on the signal. Only isolated photons with $p_T^{(1)} > 40\text{GeV}$, $p_T^{(2)} > 25\text{GeV}$ and $|\eta| < 2.5$ are accepted. The isolation criterion rejects events with hadronic energy $E_t^{\text{hard}} > 15 GeV$ in a cone of size $R = 0.4$ around the photons. The total cross-section result cannot be used directly to estimate the diphoton-signal events which pass the above cuts. Usually, a significant extrapolation is made by evaluating the ratio of the accepted to the inclusive cross-section by using leading order Monte-Carlo programs. However, the magnitude of the higher order corrections in the gluon fusion channel, indicates that such estimate is most likely unreliable. It is desirable to change this strategy and be able to implement cuts directly in the evaluation of the NNLO cross-section.

A new method [10] has been developed which solves the problem of double real radiation and permits the evaluation of fully differential cross-sections at NNLO; it has already been applied in processes [11] with diverse features, such as jet production, heavy particle decays with radiation from massive external legs, and processes with initial-state radiation. The method is based on a few simple ideas. It has been realized that the singularities of NNLO matrix-elements appear very entangled when we describe the phase-space in terms of physical variables, such as energies and angles. The new method maps the phase-space volume into a simpler geometry, the unit hypercube, replacing the momenta coordinates with variables which range from 0 to 1.

$$\{E, p_x, p_y, p_z\} \to \{0 \leq \lambda_i \leq 1\} \tag{3}$$

It turns out that, in such parameterizations, the singular regions of the phase-space are usually the corners of the hypercube. The singularities are then simplified with recursive slicing of the hypercube and remappings of the slices to the original geometry. After sufficient simplifications, the singularities are extracted in an observable independent fashion, in terms of delta and plus-distributions.

THE DIPHOTON SIGNAL

The fully differential cross-section for gluon fusion is now known [12] with NNLO accuracy. A phenomenological study of sample observables in the diphoton signal has also been made. The signal cross-section, after all cuts (p_T, pseudorapidity, and isolation) are applied, is 60% to 70% of the total cross-section (depending on the choice of parton densities and the renormalization and factorization scales).

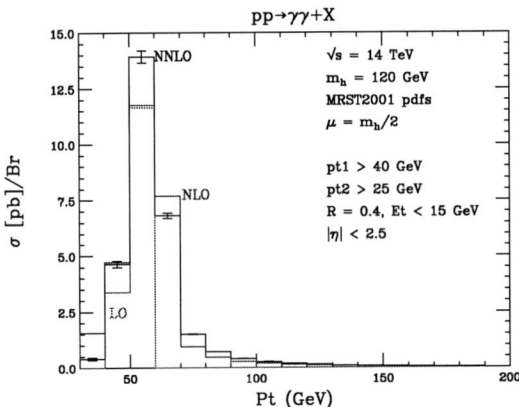

FIGURE 1. The average p_T of the two photons at NNLO

In addition to the experimentally accepted cross-section, we can now compute differential distributions for the deecay products of the Higgs boson. As an example, we present here the distribution of the average p_T of the two photons in $pp \to H + X \to \gamma\gamma + X$. At leading order, the average p_T cannot exceed half the value of the Higgs boson mass $p_T^{\text{mean}} < M_H/2$. The corresponding distribution shows a sharp edge at this kinematic boundary. At NLO and NNLO, additional radiation allows larger values of the average p_T. Potentially, the higher order contributions could smooth out the behavior at the kinematic edge. However, it turns out that the distribution remains sharply peaked at $M_H/2$. This is a very interesting feature, since the corresponding distribution for the background should not have such a peak. Therefore, there is a possibility to optimize further the cuts for the search of the Higgs boson in the diphoton signal using this information.

A more detailed description of observables for the diphoton signal at NNLO can be found in Ref. [12]. In addition, a comparison of the NNLO results and the program MC@NLO [13] was performed in Ref. [14].

CONCLUSIONS

I have presented a brief summary of a new method for NNLO calculations and results for the diphoton signal in Higgs production. I would like to thank Frank Petriello and Kirill Melnikov for a very enjoyable collaboration in this project, and the organizers of the conference for creating a very stimulating environment.

REFERENCES

1. ATLAS collaboration, report CERN/LHCC 99-15, ATLAS-TDR-15.
2. S. Dawson, Nucl. Phys. **B359**, 283 (1991).
3. A. Djouadi, M. Spira and P.M. Zerwas, Phys. Lett. **B264**, 440 (1991); D. Graudenz, M. Spira and P.M. Zerwas, Phys. Rev. Lett. **70**, 1372 (1993); M. Spira, A. Djouadi, D. Graudenz and P.M. Zerwas, Nucl. Phys. **B453**, 17 (1995).
4. J.R. Ellis, M.K. Gaillard, D.V. Nanopoulos, Nucl. Phys. **B106**, 292 (1976);
   M.A. Shifman, A.I. Vainshtein, M.B. Voloshin and V.I. Zakharov, Sov. J. Nucl. Phys. **30**, 711 (1979).
5. R. V. Harlander, Phys. Lett. B **492**, 74 (2000) [arXiv:hep-ph/0007289].
6. V. A. Smirnov, Phys. Lett. B **460**, 397 (1999); J. B. Tausk, Phys. Lett. B **469**, 225 (1999); T. Gehrmann and E. Remiddi, Nucl. Phys. **B580**, 485 (2000).
7. R. V. Harlander and W. B. Kilgore, Phys. Rev. Lett. **88**, 201801 (2002); C. Anastasiou and K. Melnikov, Nucl. Phys. B **646**, 220 (2002); V. Ravindran, J. Smith and W. L. van Neerven, Nucl. Phys. B **665**, 325 (2003).
8. S. Catani, D. de Florian, M. Grazzini and P. Nason, JHEP **0307**, 028 (2003) [arXiv:hep-ph/0306211].
9. S. Moch and A. Vogt, arXiv:hep-ph/0508265.
10. C. Anastasiou, K. Melnikov and F. Petriello, Phys. Rev. D **69**, 076010 (2004).
11. C. Anastasiou, K. Melnikov and F. Petriello, Phys. Rev. Lett. **93**, 032002 (2004); C. Anastasiou, K. Melnikov and F. Petriello, arXiv:hep-ph/0505069.
12. C. Anastasiou, K. Melnikov and F. Petriello, Phys. Rev. Lett. **93**, 262002 (2004); C. Anastasiou, K. Melnikov and F. Petriello, Nucl. Phys. B **724** 197 (2005).
13. S. Frixione and B. R. Webber, arXiv:hep-ph/0506182.
14. F. Stockli, A. G. Holzner and G. Dissertori, arXiv:hep-ph/0509130.

Perturbative QCD Corrections to $b \to s\gamma$

Luca Trentadue

Dipartimento di Fisica, Universitá di Parma,
and
Gruppo Collegato INFN di Parma, 43100 Parma, Italy

Abstract.
We present the results of a complete $O(\alpha_s)$ evaluation of transverse momentum distribution for the $b \to s\gamma$ decay. Coefficient function and remainder functions are presented in analytic form.

Keywords: Perturbative Quantum Chromodynamics, Heavy Quark Decays, Heavy Quark Effective Theory
PACS: 11.10.-z,11.15.-q,11.15.Bt,12.38.Cy, 12.39.Hg

We report results obtained in Refs.[1] and [2] concerning the evaluation and resummation of the strong radiative corrections to the $b \to s\gamma$ decay. In particular we concentrate on the transverse momentum distributions of the strange qualk with respect to the photon axis and discuss the perturbative and non perturbative sources of transverse momentum. $b \to s\gamma$ is a rare decay of the b quark (with a branching ratio of the $O(3 \cdot 10^{-4})$) widely studied in the last years both theoretically and experimentally. Theoretical efforts are dictated by many purposes: being loop mediated even at the lowest order, this process is very sensitive to new physics; it can be used as an application of the quark effective theory. The transverse momentum distribution of the s quark with respect to the photon direction is introduced and evaluated: in the chosen framework the heavy quark line is treated according to an effective field theory together with a perturbative QCD approach for the light fermion line. This quantity for this particular decay has been chosen for several reasons: from a phenomenological point of view since its kinematics is very simple, so that this can be the first step to face more difficult processes like the semileptonic decay $b \to u, c + l + \nu_l$. Moreover the presence of a single quark in the final state produce a single hadronic jet. From a theoretical point of view the transverse momentum distribution gives complementary informations to the photon spectrum and it can be a simple framework to study the singularities of the effective theory and of the perturbative expansion in processes where an heavy quark is involved. Moreover one can try to infer informations about non perturbative physics, using fragmentation functions as well as a shape function approach.

The general formula representing an improved perturbative expression, accurate in the whole phase space, is [3]

$$D(y) = K(\alpha_s)\Sigma(y;\alpha_s) + R(y;\alpha_s) \quad (1)$$

where $D(y)$ is the cumulative distribution for the variable of interest. In Ref.[2] the general theoretical approach has been introduced together with the calculation with next-to-leading accuracy of the universal function $\Sigma(y;\alpha_s)$ resumming large infrared logarithms. A detailed discussion of the singularities has been given there also. In particular a comparison between the structure of the singularities in the transverse momentum distribution and the ones in the more used threshold distribution has been presented.

In Ref.[1] the detailed complete evaluation of the $O(\alpha_S)$ corrections to the transverse momentum distribution has been performed and the results derived are presented in analytical form.

Perturbation theory, i.e. the expansion in powers of α_S, has been applied to describe decays of the beauty quark since its discovery. While the expansion parameter $\alpha_S(m_B) \sim 0.21$, being reasonably small, allows one to have confidence in the computations, it is difficult to directly compare the perturbative approach with the experimental data. As is well known, decay rates do not make good quantities to be compared with the data, because they are proportional to the fifth power of the beauty quark mass, a poorly known parameter,

$$\Gamma \propto m_b^5 \qquad (2)$$

and because they involve in principle unknown CKM matrix elements such as V_{cb}, V_{ub}, V_{ts}, etc.. By taking ratios of different widths, one can cancel the m_b^5 dependence in the observables, and, eventually, also the dependence on the CKM matrix elements. A rather good theoretical quantity is represented, for instance, by the semileptonic branching ratio:

$$B_{SL} = \frac{\Gamma_{SL}}{\Gamma_{TOT}}, \qquad (3)$$

which turns out to be marginally in agreement with present data. Inclusive quantities, $B_{inclusive}$, such as (3), have a perturbative series that involves numerical coefficients c_n of the form:

$$B_{inclusive} = \sum_{n=0}^{\infty} c_n \, \alpha_S^n(m_B). \qquad (4)$$

In less inclusive quantities, additional dynamical effects appear, due to the kinematical restrictions on the final particles, and the use of perturbation theory is, in general, less justified. In semi-inclusive quantities, $B_{semi-inclusive}$, such as threshold and transverse momentum p_t distributions, the perturbative series contains large infrared logarithms in addition to the coefficients c_n; they may be expanded as a perturbative series of the form:

$$B_{semi-inclusive} = \sum_{n=0}^{\infty} \sum_{k=0}^{2n} c_{n,k} \, \alpha_S^n \, \log^k x, \qquad (5)$$

where x represents the characteristic scale of the process as the energy or the transverse momentum. Resummation of such enhanced terms to any order in α_S can be performed in various approximations.

The simplest one, the leading logarithmic approximation, involves picking up only the terms having two powers of the logarithm for each power of the coupling, i.e. $k = 2n$. In the double-logarithmic approximation each parton is dressed with a cloud of soft and collinear gluons. Further, more refined approximations involve smaller numbers of logarithms for each power of α_S, i.e. $k = 2n-1, 2n-2, \ldots$.

In the last years, considerable effort has been devoted to the study of various spectra in B decays in the endpoint region, in the framework of resummed perturbation theory.
In order to verify the ability of the resummed perturbation theory to describe B decays in a different dynamical situation, we considered, in a previous note [2], p_t-distributions describing that of the s quark with respect to the photon direction, in the b rest frame.
In this work, [2], the following issues have been considered: the resummed p_t-distribution in the $b \to s\gamma$ decay is evaluated and both perturbative and non-perturbative sources of transverse momentum contributions discussed. The general theoretical framework for the evaluation of the corresponding matrix element defined and the strategy to evaluate leading and next-to-leading perturbative contributions is outlined, by introducing a method to treat the radiative corrections and their summation in a improved perturbative formula. The comparison of the transverse momentum distribution singularity structure with the more widely-known threshold case is also presented.
The chosen quantity manifests a clear advantage from a phenomenological point of view since, as discussed in [2], it depends only on the photon momentum in the process. Thanks to the straightforward and direct kinematics, the transverse momentum turns out to be a particularly simple variable to use to discuss the singularity structure of the perturbative expansion. The case of a possible effective theory within which to factorize these singularities can, for the transverse momentum, be considered as well.
The general formula representing the complete perturbative expression for a the resummed distribution is given by the formula

$$D(x) = K(\alpha_S)\Sigma(x;\alpha_S) + R(x;\alpha_S). \tag{6}$$

The results, already presented in [2], did concern the universal process-independent function $\Sigma(x;\alpha_S)$, resumming the infrared logarithms in exponentiated form.
Here the general perturbative expression for the whole distribution will be concisely recalled and the new entries represented by the coefficient function $K(\alpha_S)$ and by the remainder function $R(x)$ will be evaluated. Both $K(\alpha_S)$ and $R(x)$ are process-dependent and require an explicit evaluation of Feynman diagrams.
Resummation of large infrared logarithms in b decays have been studied in great detail in recent years. This scheme is justified by the fact that the double logarithm appearing to order α_S can become rather large (with respect to 1 coming from the tree level):

$$-\frac{\alpha_S C_F}{4\pi} \log^2 \frac{p_t^2}{m_b^2} \sim -0.7 \tag{7}$$

if we push the transverse momentum to such small values as $p_t \sim \Lambda_{QCD} = 300$ MeV. The single logarithm can also become rather large, having a large numerical coefficient:

$$-\frac{5\alpha_S C_F}{4\pi} \log \frac{p_t^2}{m_b^2} \sim 0.6. \tag{8}$$

The purpose of resumming classes of such terms therefore seems quite justified. If we consider running coupling effects, i.e. if the (frozen) coupling evaluated at the hard scale $Q = m_B = 5.2$ GeV is replaced by the coupling evaluated at the gluon transverse

momentum,
$$\alpha_S(m_b) \to \alpha_S(p_t) = 0.45 \quad \text{for} \quad p_t = 1 \text{ GeV}, \tag{9}$$
the logarithmic terms have sizes of order:
$$-\frac{\alpha_S(p_t)C_F}{4\pi} \log^2 \frac{p_t^2}{m_b^2} \sim -0.5 \tag{10}$$
and
$$-\frac{5\alpha_S C_F}{4\pi} \log \frac{p_t^2}{m_b^2} \sim 0.8. \tag{11}$$

The main difference with respect to resummation in Z^0 decays is a hard scale smaller by over an order of magnitude, i.e. a coupling larger by a factor 2 and infrared logarithms smaller by a factor 3.

The decay $b \to s\gamma$ is loop-mediated in the Standard Model and offers stringent tests of the latter as well as a way to extract CKM matrix elements. The relevant diagrams involve a loop with a virtual W and an up-type quark (u, c or t); the external photon can be emitted from the internal lines and from the external lines of the b or s quark.

QCD radiative corrections are affected by large logarithms of the form
$$\alpha_S^n \log^k \frac{m_W}{m_b} \quad \text{with} \quad 0 \leq k \leq n \tag{12}$$

as well as logarithms of m_t/m_W. Since the energies involved in the process are much smaller than the W or t mass, it is possible to integrate out these fields by means of an operator product expansion and write an effective low-energy hamiltonian of the form:
$$\mathcal{H}_{eff}(x) = \frac{G_F}{\sqrt{2}} V_{ts}^* V_{tb} \sum_{j=1}^{8} C_j(\mu_b) \, \hat{\mathcal{O}}_j(x;\mu_b). \tag{13}$$

With a factorization scale $\mu_b = O(m_b)$, the long-distance effects — both perturbative and non-perturbative — are factorized in the matrix elements of the operators $\hat{\mathcal{O}}_j$, while the short-distance effects are contained in the coefficient functions $C_j(\mu_b)$, calculable in perturbation theory. In particular, the large logarithms in (12) are included into the coefficient functions and can be resummed with standard renormalization group techniques.

A suitable basis for the operators $\hat{\mathcal{O}}_j$ is given by six four-quark operators, $\hat{\mathcal{O}}_1$-$\hat{\mathcal{O}}_6$ and by the penguin operators $\hat{\mathcal{O}}_7, \hat{\mathcal{O}}_8$. A similar basis was introduced at first in [4, 5]. Here we refer to an alternative choice, as given in [6].

Let us now consider the evaluation of the matrix elements of the effective hamiltonian between quark states. Only the magnetic penguin operator $\hat{\mathcal{O}}_7$ contributes in lowest order with a rate:
$$\Gamma_0 \simeq \frac{\alpha_{em}}{\pi} \frac{G_F^2 m_b^3 m_{b,\overline{MS}}^2(m_b) |V_{tb}V_{ts}^*|^2}{32\pi^3} C_7^2(\mu_b), \tag{14}$$

where m_b is the pole mass of the b quark.

Radiative QCD corrections involve gluon brehmsstrahlung. The operator $\hat{O}_7$ is affected by infrared singularities for the emission of a soft or a collinear gluon; the remaining operators $\hat{O}_1$-$\hat{O}_6$ have infrared-finite matrix elements. This implies that QCD corrections to the operator $\hat{O}_7$ only are logarithmically enhanced for $p_t \ll m_b$. We will then consider at first only the operator $\hat{O}_7$.

The process we are dealing with has a very simple kinematics: in lowest order it is the two-body decay $b \to s\gamma$. Let us define

$$x = \frac{p_t^2}{m_b^2}, \tag{15}$$

where p_t is the transverse momentum of the strange quark with respect to the photon direction, fixed as z-axis, and m_b is the mass of the heavy quark, to be identified with the hard scale of the process.

In lowest order the transverse momentum distribution then is

$$\frac{d\Gamma}{dx} = \Gamma_0\, \delta(x), \tag{16}$$

that is the strange quark and the photon are emitted in opposite directions, because of momentum conservation. Acollinearity is generated by gluon emission; in $b \to s\gamma g$, i.e. at $O(\alpha_S)$, $p_t = -k_t$ while in $b \to s\gamma g_1 \ldots g_n$, i.e. in higher orders, $p_t = -k_{t1} \ldots -k_{tn}$.

Beside the differential distribution the partially integrated distribution is also of interest

$$D(x) = \int_0^x dx'\, \frac{1}{\Gamma_0} \frac{d\Gamma}{dx'}. \tag{17}$$

Even though $\alpha_S(m_B)$ is small enough to justify a perturbative approach, the combination $\alpha_S^n(m_B) \log^k x$, with $0 \leq k \leq 2n$ can be large. A resummation, to any order in α_S, of logarithms of the same magnitude is required to obtain sensible physical results.

A partial resummation of large logarithms with next-to-leading accuracy has been performed in [2]: here we complete the calculation.

It is well known, [3], that the resummation of large logarithms is accomplished by an expression of the form in eq.(6) where

- $\Sigma(x; \alpha_S)$ is a universal, process-independent, function resumming the infrared logarithms in exponentiated form. It can be expanded in a series of functions as:

$$\log \Sigma(x; \alpha_S) = L g_1(\alpha_S L) + g_2(\alpha_S L) + \alpha_S g_3(\alpha_S L) + \ldots, \tag{18}$$

where $L = \log x$ (in general L is a large infrared logarithm). The functions g_i have a power expansion of the form

$$g_i(z) = \sum_{k=0}^{\infty} g_{i,k}\, z^k \tag{19}$$

and resum logarithms of the same size: in particular g_1 resums leading logarithms of the form $\alpha_S^n L^{n+1}$ and g_2 the next-to-leading ones $\alpha_S^n L^n$. The explicit form of $\Sigma(x; \alpha_S)$ can be found in Ref.[2];

- $K(\alpha_S)$ is a short-distance coefficient function, a process-dependent function, which can be calculated in perturbation theory:

$$K(\alpha_S) = 1 + \frac{\alpha_S C_F}{\pi} k_1 + O(\alpha_S^2). \tag{20}$$

- $R(x; \alpha_S)$ is the remainder function and satisfies the condition

$$R(x; \alpha_S) \to 0 \quad \text{for} \quad x \to 0. \tag{21}$$

It is process dependent, takes into account hard contributions and is calculable as an ordinary α_S expansion:

$$R(x; \alpha_S) = \frac{\alpha_S C_F}{\pi} r_1(x) + O(\alpha_S^2). \tag{22}$$

The result is an improved perturbative distribution, reliable in the semi-inclusive region [3], that is for small values of x, which can be matched with a fixed-order spectrum, describing the distribution for large values of x. The description of the tools used to perform the resummation of infrared logarithms can be found in Refs. [7] – [3].
Radiative corrections to transverse momentum distributions can be evaluated by taking into account the contributions of real and virtual Feynman diagrams contributions as discussed in Ref.[1]. Summing real and virtual contributions, the transverse momentum distribution for the decay $b \to s\gamma$ reads, to $O(\alpha_S)$:

$$D(x) = 1 + C_F \frac{\alpha_S}{\pi} \left[-\frac{1}{4} \log^2 x - \frac{5}{4} \log x + f + d(x) \right]. \tag{23}$$

As expected, the result contains a double logarithm and a single logarithm of x, a finite term f and a function $d(x)$ vanishing in the limit $x \to 0$.
By expanding the resummed formula to order α_S one obtains:

$$\begin{aligned} D(x) &= \left(1 + \frac{C_F \alpha_S}{\pi} k_1\right)\left(1 - \frac{A_1}{4} \alpha_S \log^2 x + B_1 \alpha_S \log x\right) + \frac{C_F \alpha_S}{\pi} r(x) \\ &= 1 - \frac{A_1}{4} \alpha_S \log^2 x + B_1 \alpha_S \log x + \frac{C_F \alpha_S}{\pi} k_1 + \frac{C_F \alpha_S}{\pi} r(x) + O(\alpha_S^2). \end{aligned} \tag{24}$$

By identifying the resummed result expanded to $O(\alpha_S)$ with the fixed-order one — matching procedure — we check the values for A_1 and B_1 evaluated in our previous paper using general properties of QCD radiation [2] and we extract the value of the coefficient function:

$$k_1 = f = -\frac{11}{4} - \frac{\pi^2}{12} + 4\log\frac{m_b}{\mu}, \tag{25}$$

as well as the remainder function $r_1(x) = d(x)$. As explained in previous sections, the remaining operators $\hat{O}_{i \neq 7}$ contribute to $D(x)$ only by finite terms $\tilde{r}_i$ and remainder functions. Since the constants $\tilde{r}_i$ come from virtual diagrams alone, we can quote their result from [11] and present an improved formula for the coefficient function, in analogy with [12]:

$$K(\alpha_S) = 1 + \frac{\alpha_S}{2\pi} \sum_{i=1}^{8} \frac{C_i^{(0)}(\mu_b)}{C_7^{(0)}(\mu_b)} \left(\Re\, \tilde{r}_i + \gamma_{i7}^{(0)} \log \frac{m_b}{\mu_b} \right) + \frac{\alpha_S}{2\pi} \frac{C_7^{(1)}(\mu_b)}{C_7^{(0)}(\mu_b)} + \mathcal{O}(\alpha_S^2) \quad (26)$$

where

$$\tilde{r}_i = r_i \quad i \neq 7$$
$$\tilde{r}_7 = \frac{8}{3}\left(f - 4\log\frac{m_b}{\mu_b}\right) = -\frac{22}{3} - \frac{2\pi^2}{9}. \quad (27)$$

Let us remark that only the coefficients related to the operators with $i = 1, 2, 7, 8$ are relevant, because the others are multiplied by very small coefficient functions and can be neglected:

$$r_1 = -\frac{1}{6}r_2$$
$$\Re\, r_2 = -4.092 - 12.78(0.29 - m_c/m_b)$$
$$r_8 = \frac{4}{27}(33 - 2\pi^2). \quad (28)$$

The analytic expressions for the coefficient functions as well as a standard numerical evaluation are given in [6]. The anomalous dimension $\gamma_{77}^{(0)}$ is derived from the coefficient of the logarithmic term in k_1. The values of $\gamma_{i7}^{(0)}$ are [6]:

$$\gamma_{i7}^{(0)} = \left(-\frac{208}{243}, \frac{416}{81}, -\frac{176}{81}, -\frac{152}{243}, \frac{6272}{81}, \frac{4624}{243}, \frac{32}{3}, -\frac{32}{9}\right). \quad (29)$$

Equation(26) is the main result of our paper and allows a complete resummation to NLO of transverse momentum logarithms.
The explicit calculation of the remainder function in (24) reads

$$r(\tau) = \frac{(\tau-1)p(\tau)}{12(\tau+1)^5(\tau^2+3\tau+1)^2} + \frac{q(\tau)}{4(\tau+1)^6(\tau^2+3\tau+1)^2}\log\tau - s(\tau)$$
$$- 2\sqrt{\tau}\arctan(\sqrt{\tau})\frac{(\tau+1)(2\tau^2+7\tau+2)}{(\tau^2+3\tau+1)^2} + \frac{\pi}{2}\sqrt{\tau}\frac{(\tau+1)(2\tau^2+7\tau+2)}{(\tau^2+3\tau+1)^2} + \frac{49}{12}$$
$$+ \frac{5}{4}\log\tau - \frac{5}{2}\log(\tau+1) + \log^2(\tau+1), \quad (30)$$

where

$$p(\tau) = 49\tau^8 + 468\tau^7 + 1797\tau^6 + 3642\tau^5$$
$$+ 4450\tau^4 + 3642\tau^3 + 1797\tau^2 + 468\tau + 49$$

$$\begin{aligned}q(\tau) &= -5 - 61\tau - 317\tau^2 - 912\tau^3 - 1622\tau^4 - 1934\tau^5 \\ &\quad - 1622\tau^6 - 912\tau^7 - 317\tau^8 - 61\tau^9 - 5\tau^{10} \\ (\tau) &= J[0,-3,\tau] + J[0,-3,1/\tau] - 2J[0,-1,\tau] \\ &\quad + J[-1,0,\tau] + J[-1,-3,\tau] - J[-1,-3,1/\tau]\end{aligned} \tag{31}$$

$$\tau = \frac{1 - \sqrt{1-4x}}{1 + \sqrt{1-4x}}. \tag{32}$$

Let us notice that τ behaves as x for small values of the transverse momentum

$$\tau(x) = x + O(x^2) \tag{33}$$

and it is a unitary variable

$$\begin{aligned}\tau &\to 0 \quad \text{for} \quad x \to 0 \\ \tau &\to 1 \quad \text{for} \quad x \to 1/4.\end{aligned}$$

The relation (32) may be inverted as

$$x = \frac{\tau}{(\tau+1)^2}. \tag{34}$$

One can easily check that $r(\tau)$ vanishes for $\tau \sim x \to 0$, by using the properties

$$J[0,-1,0] = J[0,-3,0] = J[-1,0,0] = J[-1,-3,0] = 0 \tag{35}$$

$$\lim_{\tau \to 0} J[0,-3,1/\tau] = \lim_{\tau \to 0} J[-1,-3,1/\tau] = -\frac{\pi^2}{3}. \tag{36}$$

Eq. (23), which contains the final result, represents the full evaluation to $O(\alpha_S)$ of the transverse momentum distribution. It is explicitly given in terms of an analytic expression.

Contrary to what happens in hard processes at much larger energies, at the energy scales involved here for the b decay the remainder function contribution does play a more important role.

A straightforward numerical evaluation of the remainder function $r(x)$ of eq. (24) allows us to conclude that its contribution can be safely neglected for small values of x, up to $x \simeq 0.1$, where it approaches the zero limit of $x \to 0$. For larger values of x, however, the size of its contribution increases to reach values of the order of the 10–15% of the combined leading and next-to-leading logarithmic terms.

ACKNOWLEDGMENTS

I wish to thank Ugo Aglietti for sharing with me his enthousiasm and for the unnumerous discussions and conversations along the course of our collaboration. I wish to thank Beppe Nardulli and Pietro Colangelo for the invitation to such a lively and successful meeting and the for warm hospitality extended to me in Conversano.

REFERENCES

1. U. Aglietti, R. Sghedoni and L. Trentadue, Phys. Lett. B **585**, 131 (2004) [arXiv:hep-ph/0310360].
2. U. Aglietti, R. Sghedoni and L. Trentadue, Phys. Lett. B **522**, 83 (2001) [arXiv:hep-ph/0105322].
3. S. Catani, L. Trentadue, G. Turnock and B. R. Webber, Nucl. Phys. B **407**, 3 (1993).
4. B.Grinstein, R.Springer, M.Wise, Phys. Lett. B202, 138 (1988).
5. B.Grinstein, R.Springer, M.Wise, Nucl. Phys. B339, 269 (1990).
6. K.Chetyrkin, M.Misiak, M.Munz, Phys. Lett. B400, 206 (1997),Erratum Phys. Lett. B425, 414 (1998).
7. G. Parisi and R. Petronzio, Nucl. Phys. B154, 427 (1979).
8. D.Amati, A.Bassetto, M.Ciafaloni, G.Marchesini, G.Veneziano, Nucl. Phys. B173, 429 (1980).
9. J.Kodaira, L.Trentadue, Phys. Lett. B112, 66 (1982) and SLAC-PUB-2934 (1982); L.Trentadue Phys. Lett. B151, 171 (1985).
10. S.Catani, L.Trentadue, Nucl. Phys. B327, 323 (1989); (ibidem) B353, 183 (1991).
11. C.Greub, T.Hurth, D.Wyler, Phys. Lett. B380, 385 (1996) and Phys. Rev. D54, 3350 (1996).
12. U.Aglietti, M.Ciuchini, P.Gambino, Nucl. Phys. B637, 427 (2002).

CSW Diagrams and Electroweak Vector Bosons [1]

Pierpaolo Mastrolia

Department of Physics and Astronomy, UCLA Los Angeles, CA 90095–1547, USA

Abstract. Based on the joined work performed together with Z. Bern, D. Forde, and D. Kosower [1], in this talk it is recalled the (twistor-motivated) diagrammatic formalism describing tree-level scattering amplitudes presented by Cachazo, Svrček and Witten, and it is discussed an extension of the vertices and accompaining rules to the construction of vector-boson currents coupling to an arbitrary source.

Keywords: Perturbative calculations
PACS: 11.15.Bt, 11.25.Db, 11.25.Tq, 11.55.Bq, 12.38.Bx

INTRODUCTION

The computation of amplitudes for QCD and mixed electroweak-QCD processes is an important part of a physics program at modern-day colliders, given their important role as experimentally distinctive probes of new physics.

In less than a couple of years, the progress in the evaluation of scattering processes has received a strong boost, due to a better understanding of the analytic structure of scattering amplitudes. Stimulated by Witten's realization [2] that tree-level gluon amplitudes, once transformed in twistor space [3], have a simple geometrical description, Cachazo, Svrček and Witten (CSW) [4] proposed a powerful set of new computational rules to deal with many particles scattering amplitudes in QCD. Only recently, Risager [5] has found that the CSW approach can be understood as a particular class of a more general set of *on-shell recurrence relations* for amplitudes, which meanwhile had been introduced by Britto, Cachazo, Feng and Witten (BCFW) [6].

The CSW rules are of interest in their own right for tree-level computations. Their efficiency for gluonic amplitudes, was soon extended to account for massless external fermions [7] and Higgs boson coupled to QCD via a massive top-quark loop (in the infinite-mass limit) [8], and improved when recast in a recursive form [9]. They also allow great simplification in loop calculations [10].

In this talk it is described the extension of the CSW construction to include building blocks for mixed QCD-electroweak amplitudes, by providing a construction of vector currents, which may in principle be coupled to an *arbitrary* source. We will focus on the case of coupling a process involving one quark pair and any number of gluons to one colorless off-shell vector boson. The key idea in the construction is to introduce a new set of basic vertices coupling to the off-shell vector boson, having either one or no negative-helicity gluons. The rules for combining them into new currents with additional

---

[1] Research supported by the US Department of Energy under contract DE–FG03–91ER40662.

negative-helicity legs are then in fact the same as those of CSW.

COLOR DECOMPOSITIONS

Color decompositions [11, 12] allows the disentangling of the the gauge group factors and the pure kinematical terms, namely *partial amplitudes*, in the full momentum-space amplitudes. For example, the tree-level n-gluon amplitude $\mathscr{A}_n$ has the color decomposition,

$$\mathscr{A}_n(1,2,\ldots,n) = \sum_{\sigma \in S_n/Z_n} \mathrm{Tr}(T^{a_{\sigma(1)}} \cdots T^{a_{\sigma(n)}}) A_n(\sigma(1),\ldots,\sigma(n)), \quad (1)$$

where S_n/Z_n is the group of non-cyclic permutations on n symbols, and j denotes the j-th gluon and its associated momentum. We use the color normalization $\mathrm{Tr}(T^a T^b) = \delta^{ab}$. Similar decompositions hold for cases involving quarks. In general, it is more convenient to calculate the partial amplitudes than the entire amplitude at once.

The cases in which we are interested here involve colorless vector bosons. Single massive vector boson exchange is easily obtained from pure QCD amplitudes (which are directly calculable from CSW diagrams). For example, for $e^+e^- \to \gamma^* \to q\bar{q} + n$ gluons, where γ^* represents an off-shell photon, the amplitude reduces to

$$\mathscr{A}_n(1_{e^+}, 2_{e^-}, 3_q, 4, 5, \ldots, (n-1), n_{\bar{q}}) = -2e^2 Q^q g^{n-2} \times \sum_{\sigma \in S_{n-4}} (T^{a_{\sigma(4)}} \cdots T^{a_{\sigma(n-1)}})_{i_3}^{\bar{i}_n}$$

$$\times A_n(1_{e^+}, 2_{e^-}, 3_q, \sigma(4), \ldots, \sigma(n-1), n_{\bar{q}}), \quad (2)$$

where we use an all outgoing momentum convention. The particle labels $q, \bar{q}, e^-, e^+$ stand for quarks, anti-quarks, electrons and positrons, while legs without labels, for gluons. The off-shell photon is internal to the amplitude and exchanged between the lepton pair and the quark pair.

To convert the exchanged photon to an electroweak vector boson, for describing $e^+e^- \to Z, \gamma^* \to \bar{q}q + ng$, one may adjust the coupling and modify the photon kinematic pole to account for an unstable massive particle [13, 14]. More generally, one may convert gluons to photons, purely by group theoretic rearrangements [15]. However, in general, it is not possible to then convert the photonic amplitudes to ones involving electroweak vector bosons since vector bosons have non-abelian self interactions which photons do not.

A purpose of this talk is to see how constructing an appropriate off-shell continuation so that the CSW *diagrammatricks*, originary introduced to describe pure gluon scattering, can be applied to such cases as well.

CSW DIAGRAMS

The CSW construction [4] builds amplitudes out of vertices which are off-shell continuations of the Parke–Taylor amplitudes [16, 17]. These amplitudes, with two negative-helicity gluons and any number of positive-helicity ones, are the maximally helicity-violating (MHV) non-vanishing tree-level amplitudes in a gauge theory. In the spinor helicity [18, 20, 12] notation, they are,

$$A_n(1^+,\ldots,m_1^-,\ldots,m_2^-,\ldots,n^+) = i\frac{\langle m_1 m_2\rangle^4}{\langle 12\rangle\langle 23\rangle\cdots\langle(n-1)n\rangle\langle n1\rangle}, \quad (3)$$

where the two negative-helicity gluons are labeled $m_{1,2}$. In this equation, $\langle ij\rangle = \langle k_i k_j\rangle$. We follow the standard spinor normalizations $[ij] = \text{sign}(k_i^0 k_j^0)\langle ji\rangle^*$ and $\langle ij\rangle[ji] = 2k_i \cdot k_j$. With our conventions all particle momenta are taken to be outgoing.

The remaining MHV fermionic amplitudes needed for our discussion of vector boson currents are,

$$A(1_q^+,2^+,3^+,\ldots,i^-,\ldots,(n-2)^+,(n-1)_{\bar{q}}^-,n^+) = i\frac{\langle i1\rangle\langle i, n-1\rangle^3}{\langle 12\rangle\langle 23\rangle\langle 34\rangle\cdots\langle n1\rangle}, \quad (4)$$

$$A(1_q^+,2^+,3^+,\ldots,(n-2)^+,(n-1)_{\bar{q}}^-,n^-) = i\frac{\langle n1\rangle\langle n-1, n\rangle^3}{\langle 12\rangle\langle 23\rangle\langle 34\rangle\cdots\langle n1\rangle}, \quad (5)$$

$$A(1_{\bar{q}'}^-,2_{q'}^+,3_q^+,4^+,\ldots,(n-1)^+,n_{\bar{q}}^-) = -i\frac{\langle 1n\rangle^2}{\langle 12\rangle\langle 34\rangle\langle 45\rangle\cdots\langle(n-1)n\rangle}. \quad (6)$$

The last equation gives the color-ordered amplitude appearing in eq. (2) after relabeling $q' \to e^-$ and $\bar{q}' \to e^+$.

In the CSW construction a particular off-shell continuation of these amplitudes, A, is an MHV vertex, V. The original CSW prescription for the off-shell continuation of a momentum k_j amounts to replacing $\langle jj'\rangle \longrightarrow [\eta j]\langle jj'\rangle \longrightarrow \langle \eta^+|k_j|j'^+\rangle$, where η is an arbitrary light-like reference vector, in the Parke-Taylor formula. The extra factors introduced in this off-shell continuation cancel when sewing together vertices to obtain an on-shell amplitude. As shown by CSW [4], on-shell amplitudes are in fact independent of the choice of η, implying that the sum over MHV diagrams is Lorentz invariant.

In our construction we use an alternative, but equivalent way of going off-shell [19, 9]. We instead decompose an off-shell momentum K into a sum of two massless momenta, where one is proportional to the auxiliary light-cone reference momentum η (with $\eta^2 = 0$),

$$K = K^\flat + \zeta(K)\eta. \quad (7)$$

The constraint $(K^\flat)^2 = 0$ yields $\zeta(K) = K^2/(2\eta \cdot K)$. If K goes on shell, ζ vanishes. Also, if two off-shell vectors sum to zero, $K_1 + K_2 = 0$, then so do the corresponding $k^\flat$s. The prescription for continuing MHV amplitudes or vertices off shell is to replace,

$$\langle jj'\rangle \to \langle j^\flat j'\rangle, \quad (8)$$

when k_j is taken off shell. In the on-shell limit, $\zeta(K)$ vanishes and $k_j^\flat \to k_j$. Although equivalent to the original CSW prescription, it is a bit more convenient to implement. In particular, there are no extra factors associated with going off-shell and the MHV vertices carry the same dimensions as amplitudes.

The CSW construction replaces ordinary Feynman diagrams with diagrams built out of MHV vertices and ordinary propagators. Each vertex has exactly two lines carrying

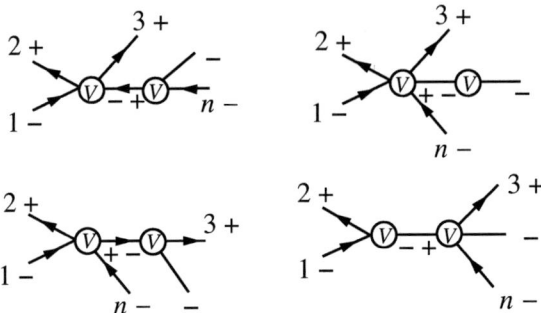

FIGURE 1. The stripped diagrams for 3 minus helicity amplitudes with vector boson exchange between two fermion pairs. Legs 1 and 2 correspond to the leptons and the legs 3 and n to the quarks. Lines with arrows represent quarks and those without arrows represent either vector bosons or gluons.

negative helicity (which may be on or off shell), and at least one line carrying positive helicity. The propagator takes the simple form i/K^2, because the physical state projector is effectively supplied by the vertices. For example, with this notation an all-gluon vertex would be,

$$V(1^+,\ldots,m_1^-,(m_1+1)^+,\ldots,n,K^-) = i\frac{\langle m_1 K^\flat\rangle^4}{\langle 1\,2\rangle\langle 2\,3\rangle\cdots\langle n K^\flat\rangle\langle K^\flat 1\rangle}. \qquad (9)$$

The CSW rules then instruct us to write down all tree diagrams with MHV vertices, subject to the constraints that each vertex has exactly two negative-helicity gluons and at least one positive-helicity gluon attached, and that each propagator connects legs of opposite helicity. For amplitudes with two negative-helicity gluons, the vertex with all legs taken on shell is then the amplitude. For each additional negative-helicity gluon, we must add a vertex and a propagator. The number of vertices is thus the number of negative-helicity gluons, less one.

As a simple example we may use the CSW rules to construct next-to-MHV (NMHV) partial amplitudes needed for the process $e^+e^- \to \gamma^*, Z, W \to q\bar{q} + ng$. The 'stripped diagrams' (where all the positive helicity gluons are not indicated) for this process are shown in fig. 1. Dressing the diagrams with the positive helicity gluon legs between q and $\bar{q}$ in the color ordering leads to

$$A(1_{q'}^-, 2_{\bar{q}'}^+, 3_q^+, 4^+, 5^+, \ldots, (n-1)^-, n_{\bar{q}}^-)$$
$$= \sum_{j=4}^{n-1} V(1_{q'}^-, 2_{\bar{q}'}^+, 3_q^+, 4^+, 5^+, \ldots, (j-1)^+, (-K_{1\ldots(j-1)})_{\bar{q}}^-) \frac{i}{K_{1\ldots(j-1)}^2}$$
$$\times V((-K_{j\ldots n})_q^+, j^+, \ldots, (n-2)^+, (n-1)^-, n_{\bar{q}}^-)$$
$$+ \sum_{j=4}^{n-2} V(1_{q'}^-, 2_{\bar{q}'}^+, 3_q^+, 4^+, 5^+, \ldots, (j-1)^+, (-K_{n1\ldots(j-1)})^+, n_{\bar{q}}^-) \frac{i}{K_{n1\ldots(j-1)}^2}$$
$$\times V((-K_{j\ldots(n-1)})^-, j^+, \ldots, (n-2)^+, (n-1)^-)$$

$$+V(1^-_{q'},2^+_{\bar q},(-K_{n,1,2})^+_q,n^-_{\bar q})\frac{i}{K^2_{n12}}V(3^+_q,4^+,\ldots,(n-2)^+,(n-1)^-,(-K_{3\ldots(n-1)})^-_{\bar q})$$

$$+V(1^-_{q'},2^+_{\bar q},(-K_{12})^-)\frac{i}{K^2_{12}}V(3^+_q,4^+,\ldots,(n-2)^+,(n-1)^-,n^-_{\bar q},(-K_{3\ldots n})^+), \tag{10}$$

where $K_{i\ldots j}=k_i+k_{i+1}+\cdots+k_j$. Renaming $\bar{q}',q' \to e^+,e^-$ gives the partial amplitudes appearing in the vector boson exchange amplitudes (2).

MHV VERTICES FOR VECTOR BOSON CURRENTS

In this section we generalize the CSW construction to allow couplings to arbitrary sources. We focus on the phenomenologically interesting case of vector boson currents, though our construction of currents is applicable more generally.

An important application of these currents is that they allow us to couple the electroweak theory to QCD, while taking full advantage of the CSW formalism on the QCD side. The currents satisfy a similar color decomposition as the photon exchange amplitude (2),

$$\mathscr{J}_\mu(1_q,2,3,\ldots,(n-1),n_{\bar q};P_V) = g_V g^n \sum_{\sigma \in S_{n-2}} (T^{a_{\sigma(2)}}T^{a_{\sigma(3)}}\ldots T^{a_{\sigma(n-1)}})^{\bar{i}_n}_{i_1}$$
$$\times J_\mu(1_q,\sigma(2),\sigma(3),\ldots,\sigma(n-1),n_{\bar q};P_V), \tag{11}$$

where g_V is the appropriate coupling for a vector boson $V=\gamma^*,Z,W$ and P_V is the momentum carried by the vector boson. Hence we need consider only the partial currents J_μ in much the same way that we need only consider color-ordered partial amplitudes.

We start by defining two currents that will serve as new basic vertices for obtaining general vector boson currents:

1. A vector-boson current with n gluon emissions, all of positive helicity

$$J^\mu(1^-_q,2^+,\ldots,(n-1)^+,n^+_{\bar q};P_V) = -\frac{i}{\sqrt{2}}\frac{\langle(-1)^-|\gamma^\mu \slashed{P}_V|(-1)^+\rangle}{\langle(-1)2\rangle\langle 23\rangle\ldots\langle(n-1)n\rangle}$$
$$= c_+\varepsilon^{(+)\mu}(P^\flat_V,\eta)+c_-\varepsilon^{(-)\mu}(P^\flat_V,\eta)$$
$$+c_L\left(P^\mu_V - \frac{P^2_V}{\eta\cdot P_V}\eta^\mu\right), \tag{12}$$

where $P_V=-K_{1\ldots n}$ by momentum conservation, where '-1' as a spinor argument denotes $-k_1$, and where

$$c_+ = -V^{\text{MHV}}(1^-_{\bar q},\ldots,n^+_q;P^-_V),$$
$$c_- = V^{\text{MHV}}(1^-_{\bar q},\ldots,n^+_q;P^-_V)\frac{\langle 1\eta\rangle^2 P^2_V}{\langle \eta P^\flat_V\rangle^2 \langle 1 P^\flat_V\rangle^2}, \tag{13}$$
$$c_L = V^{\text{MHV}}(1^-_{\bar q},\ldots,n^+_q;P^-_V)\frac{\sqrt{2}\langle 1\eta\rangle}{\langle \eta P^\flat_V\rangle\langle 1 P^\flat_V\rangle}.$$

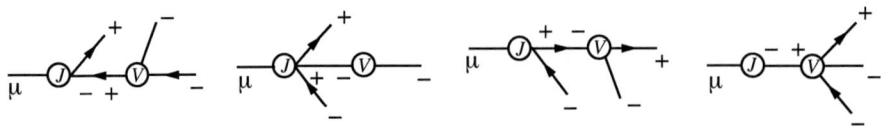

FIGURE 2. The NMHV vector boson current in terms of diagrams where positive helicity gluon lines have been stripped.

The vertex V^{MHV} is simply a CSW vertex for one photon, one quark pair, and $n-2$ gluons, obtained by fermionic phase adjustments from the amplitude in eq. (5). As with the basic CSW vertices, when any colored leg j is taken off shell, the k_j argument to all spinor products or spinor strings must be replaced by $k_j^\flat$.

2. A purely bosonic basic current emitting a single vector state,

$$J^\mu((-P_V)^-;P_V) = \frac{i}{\sqrt{2}} \frac{\langle \eta^+|\gamma^\mu|P_V^{\flat+}\rangle}{[P_V^\flat \eta]} P_V^2 = i\varepsilon^{(-)\mu}(P_V^\flat,\eta)P_V^2. \quad (14)$$

The first of these is the vector-boson current for positive helicity gluons [13]. The second is just a negative helicity polarization vector with reference momentum taken to be the CSW reference momentum.

The polarizations in the above equations are defined using the spinor helicity method and are given by [20]

$$\varepsilon_\mu^{(+)}(k,r) = \frac{1}{\sqrt{2}} \frac{\langle r^-|\gamma^\mu|k^-\rangle}{\langle rk \rangle}, \qquad \varepsilon_\mu^{(-)}(k,r) = \frac{1}{\sqrt{2}} \frac{\langle r^+|\gamma^\mu|k^+\rangle}{[kr]}, \quad (15)$$

where r is a null reference momentum.

We take the currents (12) and (14) to act as vertices, using the same CSW prescriptions (8) as used for defining vertices from MHV amplitudes.

To illustrate the construction of a current with more negative helicities, consider the NMHV vector boson current, $J_\mu(1_q^+, 2^+, \ldots, (n-2)^+, (n-1)^-, n_{\bar{q}}^-; P_V)$ where the negative helicity legs are nearest neighbors in the color ordering. The CSW diagrams for this current may be organized using the four diagrams shown in fig. 2, where the positive helicity gluon legs have all been stripped away. Inserting back the positive helicity gluon legs, leads to the following expression for this NMHV vector boson current,

$$J_\mu(1_q^+, 2^+, \ldots, (n-2)^+, (n-1)^-, n_{\bar{q}}^-; P_V)$$

$$= \sum_{j=2}^{n-1} J_\mu(1_q^+, 2^+, \ldots, (j-1)^+, (K_{j\ldots n})_{\bar{q}}^-; P_V) \frac{i}{K_{j\ldots n}^2}$$

$$\times V((-K_{j\ldots n})_q^+, j^+, \ldots, (n-2)^+, (n-1)^-, n_{\bar{q}}^-)$$

$$+ \sum_{j=2}^{n-2} J_\mu(1_q^+, 2^+, \ldots, (j-1)^+, (K_{j\ldots(n-1)})^+, n_{\bar{q}}^-; P_V) \frac{i}{K_{j\ldots(n-1)}^2}$$

$$\times V((-K_{j\ldots(n-1)})^-, j^+, \ldots, (n-2)^+, (n-1)^-)$$

$$+ J_\mu((K_{1...(n-1)})_q^+, n_{\bar{q}}^-; P_V) \frac{i}{K_{1...(n-1)}^2} V(1_q^+, 2^+, \ldots, (n-2)^+, (n-1)^-, (-K_{1...(n-1)})_{\bar{q}}^-)$$

$$+ J_\mu((K_{1...n})^-; P_V) \frac{i}{K_{1...n}^2} V(1_q^+, 2^+, \ldots, (n-2)^+, (n-1)^-, n_{\bar{q}}^-, (-K_{1...n})^+), \quad (16)$$

where the momentum of the vector boson is $P_V = -K_{1...n}$. The explicit values of the current vertices are obtained from eqs. (12) and (14) by relabeling the arguments. Other NMHV helicity configurations are only a bit more complicated. In ref. [9], Bena and two of the authors introduced a recursive reformulation of the CSW rules, useful when increasing the number of negative helicity legs. Indeed, an analogous recurrence relation also applies to the vector-boson currents considered here.

CONCLUSIONS

The twistor-inspired computational approach presented by Cachazo, Svrček, and Witten [4], and very recently demonstrated by Risager [5], is among the novel ways of computing tree amplitudes in massless gauge theories, including of course QCD. In this talk, I have discussed the main issue of [1], where, with Z. Bern, D. Forde, and D. Kosower, addressing the question of computing amplitudes containing both colored and non-colored particles, we have shown how to incorporate an additional vector leg coupling to an arbitrary source into the CSW approach. The currents we have constructed can be used directly in the computation of processes producing electroweak vector bosons. The structure of the CSW construction implies that that a similar approach can be used to build multi-Ws currents.

In outlook, novel techniques dealing directly with on-shell objects, like the CSW [4] and the BCFW [6] approaches, relying on general properties of complex analysis, and exploiting the *recursive behaviour* of scattering amplitudes, are establishing themselves as suitable tools [21, 22] for computing massless and massive multi-legs tree-level [23] and one-loop QCD (and beyond) amplitudes [24].

REFERENCES

1. Z. Bern, D. Forde, D. A. Kosower and P. Mastrolia, Phys. Rev. D **72**, 025006 (2005) [hep-ph/0412167].
2. E. Witten, Commun. Math. Phys. **252**, 189 (2004) [hep-th/0312171].
3. R. Penrose, J. Math. Phys. 8, 345 (1967).
4. F. Cachazo, P. Svrcek and E. Witten, JHEP **0409**, 006 (2004) [hep-th/0403047].
5. K. Risager, hep-th/0508206.
6. R. Britto, F. Cachazo and B. Feng, Nucl. Phys. B **715**, 499 (2005) [hep-th/0412308]. R. Britto, F. Cachazo, B. Feng and E. Witten, Phys. Rev. Lett. **94**, 181602 (2005) [hep-th/0501052].
7. G. Georgiou and V. V. Khoze, JHEP **0405**, 070 (2004) [hep-th/0404072]. J. B. Wu and C. J. Zhu, JHEP **0409**, 063 (2004) [hep-th/0406146]. G. Georgiou, E. W. N. Glover and V. V. Khoze, JHEP **0407**, 048 (2004) [hep-th/0407027].
8. L. J. Dixon, E. W. N. Glover and V. V. Khoze, JHEP **0412**, 015 (2004) [hep-th/0411092]. S. D. Badger, E. W. N. Glover and V. V. Khoze, JHEP **0503**, 023 (2005) [hep-th/0412275].
9. I. Bena, Z. Bern and D. A. Kosower, Phys. Rev. D **71**, 045008 (2005) [hep-th/0406133]
10. F. Cachazo, P. Svrček and E. Witten, JHEP **0410**, 074 (2004) [hep-th/0406177] A. Brandhuber, B. Spence and G. Travaglini, Nucl. Phys. B **706**, 150 (2005) [hep-th/0407214] F. Cachazo, P. Svrček and E. Witten, JHEP **0410**, 077 (2004) [hep-th/0409245]. I. Bena, Z. Bern, D. A. Kosower and

R. Roiban, Phys. Rev. D **71**, 106010 (2005) [hep-th/0410054]. F. Cachazo, hep-th/0410077. R. Britto, F. Cachazo and B. Feng, Phys. Rev. D **71**, 025012 (2005) [hep-th/0410179]; Phys. Lett. B **611**, 167 (2005) [hep-th/0411107]; Nucl. Phys. B **725**, 275 (2005) [hep-th/0412103]. J. Bedford, A. Brandhuber, B. Spence and G. Travaglini, Nucl. Phys. B **706**, 100 (2005) [hep-th/0410280]; Nucl. Phys. B **712**, 59 (2005) [hep-th/0412108]. C. Quigley and M. Rozali, JHEP **0501**, 053 (2005) [hep-th/0410278] Z. Bern, V. Del Duca, L. J. Dixon and D. A. Kosower, Phys. Rev. D **71**, 045006 (2005) [hep-th/0410224]. S. J. Bidder, N. E. J. Bjerrum-Bohr, L. J. Dixon and D. C. Dunbar, Phys. Lett. B **606**, 189 (2005) [hep-th/0410296]. S. J. Bidder, N. E. J. Bjerrum-Bohr, D. C. Dunbar and W. B. Perkins, Phys. Lett. B **608**, 151 (2005) [hep-th/0412023].
11. J. E. Paton and H. M. Chan, Nucl. Phys. B10, 516 (1969). P. Cvitanovic, P. G. Lauwers and P. N. Scharbach, Nucl. Phys. B186, 165 (1981). F. A. Berends and W. Giele, Nucl. Phys. B294, 700 (1987). D. Kosower, B. H. Lee and V. P. Nair, Phys. Lett. B201, 85 (1988). M. L. Mangano, S. J. Parke and Z. Xu, Nucl. Phys. B298, 653 (1988). Z. Bern and D. A. Kosower, Nucl. Phys. B362, 389 (1991). V. Del Duca, L. J. Dixon and F. Maltoni, Nucl. Phys.B, 571, 51 (2000) [hep-ph/9910563].
12. M. L. Mangano and S. J. Parke, Phys. Rept. 200, 301 (1991). L. J. Dixon, in *QCD & Beyond: Proceedings of TASI '95*, ed. D. E. Soper (World Scientific, 1996) [hep-ph/9601359].
13. F. A. Berends, W. T. Giele and H. Kuijf, Nucl. Phys. B321, 39 (1989).
14. L. J. Dixon, Z. Kunszt and A. Signer, Nucl. Phys. B531, 3 (1998) [hep-ph/9803250].
15. Z. Bern, L. J. Dixon and D. A. Kosower, Nucl. Phys. B437, 259 (1995) [hep-ph/9409393].
16. S. J. Parke and T. R. Taylor, Phys. Rev. Lett. 56, 2459 (1986).
17. F. A. Berends and W. T. Giele, Nucl. Phys. B306, 759 (1988).
18. P. De Causmaecker, R. Gastmans, W. Troost and T. T. Wu, Phys. Lett. B**105**, 215 (1981). F. A. Berends, R. Kleiss, P. De Causmaecker, R. Gastmans, W. Troost and T. T. Wu, Nucl. Phys. B206, 61 (1982);
P. De Causmaecker, R. Gastmans, W. Troost and T. T. Wu, Nucl. Phys. B206, 53 (1982);
J. F. Gunion and Z. Kunszt, Phys. Lett. B161, 333 (1985).
19. D. A. Kosower, hep-th/0406175.
20. Z. Xu, D. H. Zhang and L. Chang, preprint TUTP-84/3-TSINGHUA, unpublished; Nucl. Phys. B **291**, 392 (1987).
21. F. Cachazo and P. Svrcek, hep-th/0504194.
22. L. Dixon, hep-ph/0507064.
23. J. Bedford, A. Brandhuber, B. J. Spence and G. Travaglini, Nucl. Phys. B **721**, 98 (2005) [hep-th/0502146]. T. G. Birthwright, E. W. N. Glover, V. V. Khoze and P. Marquard, JHEP **0505**, 013 (2005) [hep-ph/0503063]. T. G. Birthwright, E. W. N. Glover, V. V. Khoze and P. Marquard, JHEP **0507**, 068 (2005) [hep-ph/0505219]. N. E. J. Bjerrum-Bohr, D. C. Dunbar, H. Ita, W. B. Perkins and K. Risager, hep-th/0509016. K. J. Ozeren and W. J. Stirling, hep-th/0509063. M. x. Luo and C. k. Wen, JHEP **0503**, 004 (2005) [hep-th/0501121]. M. x. Luo and C. k. Wen, Phys. Rev. D **71**, 091501 (2005) [hep-th/0502009]. R. Britto, B. Feng, R. Roiban, M. Spradlin and A. Volovich, Phys. Rev. D **71**, 105017 (2005) [hep-th/0503198]. S. D. Badger, E. W. N. Glover, V. V. Khoze and P. Svrcek, JHEP **0507**, 025 (2005) [hep-th/0504159].
24. Z. Bern, L. J. Dixon and D. A. Kosower, Phys. Rev. D **72**, 045014 (2005) [hep-th/0412210]. Z. Bern, L. J. Dixon and D. A. Kosower, Phys. Rev. D **71**, 105013 (2005) [hep-th/0501240]. S. J. Bidder, N. E. J. Bjerrum-Bohr, D. C. Dunbar and W. B. Perkins, Phys. Lett. B **612**, 75 (2005) [hep-th/0502028]. Z. Bern, L. J. Dixon and D. A. Kosower, hep-th/0505055. A. Brandhuber, S. McNamara, B. J. Spence and G. Travaglini, hep-th/0506068. E. I. Buchbinder and F. Cachazo, hep-th/0506126. Z. Bern, L. J. Dixon and D. A. Kosower, hep-ph/0507005. Z. Bern, N. E. J. Bjerrum-Bohr, D. C. Dunbar and H. Ita, hep-ph/0507019. D. Forde and D. A. Kosower, hep-ph/0509358.

QCD Effective Couplings in Minkowskian and Euclidean Domains

D.V. Shirkov

Bogoliubov Lab. of Theor. Physics, JINR, Dubna, 141980, Russia

Abstract. We argue for essential upgrading of the defining equations (9.5) and (9.6) in Section 9.2. "The QCD coupling ... " of PDG review and their use for data analysis in the light of recent development of the QCD theory. Our claim is twofold. First, instead of universal expression (9.5) for $\bar{\alpha}_s$, one should use various ghost-free couplings $\alpha_E(Q^2)$, $\alpha_M(s)$ ... specific for a given physical representation. Second, instead of power expansion (9.6) for observable, we recommend to use nonpower functional ones over particular functional sets $\{\mathscr{A}_k(Q^2)\}$, $\{\mathfrak{A}_k(s)\}$ ... related by suitable integral transformations. We remind that use of this modified prescription results in a better correspondence of reanalyzed low energy data with the high energy ones.

Keywords: Non-perturbative QCD, non-power expansion
PACS: 11.10.Hi, 12.38.Cy, 12.38.Lg

1. PREAMBLE

The main message consists of two statements:

A: Instead of common effective QCD coupling $\bar{\alpha}_s$, (with its ghost defect) as, e.g., it is implicitly mentioned by eq.(9.5) of PDG review [1], one should use (at least) two different **ghost-free forms for QCD effective coupling** $\alpha_E(Q^2)$ in the Euclidean and $\alpha_M(s)$ in the Minkowskian (and, possibly, some others) pictures;

B: The RG-invariant perturbative expansions for observables, like eq.(9.6) in PDG,

$$O(\xi) = o_1\bar{\alpha}_s(\xi) + o_2\bar{\alpha}_s^2(\xi) + o_3\bar{\alpha}_s^3(\xi) + ...,$$

– in powers *of the same* $\bar{\alpha}_s$ in different pictures, Euclidean ($\xi = Q^2$) or Minkowskian ($\xi = s$) are neither based theoretically, nor adequate practically to low-energy QCD. Instead, one should use **diverse nonpower functional expansions**

$$d(Q^2) = \sum_{i\geq 1} d_i \mathscr{A}_i(Q^2), \quad r(s) = \sum_{i\geq 1} d_i \mathfrak{A}_i(s),$$

(each particular one for a given representation) over nonpower sets of ghost-free functions like $\{\mathscr{A}_k(Q^2)\}$ in Euclidean and $\{\mathfrak{A}_k(s)\}$ in Minkowskian, mutually related by suitable integral transformations.

Below we demonstrate that a reasonable revising of the above mentioned PDG Eqs. essentially modifies the results of the analysis of some low energy data like GLM and Bjorken sum-rules, τ-lepton and Ypsilon decays and $e + e-$ inclusive cross-sections (Sections 9.3., 9.4. and 9.6 in PDG).

As a result, new overall fit for Euclidean data in terms of $\alpha_E(Q^2)$ and Minkowskian data in $\alpha_M(s)$ results in (see our recent review [2]) $\bar{\alpha}_s(M_Z^2) = 0.123$ with an essentially smaller χ^2 than the commonly accepted one.

2. THE APT ESSENCE AND STRUCTURE

2a. Minkowskian And Euclidean Couplings α_M and α_E

RG defined invariant coupling $\bar{\alpha}(Q^2)$ is a real function of space-like argument Q^2. It effectively sums up UV logs into an expression with ghost. In the 1-loop QCD case

$$\bar{\alpha}_s^{(1)}(Q^2) = \frac{\alpha_s}{1 + \beta_0 \alpha_s \ln(Q^2/\mu^2)} = \frac{1}{\beta_0 \ln(Q^2/\Lambda^2)}$$

Instead, in the APT scheme[2], we deal with differing ghost-free couplings

Minkowskian: $\quad \alpha_M^{(1)}(s) = \frac{1}{\pi \beta_0} \arccos \frac{L}{\sqrt{L^2 + \pi^2}} \Big|_{L>0} = \frac{1}{\pi \beta_0} \arctan \frac{\pi}{L}; \quad$ and $\quad$ (1)

Euclidean: $\quad \alpha_E^{(1)}(Q^2) = \frac{1}{\beta_0} \left[\frac{1}{\ell} - \frac{\Lambda^2}{Q^2 - \Lambda^2} \right]; \quad \ell = \ln \frac{Q^2}{\Lambda^2}, \quad L = \ln \frac{s}{\Lambda^2}.$ (2)

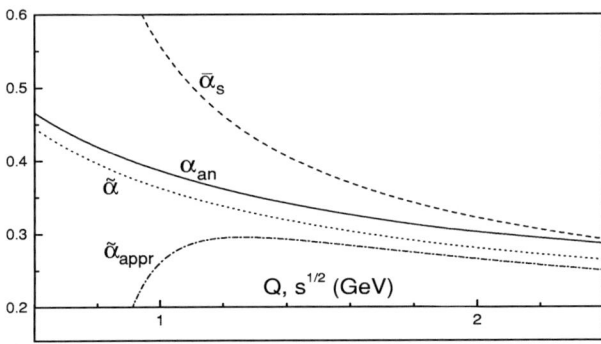

FIGURE 1. Comparison of usual QCD coupling $\bar{\alpha}_s$ with Euclidean $\alpha_{an} = \alpha_E$ and Minkowskian one $\tilde{\alpha} = \alpha_M$ in a few GeV region.

On Fig.1 one can see[1] the comparison of $\bar{\alpha}_s$ with α_E and α_M in the 1-2 GeV region. Transition to the "s picture" performed first by contour integration by Radyushkin[3], Krasnikov and Pivovarov [4], (see also [5])

$$\bar{\alpha}_s(Q^2) \rightarrow \frac{i}{2\pi} \int_{s-i\varepsilon}^{s+i\varepsilon} \frac{dz}{z} \bar{\alpha}_s(-z) = \bar{\alpha}_M(s) \equiv \mathbb{R}[\bar{\alpha}_s](s) \quad (3)$$

results in a ghost-free expression with π^2 terms summed.

Reverse transformation $\mathbb{D} = [\mathbb{R}]^{-1}$ [6, 7] yields[2] a ghost-free expression in the Q^2 picture with subtracted singularity; see below eqs.(7) and (8).

---

[1] In this figure taken from our previous papers, a bit different notation $\alpha_{an} = \alpha_E$, $\tilde{\alpha} = \alpha_M$ is used. Here $\tilde{\alpha}_{appr} = \bar{\alpha}_s - \frac{\pi^2 \beta_0^2}{3} \bar{\alpha}_s^3$. All the curves are given in the 2-loop approximation for $\Lambda = 350$ GeV.
[2] For its explicit form see below eq.(10)

2b. Minkowskian: π^2 Summation

Summation of π^2-terms by contour integration (3) for the 1-loop case results in

$$\bar{\alpha}_s^{(1)}(Q^2) = \frac{1}{\beta_0 L} \to \bar{\alpha}_M^{(1)}(s) = \frac{1}{\pi \beta_0} \arccos \frac{L}{\sqrt{L^2+\pi^2}} \equiv \mathfrak{A}_1^{(1)}(s), \quad (4)$$

$$\left[\mathfrak{A}_1^{(1)}(s)\right]_{L>0} = \frac{1}{\pi \beta_0} \arctan \frac{\pi}{L}; \quad L = \ln \frac{s}{\Lambda^2}. \quad (5)$$

This expression was first obtained by Radyushkin[3] in the form (5). Later on, Jones and Solovtsov [8] considered the region $Q^2 \leq \Lambda^2$ and proposed treating expression (4) as a ghost-free Minkowskian effective coupling.

At the same time, the procedure (3) transforms square and cube of $\bar{\alpha}_s^{(1)}$ into ghost-free forms[4]

$$\mathfrak{A}_2^{(1)} = \frac{1}{\beta_0^2 [L^2+\pi^2]}, \quad \mathfrak{A}_3^{(1)} = \frac{L}{\beta_0^3 [L^2+\pi^2]^2},$$

which are not powers of $\alpha_M^{(1)}(s)$. They are rather connected with (4) by the iterative differential relation

$$\mathfrak{A}_{k+1}(s) = -\frac{1}{k \beta_0} \frac{d \mathfrak{A}_k(s)}{d \ln s}. \quad (6)$$

2c. Euclidean: Källen-Lehmann Analyticity

APT uses imperative of the Q^2 analyticity[9] in the form of the Källen–Lehmann spectral representation [3]. Being applied to the QCD one-loop case, it gives

$$\bar{\alpha}_s^{(1)} = \frac{1}{\beta_0 \ell} \Rightarrow \mathbb{A}\left[\bar{\alpha}_s^{(1)}\right] = \alpha_E^{(1)}(Q^2) = \frac{1}{\beta_0}\left[\frac{1}{\ell} - \frac{\Lambda^2}{Q^2-\Lambda^2}\right] = \mathscr{A}_1^{(1)}(Q^2). \quad (7)$$

For coupling $\bar{\alpha}_s^{(1)}$ squared

$$\mathbb{A}\left[\frac{1}{\ell^2}\right] = \frac{1}{\ln^2(Q^2/\Lambda^2)} + \frac{Q^2 \Lambda^2}{(Q^2-\Lambda^2)^2} = \beta_0^2 \, \mathscr{A}_2^{(1)}(Q^2) \neq \left(\beta_0 \, \mathscr{A}_1^{(1)}(Q^2)\right)^2. \quad (8)$$

The Minkowskian and Euclidean ghost-free functions are related [10], [2] by $\mathbb{D}$ and $\mathbb{R}$ transformations: $\mathscr{A}_k(Q^2) = \mathbb{D}[\mathfrak{A}_k]$, $\mathfrak{A}_k(s) = \mathbb{R}[\mathscr{A}_k]$. Accordingly,

$$\mathbb{D}\left[R(s) = \sum_k d_k \mathfrak{A}_k(s)\right] \Rightarrow D(Q^2) = \sum_k d_k \mathscr{A}_k(Q^2).$$

---

[3] In the form of the first of Eqs.(9) For detail see Refs.[6, 7].

2d. Sketch Of The Global APT Algorithm

The most convenient form of the APT formalism uses a spectral density $\rho(\sigma) = Im\bar{\alpha}_s(-\sigma)$ taken from the perturbative input

$$\mathscr{A}_k = \frac{1}{\pi}\int_0^\infty \frac{\rho_k(\sigma)\,d\sigma}{\sigma+Q^2}\,, \quad \mathfrak{A}_k = \frac{1}{\pi}\int_s^\infty \frac{d\sigma}{\sigma}\rho_k(\sigma)\,. \tag{9}$$

In the 1-loop case

$$\rho_1^{(1)} = \frac{1}{\beta_0[L_\sigma^2+\pi^2]}\,; \quad L_\sigma = \ln\frac{\sigma}{\Lambda^2}\,; \quad \rho_{k+1}^{(1)}(\sigma) = -\frac{1}{k\beta_0}\frac{d\rho_k^{(1)}(\sigma)}{dL_\sigma}$$

These expressions were generalized for a higher-loop case and for real QCD with transitions across quark thresholds. This *global* APT was successively used for fitting of various data, e.g. for describing mass spectrum of light mesons [11] and for description of pion formfactor [12]. Logic of the APT scheme is displayed[4] in Fig.2.

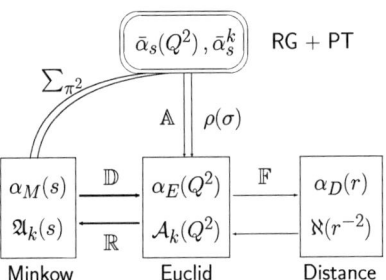

FIGURE 2. Logic of the APT scheme.

3. THE APT RESUME

3a. Non-Power Ghost-Free Sets $\{\mathscr{A}_k\}$, $\{\mathfrak{A}_k\}$

By construction, all APT expansion functions $\mathscr{A}_k$ and $\mathfrak{A}_k$ (for 2-loop etc. as well) are free of unphysical singularities and at weak-coupling limit tend to powers $\bar{\alpha}_s^k$ of common QCD coupling. On Fig.3 we demonstrate the behavior of the first three functions.

Their more detailed properties can be described as follows:

I. First ones, new couplings, α_E, α_M:
◇ are monotonic and IR finite, $\alpha_E(0) = \alpha_M(0 = 1/\beta_0 \simeq 1.4$
◇ in the UV limit $\sim 1/\ln x \sim \bar{\alpha}_s(x)$.

---

[4] Here, the distance picture with functions α_D and $\{\aleph_k\}$ is mentioned. It is related with Q^2 picture by Fourier transfirmation $\mathbb{R}$. For detail, see Ref.[13].

II. All the other functions $(k \geq 2)$:
♡ start from zero $\mathscr{A}_k(0), \mathfrak{A}_k(0) = 0$;
♡ in the UV limit $\sim 1/(\ln x)^k \sim \bar{\alpha}_s^k(x)$.
♡ 2nd ones, $\mathscr{A}_2, \mathfrak{A}_2$ obey max at $\sim \Lambda^2$.
♡ Higher ones, $\mathscr{A}_{k>3}$; $\mathfrak{A}_{k>3}$ oscillate near Λ^2 with $k-1$ zeroes.

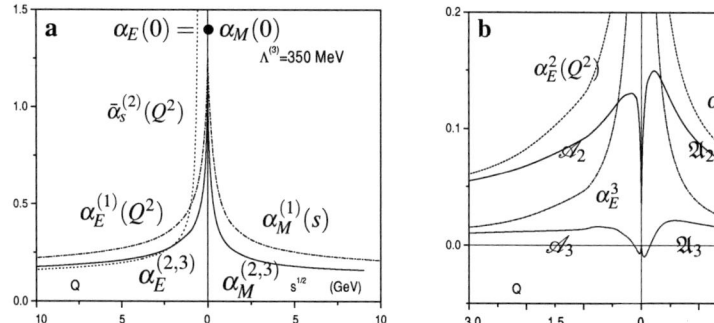

FIGURE 3. **a** – *Space-like and time-like APT couplings for 1-,2- and 3-loop case in a few GeV domain.* **b** – *"Distorted mirror symmetry" for global expansion functions. All the solid curves here correspond to exact two–loop solutions $\mathscr{A}_{2,3}$ and $\mathfrak{A}_{2,3}$. expressed in terms of the Lambert function. They are compared with powers of APT couplings α_E and α_M depicted by dotted lines.*

The last property results [12, 14] in the reduced renormalization-scheme and higher loop sensitivity and better convergence [15] in the low-energy region, see below Sect.3b.

3b. Non-Power Expansions: Quick Loop Convergence

New effective couplings are related by integral transformations (3) and

$$\alpha_E(Q^2) = Q^2 \int_0^\infty \frac{\alpha_M(s)\,ds}{(s+Q^2)^2} \equiv \mathbb{D}[\bar{\alpha}_M](Q^2). \tag{10}$$

The same transformations induce a nonpower structure

$$\mathfrak{A}_k(s) \to \mathscr{A}_k(Q^2) \equiv \mathbb{D}[\mathfrak{A}_k](Q^2)$$

of expansion functions for observables.

Due to this, instead of the "PDG–recommended" universal power-in-$\bar{\alpha}_s$ expansion,

$$d_{pt}(Q^2/s) = d_1\bar{\alpha}_s(Q^2/s) + d_2\,\bar{\alpha}_s^2 + \mathbf{d_3}\,\bar{\alpha}_s^3 \tag{11}$$

one should use non-power expansions

$$d_{\mathrm{an}}(Q^2) = d_1\,\alpha_E(Q^2) + d_2\,\mathscr{A}_2(Q^2) + \mathbf{d_3}\,\mathscr{A}_3(\mathbf{Q^2}) + \ldots, \tag{12}$$

$$r_\pi(s) = d_1\,\alpha_M(s) + d_2\,\mathfrak{A}_2(s) + \mathbf{d_3}\,\mathfrak{A}_3(s) + \ldots. \tag{13}$$

The numerical effect of this change is demonstrated in the Table 1. There, relative contributions in per cent for usual, PT 3-loop power-in-α_s expansions (11) are confronted with the APT ones (12) and (13). Besides, they are compared with the experimental error given in the last column in the same (i.e., in α_s/π) units.

Table 1. Contributions in %% of 1-, 2-,3-loop terms and data errors

Process	Energy	PT (11)			APT (12)/(13)			Exp Errors
Bjorken SR	1.6 GeV	55	26	**19**	80	19	**1**	±14
GLS SRule	1.7 GeV	65	24	**11**	75	21	**4**	±20
Incl τ-decay	1.8 GeV	55	29	**16**	88	11	**1**	±8
$e^+e^- \to$ hadr	10 GeV	96	8	**-4**	92	7	**.5**	±27
$Z_o \to$ hadr.	91 GeV	99	3.7	**-2.3**	97	3.5	**-.4**	±4

It follows that APT expansion converges much better than common PT one. Besides, the APT 3-loop term contribution is much less than data errors. Effective suppression of higher-loop terms yields also a reduced scheme [12] and loop dependence.

All these nice features of APT are connected with due account for nonanalyticity with respect to usual expansion parameter, the coupling constant at $\alpha = 0$.

3c. The QFT Nonanalyticity In Coupling

Here, we shortly remind a few general arguments on this non-analyticity.

- **General Dyson [16] argument in QED.** Transition $\alpha \to -\alpha$ corresponds to $e \to ie$; it destroys Hermiticity of Lagrangian and the S-matrix unitarity. Hence, the origin $\alpha = 0$ in the complex α plane can not be a regular point.

- **RG + Q^2-analyticity arguments.** Combining the Q^2 analyticity for a photon propagator in QED with RG invariance, one could define [17] the type of essential singularity at $\alpha = 0$ as $\sim e^{-1/\alpha}$.

- **Functional integral reasoning.** By the method of functional-integral steepest descent for propagators, it was shown[18] that expansion coefficients $c_n \alpha^n$ at $n \gg 1$ behave like $c_n \sim n!\, n^m$ which corresponds[19] to the same singularity $\sim e^{-1/\alpha}$.

3d. Analytic approximations for 2-, 3-loop $\mathfrak{A}_k$ and $\mathcal{A}_k$

Analytic expressions for 2-,3-loop APT Minkowskian $\mathfrak{A}_k$ and Euclidean $\mathcal{A}_k$ couplings involving a special Lambert function W_{-1} are rather cumbersome. Due to this, several analytic approximations for them were devised[20, 15]. In addition, we can mention[21] very simple "1-loop-like" model expressions with "two-loop effective logs" ℓ_2, L_2:

$$\mathfrak{A}_1^*(s) = \frac{1}{\pi \beta_0} \arctan \frac{\pi}{L_2}, \quad \mathfrak{A}_2^* = \frac{1}{\beta_0^2 \left[L_2^2 + \pi^2\right]} \ldots; \quad L_2 = L + b \ln L, \quad b = \frac{\beta_1}{\beta_0},$$

$$\mathscr{A}_1^{appr} = \frac{1}{\beta_0}\left(\frac{1}{l_2} - \frac{1}{\exp(l_2)-1}\right), \ldots ; \quad l_2 = \ell + b\ln\ell$$

and modified parameter $\Lambda \to \Lambda_* = f(\Lambda)$. Such analytic approximations, typically, could provide us with accuracy at the level of few %% quite adequate to practical need.

4. CONCLUSION

1. Numerous non-perturbative data (lattice simulations, Schwinger-Dyson eqs solution) reveal the ghost-free $\bar{\alpha}_s$ behavior in low energy region with finite $\bar{\alpha}_s(0)$ value.
2. The "representation invariance" implies that functional expansions – even in powers of some non-singular $\bar{\alpha}(Q^2/s)$ – are not natural and should be changed for non-power perturbative-inspired expansions; this is essential in a few GeV region.
3. Hence, in this region:
   * the notion of a single universal effective charge $\bar{\alpha}_s$ is not adequate,
   ** to correlate data, one needs two effective couplings $\alpha_E(Q^2)$ and $\alpha_M(s)$.
4. Instead of expansion (9.6) of PDG, one should use APT expansions eqs.(12) and (13) over sets of nonpower functions $\{A_k(Q^2)\}$ and $\{\mathfrak{A}_k(s)\}$..

ACKNOWLEDGEMENTS

The author is indebted to A.Bakulev, S.Mikhailov, A. Sidorov, N.Stefanis, O.Teryaev and A.Zayakin for useful discussions. This investigation was partially supported by grants RFBR 05-01.00992 and Sc.Sch 2339.2003.2.

REFERENCES

1. S.Eidelman et al., *Phys. Lett. B* (2004) **592**.
2. D V Shirkov, *Europ. Phys.J.* **C 22** 331-40 (2001); hep-ph/0107282.
3. A. Radyushkin, JINR preprint E2-82-159 (1982); also *JINR Rapid Comm.* No.4[78]-96 (1996) pp 9-15 and hep–ph/9907228.
4. N.V. Krasnikov, A.A. Pivovarov, *Phys. Lett.* **116 B** (1982) 168–170.
5. J.D. Bjorken, "Two topics in QCD", Preprint SLAC-PUB-5103 (Dec 1989); in *Proceed. Cargese Summer Institute*, eds. M.Levy et al., Nato Adv.Inst., Serie B, vol.223, Plenum, N.Y., 1990.
6. D.V. Shirkov and I.L. Solovtsov, *JINR Rapid Comm.* No.2[76]-96, pp 5-10; hep-ph/9604363.
7. D.V. Shirkov and I.L. Solovtsov, *Phys. Rev. Lett.* **79** 1209-12 (1997); hep-ph/9704333.
8. H.F. Jones and I.L. Solovtsov, *Phys. Let.* **B 349** 519-525 (1995).
9. N.N. Bogoliubov, A.A. Logunov and D.V. Shirkov, Sov. Phys. JETP, **37(10)** 574 (1960).
10. K.A. Milton and I.L. Solovtsov,*Phys. Rev.* **D 55**, 5295-5298 (1997); hep-ph/9611438.
11. M. Baldicchi, G. M. Prosperi, AIP Conf.Proc.756:152-161,2005; hep-ph/0412359
12. A. P. Bakulev, A. I. Karanikas and N. G. Stefanis, *Phys. Rev.* **D 72**, 074015 (2005); hep-ph/0504275
13. D.V. Shirkov, "Nonpower Expansions for QCD Observables at Low Energy", hep-ph/0408272; to appear in *Nucl.Phys. (Proc. Suppl)* (2005).
14. D.V. Shirkov and I.L. Solovtsov, *Phys.Lett.* **442** (1998) 344-348; hep-ph/9711251.
15. A. P. Bakulev, S. V. Mikhailov and N. G. Stefanis, *Phys. Rev.* **D 72**, 074014 (2005); hep-ph/0506311
16. F.J. Dyson, *Phys. Rev.* **85** 631 (1952).
17. D.V. Shirkov, *Lett. Math. Phys.* **1** 179 (1976).
18. L.N Lipatov, *Sov. Fiz ZHETP*, **72** 411 (1977)
19. E.B. Bogomolny, *Phys. Lett.* **67 B** 193 (1977)
20. K.A. Milton, I.L. Solovtsov, O.P. Solovtsova, Phys. Rev. **D 65**, 076009 (2002);hep-ph/0111197].
21. D.V. Shirkov and A.V. Zayakin, in preparation.

Deconstructed Higgsless Models

Roberto Casalbuoni

Dipartimento di Fisica dell' Universita' di Firenze and Sezione INFN, Via G. Sansone 1, 50019 Sesto Fiorentino (Firenze), Italy. E-mail: casalbuoni@fi.infn.it

Abstract. We consider the possibility of constructing realistic Higgsless models within the context of deconstructed or moose models. We show that the constraints coming from the electro-weak esperimental data are very severe and that it is very difficult to reconcile them with the requirement of improving the unitarity bound of the Higgsless Standard Model. On the other hand, with some fine tuning, a solution is found by delocalizing the standard fermions along the lattice line, that is allowing the fermions to couple to the moose gauge fields.

Keywords: Higgsless Models, Higher dimensional theories, Kaluza-Klein modes
PACS: 12.60.Cn, 11.25.Mj, 12.39.Fe

1. HIGHER DIMENSIONAL GAUGE THEORIES

In the past few years a renewal of interest in higher dimensional theories came out of the possibility of sub-millimiter extra dimensions due to the softening of gravitational theories in a subspace [1, 2]. In this way a strong gravitational interaction in D space-time dimensions ($D > 4$) might give rise to a weak gravitational interaction in the usual 4 dimensions. If the extra dimensions, $d = D - 4$, are compactified, one gets a relation between the Planck scale M_D in D dimensions and the four-dimensional one, M_P

$$M_P^2 = R^d M_D^{2+d}, \tag{1}$$

with R the compactification radius. By choosing $R \gg M_D^{-1}$ one can make $M_D^2 \ll M_P^2$. As an example, with $M_D = 1\ TeV$ and $d = 2$ one gets $R \approx 0.1\ mm$.

On the other hand, gauge theories in higher dimensional spaces offer extra bonus as the possibility of realizing a geometrical Higgs mechanism. As an example we consider an abelian gauge theory in 4+1 dimensions:

$$\mathscr{L} = -\frac{1}{2g_5^2}F_{AB}F^{AB} = -\frac{1}{2g_5^2}F_{\mu\nu}F^{\mu\nu} - \frac{1}{g_5^2}F_{\mu 5}F^{\mu 5}. \tag{2}$$

Here g_5 is the gauge coupling in 5D having dimensions of $M^{-1/2}$, A, B are the space-time indices in D dimensions, and μ, ν the usual 4-dimensional indices. Furthermore

$$F_{AB} = \partial_A A_B - \partial_B A_A, \tag{3}$$

Performing the gauge transformation (with the understanding that we omit the zero mode of the operator ∂_5)

$$A_B \to A_B - (\partial_5)^{-1}(\partial_B A_5), \tag{4}$$

CP806, *QCD@Work 2005: International Workshop on Quantum Chromodynamics*
edited by P. Colangelo, F. De Fazio, E. Nappi, and G. Nardulli
© 2006 American Institute of Physics 0-7354-0302-3/06/$23.00

we get
$$A_5 = 0 \Rightarrow F_{\mu 5} = -\partial_5 A_\mu. \tag{5}$$

If the fifth dimension is compactified on a circle S^2 of length $2\pi R$, the non zero eigenmodes A_μ^n of the fields A_μ acquire a mass $M_n = n/R$ since in this case

$$A_\mu(x_\mu, x_5) = \sum_n e^{inx_5/R} A_\mu^n(x_\mu). \tag{6}$$

However the zero mode remains massless and a GB is present. This zero mode can be eliminated compactifying the model on an orbifold, that is on the coset S^2/Z, Z being the discrete group of reflections along the fifth dimension:

$$Z: \quad x_5 \to -x_5. \tag{7}$$

This allows to define fields as eigenstates of Z

$$A_B(x_\mu, x_5) = \pm A_B(x_\mu, -x_5). \tag{8}$$

In this way various possibilities open up. As an instance, by taking the odd eigenstates no zero modes are in the spectrum and one gets only massive gauge bosons. In other words we have obtained massive gauge bosons in the framework of a gauge theory without Higgs fields. If the extra dimension is discretized [3, 4] one gets a so-called deconstructed gauge theory [5, 6]. In this construction the connection field along the fifth dimension, A_5, gives rise to a non-linear σ–field. In fact a gauge field is nothing but a connection, that is a way of relating the phases of fields at nearby points. Once the space is discretized the connection goes naturally into a link variable realizing the parallel transport between two lattice sites. The link variable $\Sigma_i = e^{-iaA_5^{i-1}}$ satisfies the condition $\Sigma\Sigma^\dagger = 1$ and it can be identified with a chiral field. In fact, if we consider a non-abelian gauge theory acting on the five-dimensional space, through discretization of the fifth dimension we get a discrete infinity of four-dimensional gauge theories each of them acting at a particular lattice site. It can be easily seen that the Σ_i fields transform according to

$$\Sigma_i \to U_{i-1} \Sigma_i U_i^\dagger, \tag{9}$$

with U_{i-1} and U_i group transformations belonging to the gauge group G located at the lattice sites $i-1$ and i respectively. Then the covariant derivatives of the chiral fields can be connected with the field strengths $F_{\mu 5}$ by

$$D_\mu \Sigma_i = \partial_\mu \Sigma_i - iA_\mu^{i-1} + i\Sigma_i A_\mu^i \approx -iaF_{\mu 5}^{i-1}, \tag{10}$$

where a is the lattice size. In this way the discretized version of our original 5-dimensional gauge theory is substituted by an infinite collection of four-dimensional gauge theories with gauge interacting chiral fields Σ_i

$$S = \int d^4x \frac{a}{g_5^2} \left(-\frac{1}{2} \sum_i \text{Tr}\left[F_{\mu\nu}^i F^{\mu\nu i}\right] + \frac{1}{a^2} \text{Tr}\left[(D_\mu \Sigma_i)(D^\mu \Sigma_i)^\dagger\right] \right). \tag{11}$$

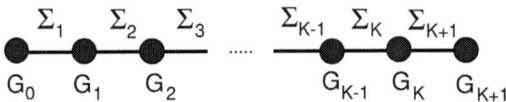

FIGURE 1. The diagram illustrates a deconstructed theory described by the gauge groups G_i and by the chiral fields Σ_i.

The theory obtained in this way is just an example of a larger set of theories generically called "deconstructed theories" [5] synthetically described by a moose diagram (see Fig. 1).

2. BREAKING THE EW SYMMETRY WITHOUT HIGGS FIELDS

As we have seen in the previous Section, abstracting from the 5-dimensional example one can study more general moose geometries. The general structure will consist in many copies of the gauge group G intertwined by link variables Σ. Now suppose that we want to describe the electro-weak (EW) symmetry breaking in this context. The condition we have to satisfy is that, before the EW gauge group is introduced, **3 massless Goldstone bosons should be present (to give masses to $W^\pm$ and Z) and all the moose gauge fields should be massive**. In the simplest case we take all the moose gauge groups equal to $SU(2)$. Then, each Σ field is an $SU(2)$ matrix

$$\Sigma_i = e^{i\vec{\pi}\cdot\vec{\tau}/(2f_i)}, \tag{12}$$

with $\vec{\tau}$ the Pauli matrices. Therefore each Σ_i describes three spin zero fields (π_i). In a connected moose diagram any site (containing three gauge fields) may absorb one link (the 3 Goldstones π_i) giving rise to three massive gauge bosons. Therefore our condition translates into

$$number\ of\ links = number\ of\ sites + 1. \tag{13}$$

The simplest of these moose is the "linear moose" whose diagram is given in Fig. 2. The corresponding action is

$$S_{moose} = \int d^4x \left(-\sum_{i=1}^{K} \frac{1}{2} \text{Tr}\left[F^i_{\mu\nu}F^{\mu\nu i}\right] + \sum_{i=1}^{K+1} f_i^2 \text{Tr}\left[(D_\mu \Sigma_i)(D^\mu \Sigma_i)^\dagger\right] \right). \tag{14}$$

We have now K gauge groups $SU(2)$ and $K+1$ chiral fields. Notice that the model has two global symmetries G_L and G_R associated to the chiral fields Σ_1 and Σ_{K+1}

$$\Sigma_1 \to U_L \Sigma_1, \quad \Sigma_{K+1} \to \Sigma_{K+1} U_R^\dagger. \tag{15}$$

As such they have been associated to the ends of the moose in Fig. 2. It is this global symmetry, $G_L \otimes G_R = SU(2)_L \otimes SU(2)_R$, that is gauged by the standard group $SU(2)_L \otimes U(1)$, in order to give the standard massive gauge bosons $W^\pm$ and Z and the

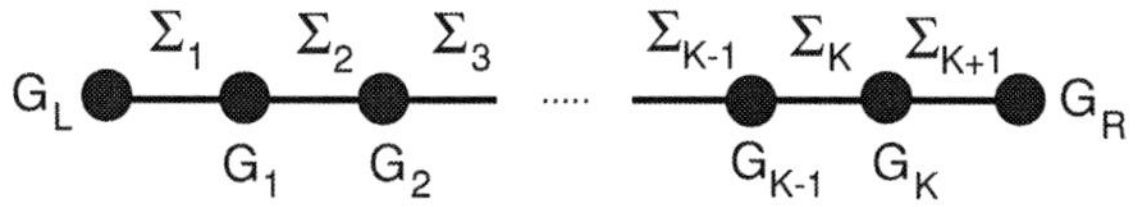

FIGURE 2. The simplest moose diagram for the Higgsless breaking of the EW symmetry.

massless photon. In fact, the three Goldstones remaining after that the moose gauge fields have eaten up the chiral fields are just the ones necessary for the breaking of the EW symmetry. Prototypes of this theory are the BESS model for $K = 1$ [7] and its generalizations [8].

3. EW CORRECTIONS FOR THE LINEAR MOOSE

If the moose vector fields are heavy enough it is possible to derive an effective action describing only the Standard Model (SM) fields. By denoting the typical mass of the moose vector fields by M_V, at the leading order in $(M_W/M_V)^2$ one gets the usual SM relations

$$M_W^2 = \frac{v^2}{4}g^2, \quad M_Z^2 = \frac{M_W^2}{c_\theta^2}, \quad e = gs_\theta = g'c_\theta, \quad (16)$$

with ($v \approx 250\ GeV$)

$$\frac{4}{v^2} \equiv \frac{1}{f^2} = \sum_{i=1}^{K+1} \frac{1}{f_i^2}. \quad (17)$$

In this class of models all the corrections from new physics arise from mixing of the SM vector bosons with the moose vector fields and therefore are oblique corrections. As well known the oblique corrections are completely captured by the parameters S, T and U [9, 10] or, equivalently by the parameters ε_i, $i = 1,2,3$ [11, 12]. For the linear moose, the existence of the global symmetry (custodial) $SU(2)_V$ ensures that

$$\varepsilon_1 = \varepsilon_2 = 0, \quad (18)$$

or, equivalently $U = T = 0$.

To compute the new physics contribution to the electroweak parameter ε_3 [11] we will make use of the dispersive representation given in Refs. [9, 10] for the related parameter S ($\varepsilon_3 = g^2 S/(16\pi)$)

$$\varepsilon_3 = -\frac{g^2}{4\pi} \int_0^\infty \frac{ds}{s^2} Im\left[\Pi_{VV}(s) - \Pi_{AA}(s)\right], \quad (19)$$

where g is the $SU(2)_L$ gauge coupling and $\Pi_{VV}(AA)$ is the current-current correlator

$$\int d^4x e^{-iq\cdot x} \langle J^\mu_{V(A)} J^\nu_{V(A)} \rangle = ig^{\mu\nu}\Pi_{VV(AA)}(q^2) + (q^\mu q^\nu\ \text{terms}). \quad (20)$$

$J_{V/A\mu}$ are the vector and axial currents associated to the global symmetry $SU(2)_L \otimes SU(2)_R$, getting the following contributions from the moose vector fields

$$J^a_{V\mu}\Big|_{\text{vector mesons}} = f_1^2 g_1 A^{1a}_\mu + f_{K+1}^2 g_K A^{Ka}_\mu,$$
$$J^a_{A\mu}\Big|_{\text{vector mesons}} = f_1^2 g_1 A^{1a}_\mu - f_{K+1}^2 g_K A^{Ka}_\mu. \quad (21)$$

It should be noticed that the ε_3 parameter is evaluated with reference to the SM, and therefore the corresponding contributions should be subtracted. For instance the contribution of the pion pole to Π_{AA}, that is of the Goldstone particles giving mass to the W and Z gauge bosons, does not appear in ε_3. As described previously, in the model described by the action (14) all the new physics contribution comes from the new vector bosons (we are assuming the standard couplings for the fermions to $SU(2)_L \otimes U(1)$). Therefore from

$$Im\Pi_{VV(AA)} = -\pi \sum_{Vn,An} g^2_{nV,nA} \delta(s - m_n^2), \quad (22)$$

we get

$$\varepsilon_3 = \frac{g^2}{4} \sum_n \left(\frac{g^2_{nV}}{m_n^4} - \frac{g^2_{nA}}{m_n^4} \right), \quad (23)$$

where $g_{nV/A}$ are the decay coupling constants of the moose vector fields defined by

$$\langle 0|J^a_{V\mu}|\tilde{A}^n_b(p,\varepsilon)\rangle = g_{nV}\delta^{ab}\varepsilon_\mu, \quad \langle 0|J^a_{A\mu}|\tilde{A}^n_b(p,\varepsilon)\rangle = g_{nA}\delta^{ab}\varepsilon_\mu, \quad (24)$$

and $\tilde{A}^n_b(p,\varepsilon)$ are the mass eigenstates of the moose vector bosons. As shown in [13] we can express ε_3 in two equivalent ways (see also [14, 15])

$$\varepsilon_3 = g^2 g_1 g_K f_1^2 f_{K+1}^2 (M_2^{-2})_{1K} = g^2 \sum_{i=1}^K \frac{(1-y_i)y_i}{g_i^2}, \quad (25)$$

where M_2 is the matrix of the square masses of the moose vector bosons, and

$$y_i = \sum_{j=1}^i x_j, \quad x_i = \frac{f^2}{f_i^2}, \quad \frac{1}{f^2} = \sum_{i=1}^{K+1} \frac{1}{f_i^2} \Rightarrow \sum_{i=1}^{K+1} x_i = 1. \quad (26)$$

Since $0 \leq y_i \leq 1$ it follows $\varepsilon_3 \geq 0$ (see also [14, 16, 17]). As an example, let us take all the link couplings f_i equal to a common value f_c, and the same for the gauge couplings $g_i = g_c$. Then (see also [18])

$$\varepsilon_3 = \frac{1}{6}\frac{g^2}{g_c^2}\frac{K(K+2)}{(K+1)}. \quad (27)$$

If we want to be compatible with the experimental data we need to get $\varepsilon_3 \approx 10^{-3}$. For $K = 1$ this would require $g_c \geq 15.8g$, implying a strong interacting gauge theory in the moose sector. Notice also that insisting on a weak gauge theory would imply g_c of the order of g, let us say $g_c \approx 2 \div 5g$. Then the natural value of ε_3 would be of the order $10^{-1} - 10^{-2}$, incompatible with the experimental data.

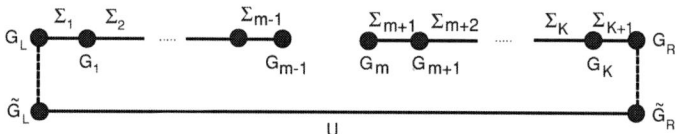

FIGURE 3. The diagram illustrates how a cut link model is generated from a linear moose.

Possible ways of evading the ε_3 problem have been considered in [13]. A way is to cut a link, that is to assume one of the link couplings, say f_m, equal to zero. In this case the matrix of the mass square of the moose vector bosons becomes block diagonal and, as a consequence, the same happens for M_2^{-2}. Therefore $(M_2^{-2})_{1K} = 0$ and $\varepsilon_3 = 0$. Since, suppressing a link amounts to eliminate three scalar fields, we need a way to reintroduce them. The Σ_m field can be reintroduced through a discretized version of a Wilson line

$$U = \Sigma_1 \Sigma_2 \cdots \Sigma_K \Sigma_{K+1}, \tag{28}$$

and inserting in the lagrangian a term

$$f_0^2 Tr[\partial_\mu U^\dagger \partial^\mu U]. \tag{29}$$

This term has a global invariance $\tilde{G}_L \otimes \tilde{G}_R = SU(2)_L \otimes SU(2)_R$ originating from a transformation $U \to \tilde{U}_L U \tilde{U}_R$. This invariance is different from the original $G_L \otimes G_R$ before the EW gauging. As a consequence the model has an enhanced custodial symmetry $[SU(2)_L \otimes SU(2)_R]$ which is enough to ensure $\varepsilon_3 = 0$ [19]. A particular example of this model, for $K = 2$ ($D - BESS$), was studied in [20, 21] (originally introduced in [8]).

Another possibility [13] is to suppress a link, that is to assume a hierarchy among the links. As an example assume an exponential behavior

$$f_i = \bar{f} e^{c(i-1)}, \quad g_i = g_c. \tag{30}$$

From Fig. 4 we see that there is a big suppression factor, of order 10^{-2} already for $c = 2$. In fact expanding at the leading order for large c it is easily seen that

$$\varepsilon_3 \to \frac{g^2}{g_c^2} e^{-2c}. \tag{31}$$

However, lowering or cutting the links may give rise to unitarity problems. For instance, in the cut model f_0 must be of the order of the v.e.v. of the Higgs field in the SM making the unitarity limit of these class of model the same as in the SM without Higgs. We will study the unitarity limits of the moose models in the next Section.

4. UNITARITY BOUNDS FOR THE LINEAR MOOSE

The worst high-energy behavior of the moose models arises from the scattering of longitudinal vector bosons. To simplify the calculation we will make use of the equivalence

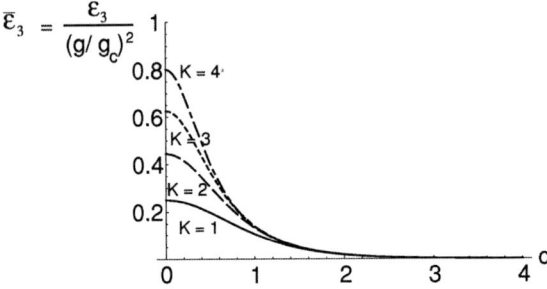

FIGURE 4. The behavior of ε_3 (normalized to $(g/g_c)^2$) as a function of c in the exponentially suppressed linear model for different values of K.

theorem, that is of the possibility of evaluating this amplitude in terms of the scattering amplitude of the corresponding Goldstone bosons [22]. However this theorem holds in the approximation where the energy of the process is much higher of the mass of the vector bosons. We will consider two situations. In the first one we assume that all the moose vectors have a mass, M_{V_i}, much higher than the SM vector boson masses, in such a way that we can evaluate the amplitude for the SM W and Z at energies $M_{W/Z} \ll E \ll M_{V_i}$. The only Goldstone bosons of interest here are the ones giving mass to W and Z. The unitary gauge for these bosons is given by the choice

$$\Sigma_i = e^{if\vec{\pi}\cdot\vec{\tau}/(2f_i^2)}, \tag{32}$$

with f given in eq. (17). The resulting four-pion amplitude is

$$A_{\pi^+\pi^-\to\pi^+\pi^-} = -\frac{f^4 u}{4}\sum_{i=1}^{K+1}\frac{1}{f_i^6} + \frac{f^4}{4}\sum_{i,j=1}^{K} L_{ij}\left((u-t)(s-M_2)_{ij}^{-1} + (u-s)(t-M_2)_{ij}^{-1}\right), \tag{33}$$

with

$$L_{ij} = g_i g_j \left(\frac{1}{f_i^2} + \frac{1}{f_{i+1}^2}\right)\left(\frac{1}{f_j^2} + \frac{1}{f_{j+1}^2}\right). \tag{34}$$

This expression reproduces correctly the low-energy limit, $E \ll M_{V_i}$:

$$A_{\pi^+\pi^-\to\pi^+\pi^-} \to -\frac{f^4 u}{4}\left(\sum_{i=1}^{K+1}\frac{1}{f_i^2}\right)^3 = -\frac{u}{4f^2} = -\frac{u}{v^2}, \tag{35}$$

whereas in the high-energy limit, where we can neglect the second term,

$$A_{\pi^+\pi^-\to\pi^+\pi^-} = -\frac{f^4 u}{4}\sum_{i=1}^{K+1}\frac{1}{f_i^6}. \tag{36}$$

The best unitarity limit is obtained for all the f_i's being equal to a common value f_c. In this case

$$A_{\pi^+\pi^-\to\pi^+\pi^-} = -\frac{u}{(K+1)v^2}, \qquad (37)$$

leading to the unitarity bound

$$\Lambda_{moose} = (K+1)\Lambda_{HSM} \approx 1.2(K+1)\ TeV, \qquad (38)$$

where Λ_{HSM} is the unitary bound for the Higgsless SM. In this case it is possible to improve as much as we like the unitarity bound of the SM increasing K. However this would lead to contradictions with the experimental bounds on ε_3.

As a second instance we consider an energy much higher than all the masses of the vector bosons. In this case to determine the unitarity bounds one has to consider the eigenchannel amplitudes corresponding to all the possible four-longitudinal vector bosons. But, since the unitary gauge for all the vector bosons is simply given by the expression (12), the amplitudes are already diagonal, and the result at high energy is simply

$$A_{\pi^+\pi^-\to\pi^+\pi^-} \to -\frac{u}{4f_i^2}. \qquad (39)$$

We see that the unitarity limit is determined the smallest link coupling. Therefore in the exponentially suppressed model the unitarity bound is essentially the same as in the SM, since in order to respect the constraint given by the first equality in eq. (17), the lowest coupling must be of order v. Also in this case the best unitarity limit is for all the link couplings being equal $f_i = f_c$. Then (for similar results see [23])

$$\Lambda_{moose} = \sqrt{K+1}\Lambda_{HSM} \approx 1.2\sqrt{K+1}\ TeV. \qquad (40)$$

However, in order our approximation is correct we have to require $M_{V_i}^{max} \ll \Lambda_{moose}$, and since we expect roughly (assuming $g_c \approx g$) $M_{V_i}^{max} \approx KM_W$, we get a bound $\sqrt{K} \ll 14$. By taking $\sqrt{K}$ of order $2 \div 3$ one could improve of the same factor the SM unitarity bound, but again this would be hardly compatible with the electro-weak experimental data.

5. DELOCALIZING FERMIONS

As we have seen, it is not possible to satisfy at the same time the experimental bounds on ε_3 and improve in a sensible way the unitarity limit. A way out has been considered in [24, 25, 26] allowing delocalized couplings of the SM fermions to the moose gauge fields and some amount of fine tuning. In fact, the SM fermions can be coupled to any of the gauge fields staying at the lattice sites by means of a Wilson line. However, we will consider only left-handed fermions, since analogous interactions for the right-handed ones are very much constrained [7, 27]. Define

$$\chi_L^i = \Sigma_i^\dagger \Sigma_{i-1}^\dagger \cdots \Sigma_1^\dagger \psi_L. \qquad (41)$$

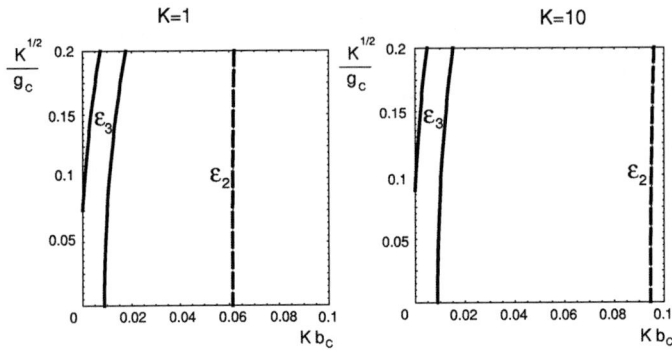

FIGURE 5. The 95% C.L. allowed region in the plane $(Kb_c, \sqrt{K}/g_c)$ is the region on the left delimited by the two continuous lines coming form the bounds on ε_3. The dashed line comes from a bound on ε_2, whereas the other bound form ε_2 and the bounds from ε_1 are out of the figure. The radiative corrections have been assumed as in the SM with $m_H = 1\ TeV$ and $m_{top} = 178\ GeV$.

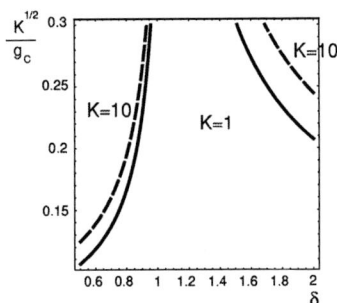

FIGURE 6. The 95% C.L. allowed regions in the plane $(\delta, \sqrt{K}/g_c)$ are the ones at the interior of the lines corresponding to $K = 1$ (continuous) and $K = 10$ (dashed). The radiative corrections have been chosen as in Fig. 5.

Then, under a gauge transformation, $\chi_L^i \to U_i \chi_L^i$, with $U_i \in G_i$. We see that at each site we can introduce a gauge invariant coupling given by

$$b_i \bar{\chi}_L^i \gamma^\mu \left(\partial_\mu + i g_i A_\mu^i + \frac{i}{2} g'(B-L) Y_\mu \right) \chi_L^i. \tag{42}$$

The expressions for the parameters ε_i are modified, and at first order in the couplings b_i we get

$$\varepsilon_1 \approx 0, \quad \varepsilon_2 \approx 0, \quad \varepsilon_3 \approx \sum_{i=1}^{K} y_i \left(\frac{g^2}{g_i^2}(1-y_i) - b_i \right). \tag{43}$$

Therefore, with some amount of fine tuning is possible to agree with the electro-weak experimental data. To show it, let us take again all the link couplings equal to f_c and the gauge couplings equal to g_c. We have considered two possibilities. In the first one we take also the b_i equal to a common value b_c. Then the allowed region in the space $(Kb_c, \sqrt{K}/g_c)$ (we have chosen these parameters due to the scaling properties of g_c and b with K) is given in Fig. 5.

In the second case we require a sort of local cancelation, assuming (again $g_i = g_c$, $f_i = f_c$)

$$b_i = \delta \frac{g^2}{g_i^2}(1 - y_i) = \delta \frac{g^2}{g_c^2}\left(1 - \frac{i}{K+1}\right). \tag{44}$$

The allowed region in the space $(\delta, \sqrt{K}/g_c)$ is given in Fig. 6. In this way it is possible to satisfy the EW constraints and improve the unitarity bound of the Higgsless SM at the same time.

REFERENCES

1. N. Arkani-Hamed, S. Dimopoulos, and G. R. Dvali, Phys. Lett. **B429**, 263 (1998), hep-ph/9803315.
2. I. Antoniadis, N. Arkani-Hamed, S. Dimopoulos, and G. R. Dvali, Phys. Lett. **B436**, 257 (1998), hep-ph/9804398.
3. C. T. Hill, S. Pokorski, and J. Wang, Phys. Rev. **D64**, 105005 (2001), hep-th/0104035.
4. H.-C. Cheng, C. T. Hill, S. Pokorski, and J. Wang, Phys. Rev. **D64**, 065007 (2001), hep-th/0104179.
5. N. Arkani-Hamed, A. G. Cohen, and H. Georgi, Phys. Rev. Lett. **86**, 4757 (2001), hep-th/0104005.
6. N. Arkani-Hamed, A. G. Cohen, and H. Georgi, Phys. Lett. **B513**, 232 (2001), hep-ph/0105239.
7. R. Casalbuoni, S. De Curtis, D. Dominici, and R. Gatto, Phys. Lett. **B155**, 95 (1985).
8. R. Casalbuoni, S. De Curtis, D. Dominici, F. Feruglio, and R. Gatto, Int. J. Mod. Phys. **A4**, 1065 (1989).
9. M. E. Peskin and T. Takeuchi, Phys. Rev. Lett. **65**, 964 (1990).
10. M. E. Peskin and T. Takeuchi, Phys. Rev. **D46**, 381 (1992).
11. G. Altarelli and R. Barbieri, Phys. Lett. **B253**, 161 (1991).
12. G. Altarelli, R. Barbieri, and F. Caravaglios, Int. J. Mod. Phys. **A13**, 1031 (1998), hep-ph/9712368.
13. R. Casalbuoni, S. De Curtis, and D. Dominici, Phys. Rev. **D70**, 055010 (2004), hep-ph/0405188.
14. J. Hirn and J. Stern, Eur. Phys. J. **C34**, 447 (2004), hep-ph/0401032.
15. H. Georgi, Phys. Rev. **D71**, 015016 (2005), hep-ph/0408067.
16. R. Barbieri, A. Pomarol, and R. Rattazzi, Phys. Lett. **B591**, 141 (2004), hep-ph/0310285.
17. R. S. Chivukula, M. Kurachi, and M. Tanabashi, JHEP **06**, 004 (2004), hep-ph/0403112.
18. R. Foadi, S. Gopalakrishna, and C. Schmidt, JHEP **03**, 042 (2004), hep-ph/0312324.
19. T. Inami, C. S. Lim, and A. Yamada, Mod. Phys. Lett. **A7**, 2789 (1992).
20. R. Casalbuoni et al., Phys. Lett. **B349**, 533 (1995), hep-ph/9502247.
21. R. Casalbuoni et al., Phys. Rev. **D53**, 5201 (1996), hep-ph/9510431.
22. J. M. Cornwall, D. M. Levin, and G. Tiktopoulos, Phys. Rev. **D11**, 1145 (1974).
23. R. S. Chivukula and H.-J. He, Phys. Lett. **B532**, 121 (2002), hep-ph/0201164.
24. R. Casalbuoni, S. De Curtis, D. Dolce, and D. Dominici, Phys. Rev. **D71**, 075015 (2005), hep-ph/0502209.
25. R. S. Chivukula, E. H. Simmons, H.-J. He, M. Kurachi, and M. Tanabashi, Phys. Rev. **D71**, 115001 (2005), hep-ph/0502162.
26. R. S. Chivukula, E. H. Simmons, H.-J. He, M. Kurachi, and M. Tanabashi, Phys. Rev. **D72**, 015008 (2005), hep-ph/0504114.
27. R. Casalbuoni, S. De Curtis, D. Dominici, and R. Gatto, Nucl. Phys. **B282**, 235 (1987).

COULD SPIN-CHARGE SEPARATION BE THE SOURCE OF CONFINEMENT?

Antti J. Niemi

Department of Theoretical Physics, Uppsala University P.O. Box 803, S-75108, Uppsala, Sweden;
Laboratoire de Mathematiques et Physique Theorique CNRS UMR 6083, Universite de Tours,
Parc de Grandmont, F37200, Tours, France

Abstract. Yang-Mills gauge field with gauge group $SU(2)$ decomposes into a single charge neutral complex vector, and two spinless charged scalar fields. At high energies these constituents are tightly confined into each other by a compact $U(1)$ interaction, and the Yang-Mills Lagrangian describes the dynamics of asymptotically free massless gauge vectors. But in a low energy and finite density environment the interaction between the constituents can become weak, and a spin-charge separation may occur. We suggest that the separation between the spin and charge with the ensuing condensation of the charged scalars takes place when the Yang-Mills theory enters confinement. The confining phase becomes then surprisingly similar to the superconducting phase of a high-T_c superconductor.

Keywords: confinement
PACS: 11.15.-q, 12.38.Aw

INTRODUCTION

According to popular folklore color (quark) confinement follows from an electric version of the BCS mechanism. This proposal is based on an assumption that the confining string is an electric version of the Abrikosov vortex [1]

The Abrikosov vortex is present in a type-II superconductor, where electrons condense into Cooper pairs. It is a static string-like configuration along which an undamped magnetic field line penetrates into the superconducting material. When such a magnetic vortex line forms between static particles with opposite magnetic charges (if such particles exist) it leads to a confining force that increases linearly in distance between the particles.

But a magnetic vortex line does not lead to a confining force between static, electrically charged particles. For the confinement of electrically charged particles such as electrons, one needs vortex lines that conduct an electric field. However, the observation that magnetically charged point particles are confined by magnetic Abrikosov vortices provides an attractive picture for explaining the confinement of quarks: Suppose the confining string is an analog of an Abrikosov vortex and suppose the quarks have a charge which couples to the component of the Yang-Mills field that is conducted along the string. Then quark confinement can be explained in the same way as the confinement of magnetic point charges in type-II superconductors is explained by (magnetic) Abrikosov vortices.

The quarks couple to the Yang-Mills field minimally, in the same manner as electrons couple to Maxwell's field in QED. As a consequence the confining string must couple to quarks in a manner which is different from the coupling between an electron and an Abrikosov vortex. Instead of a (nonabelian) magnetic field, the confining string must be a carrier of a (nonabelian) electric field, it must be an electric dual version of the Abrikosov vortex.

The BCS picture of quark confinement is consistent with the structure of N=2 and N=1 supersymmetric Yang-Mills theories [2]. In these theories we have elementary Higgs fields that can describe the Cooper pairing and condensation of magnetic monopoles. This leads to an electric dual version of the Meissner effect and to the ensuing confinement of (nonabelian) electrically charged particles such as quarks. This supersymmetry approach to confinement is intimately based on the existence and properties of the elementary Higgs fields, and confinement is basically a consequence of a relatively straightforward extension of the BCS theory.

But in order to implement the BCS picture in a pure Yang-Mills theory we first need to understand how to describe vortices in an appropriate magnetic condensate.

In all known physical scenarios where vortices are present, vorticity is supported by some kind of a medium. In ordinary liquids such as helium superfluids or water, a vortex is formed in a concrete material environment. In a spontaneously broken (gauge) theory vorticity is supported by a (material) condensation of the relevant order parameter.

But in a pure Yang-Mills theory there is no apparent medium, no elementary Higgs field that could condense. Since there are no known vortex configurations that are formed in the absence of a supporting medium, we have a fundamental problem in pure non-supersymmetric Yang-Mills theory: The formation of a confining string between quarks necessitates the introduction of a medium that carries vorticity. But there is no known mechanism how a medium could be constructed or described in a pure Yang-Mills theory.

In order to characterize a material environment that can support vorticity, we need some kind of a fundamental or effective (Higgs-like) field that can condense. In a pure Yang-Mills theory, the emergence of an effective Higgs field would mean that we can introduce some kind of a mechanism that leads to the formation of a condensation that consists of gluons. Since no such gluonic version of Cooper pair formation is known, we then either need to develop new concepts and structures for describing vorticity, or alternatively we need to explain how an effective Higgs field could arise from outside of the pure Yang-Mills theory.

The Abrikosov vortex in a type-II superconductor is supported by a condensate that consists of Cooper pairs of electrons. As a consequence it does not confine electrons, even though it can confine magnetically charged point particles. Thus it is unlikely that the Cooper pairing of quarks can lead to a confining force between quarks. In order to explain quark confinement by a version of the BCS formalism, one needs instead a Cooper pairing of (nonabelian) magnetically charged particles. This means the confining flux tube must arise from the Yang-Mills field, and it receives no contribution from the condensation of quarks into (colored) Cooper pairs.

In the wider context of the Standard Model it is intrinsically possible, but highly unlikely, that the Higgs field of the electro-weak sector could provide a condensate that also supports the confining string in the strong sector of the theory. At the moment there are no theoretical arguments that anything like this could happen. The confinement of quarks appears to be an intrinsic property of the strong sector of the theory, with no contribution from the electroweak sector. Furthermore, at the moment we do not even have any experimental evidence that a fundamental electroweak Higgs exists. If it can not be found, we may well have a very similar problem in both the strong and electroweak sector of the standard model, the absence of a fundamental Higgs field that describes a condensate.

In a lattice formulation of Yang-Mills theory the problem of a fundamental Higgs field can be avoided, by placing a singular vortex line between the lattice sites. The finiteness of the lattice site then ensures the absence of singularities in the theory, at least as long as the lattice site is finite. But it remains to be explained how anything like this could be implemented in the continuum limit of the theory.

Finally, it could be that instead of a material vortex structure the confining string has an intrinsic string theory description. But in order to describe an intrinsic string, it is necessary to introduce additional structures that are beyond a pure Yang-Mills theory: The intrinsic string approach to confinement would involve hypothetical properties of the space-time that are at the moment unknown, besides that the pure Yang-Mills theory should emerge as a particular limit of the description.

Maybe 30 years of intense but unsuccesfull efforts by the theoretical community to construct a magnetic Cooper pair condensate in a pure Yang-Mills theory should be viewed as evidence that quark confinement can not be explained by the BCS formalism. In fact, we propose that there is no *a priori* reason why any version of the BCS formalism should explain confinement in a pure Yang-Mills theory, there is no evidence of any kind of magnetic Cooper pair formation. It could be that confinement in a pure, non-supersymmetric Yang-Mills theory is due to an as yet unidentified mechanism which is quite different from the BCS picture.

Curiously, a very similar problem is also present in high temperature superconductivity where the implementation of the BCS formalism has thus far also failed: there is no theoretical or experimental evidence that the electrons form Cooper pairs in superconducting cuprates [3]. While the Cooper pair formation can not be definitely excluded, and there may even be some experimental support for a Cooper pair formation, the lack of any clear evidence for electron condensation into Cooper pairs has led to new ways for describing high-T_c superconductivity. Curiously, the situation there is surprisingly similar to that in strong interaction physics:

In the case of strong interaction physics, Yang-Mills theory is widely accepted. Similarly, in the case of high temperature superconductivity there is a consensus that the materials can be described by a definite theory, the $t-J$ model. In analogy to Yang-Mills theory, in this model there are no fundamental or effective Higgs fields that could support vortex structures with the ensuing Meissner effect. Consequently, at the moment, there is no theoretical understanding how BCS formalism could be implemented to explain high-T superconductivity. This has led to speculations that maybe high temperature superconductivity is due to a mechanism which is fundamentally different from the BCS formalism.

Could it then be, that high temperature superconductivity in $t-J$ model has an origin which is similar to the origin of quark confinement in a Yang-Mills theory?

The lack of a Cooper pair in the $t-J$ model has led to a very interesting theoretical proposal which, if correct, has far reaching consequences to our understanding of the fundamental structure of Matter. This proposal is based on the very radical idea [4], [3] that in the strongly correlated environment of cuprate superconductors an electron ceases to be a fundamental particle. Instead an electron is a bound state of two other particles, which are called spinon and holon. The spinon is a fermion that carries the spin degree of freedom of the electron. It does not directly couple to Maxwell's electrodynamics. The holon is a spinless, complex boson and it carries the electric charge of the electron. Under normal circumstances the spinon and holon are tightly bound into each other by a confining force, consistent with the observational fact that at high energies an electron behaves as a structureless point particle. But in the strongly correlated environment of cuprate superconductors the force between the spinon and holon could become weak, and a spin-charge separation may take place. A holon condensation can then provide a material environment that support vorticity, leading to the Meissner effect and an explanation of superconductivity [3].

FERMIONS

In order to outline the slave-boson decomposition of an electron we start from a four-dimensional Dirac spinor ψ_D^a. Here $a = 1,...,4$ label its four anticommuting components that obey the (graded) Poisson bracket

$$\{\psi_D^{a\dagger}(x), \psi_D^b(y)\} = \delta^{ab}(x-y) \tag{1}$$

We select the Weyl basis of the γ-matrices,

$$\gamma^\mu = \begin{pmatrix} 0 & \sigma^\mu \\ \bar{\sigma}^\mu & 0 \end{pmatrix}$$

where $\sigma^0 = \bar{\sigma}^0$ is the 2×2 unit matrix, and $\sigma^i = -\bar{\sigma}^i$ ($i = 1,2,3$) are the standard Pauli matrices. In this basis we represent the Dirac fermion as

$$\psi_D = \begin{pmatrix} \xi_\alpha \\ \chi^{\dagger\dot\alpha} \end{pmatrix}$$

where ξ_α and $\chi^{\dagger\dot\alpha}$ (with $\alpha, \dot\alpha = 1,2$) are two-component Weyl fermions. The spinor indices are raised and lowered using the antisymmetric tensors $\varepsilon_{\alpha\beta}$ and $\varepsilon^{\dot\alpha\dot\beta}$ with non-vanishing components determined by setting $\varepsilon^{12} = \varepsilon_{21} = 1$. Explicitely, we have e.g. $\xi^\alpha = \varepsilon^{\alpha\beta}\xi_\beta$ and $\chi^\dagger_{\dot\alpha} = \varepsilon_{\dot\alpha\dot\beta}\chi^{\dagger\dot\beta}$. Furthermore, when we introduce the conjugate variables $\chi^\dagger_{\dot\alpha} = (\sigma_0)_{\alpha\dot\beta}\chi^{\dagger\dot\beta}$ and $\xi^\alpha = (\bar\sigma_0)^{\dot\alpha\beta}\xi_\beta$ we get the graded Poisson brackets

$$\{\chi^\alpha(x), \chi^\dagger_\beta(y)\} = \delta^\alpha{}_\beta(x-y) \quad \& \quad \{\xi^\dagger_{\dot\alpha}(x), \xi^{\dot\beta}(y)\} = \delta_{\dot\alpha}{}^{\dot\beta}(x-y) \tag{2}$$

The relativistic version of the slave-boson decomposition is obtained by setting

$$\chi^\alpha = b^\dagger \cdot f^\alpha + \varepsilon^{\alpha\gamma} f^\dagger_\gamma \cdot d \tag{3}$$

For the right-handed Weyl spinor $\xi_{\dot\alpha}^\dagger$ we introduce an analogous decomposition, but here we do not need to display it explicitly. Here b and d are bosonic fields, they are the *holons* and subject to the Poisson brackets

$$\{b^\dagger(x), b(y)\} = \{d^\dagger(x), d(y)\} = \delta(x-y)$$

The f_α is an anticommuting (left-handed) Weyl spinor. It is the *spinon* and it obeys the graded Poisson bracket

$$\{f^\alpha(x), f_\beta^\dagger(y)\} = \delta^\alpha{}_\beta(x-y)$$

As a consequence, when we substitute the slave-boson decomposition (3) in (2), we find that the decomposed Weyl fermion χ^α obeys the graded Poisson bracket

$$\{\chi^\alpha(x), \chi_\beta^\dagger(y)\} == \delta^\alpha{}_\beta(x-y) \cdot \{f^\gamma f_\gamma^\dagger + b^\dagger b + d^\dagger d\}$$

We also verify that

$$\{\chi^\alpha(x), \chi^\beta(y)\} = 0$$

Thus the decomposed field (3) reproduces the entire Poisson bracket structure of the original Weyl fermion χ^α provided we introduce the constraint

$$N = f^\gamma f_\gamma^\dagger + b^\dagger b + d^\dagger d = 1 \tag{4}$$

With this constraint, the decomposition (3) then becomes an operator identity.

More generally, we can set

$$N = f^\gamma f_\gamma^\dagger + b^\dagger b + d^\dagger d = \mu \tag{5}$$

where μ is some function. It can be selected arbitrarily, with the sole condition that $\mu(x)$ is non-vanishing for *all* x. This ensures that the resulting Poisson brackets of the decomposed fermion χ^α continue to define a graded symplectic two-form. The only difference between (4) and (5) is, that when $\mu \neq 1$ the decomposed fermions are graded canonical variables which are not of the Darboux form.

The condition (4), and its more general version (5), can be interpreted as the statement that for a separation between spin and charge, the fermionic system must be in a physical environment with a finite density, and the density is determined by the function $\mu(x)$. If this density vanishes for some x, the Poisson brackets of the decomposed fermion fail to reproduce the symplectic structure of the original fermion, and a spin-charge separation can not occur. In particular, for all fields b, d, f to have well defined Poisson brackets so that they can be dynamical, each of the number densities $b^\dagger b$, $d^\dagger d$, $f^\alpha f_\alpha^\dagger$ must be nonvanishing: An isolated electron can not become decomposed into its spin and charge consituents, for a separation we need a material finite density environment.

Both the holons b and d and the spinon f_α are complex fields. Consequently the decomposition (3) has an internal local $U(1)$ symmetry, the Weyl fermion χ^α in (3) remains intact when we send

$$b \to e^{i\theta} b \quad \& \quad d \to e^{i\theta} d \quad \& \quad f^\alpha \to e^{i\theta} f^\alpha \tag{6}$$

We note that this symmetry is generated by the canonical Poisson bracket action of the number operator N in (4). It is a *compact* $U(1)$ symmetry, that leads to an interaction between the holons and spinons. For a large value of its coupling, a compact $U(1)$ interaction is known to be confining. Thus we expect that (6) in general leads to an interaction between the spinons and holons which in a non-material environment where μ vanishes confines them into the (pointlike) fermion.

Conventionally, we couple Maxwell's eletromagnetism to the canonical charge operator defined by

$$Q = \chi^\alpha \chi^\dagger_\alpha$$

When we compute the canonical Poisson bracket action of Q on the Weyl spinor χ^α using the decomposed representation (3), we get from (5)

$$\{Q(x), \chi^\alpha(y)\} = N(x) \cdot \chi^\alpha(x)\delta(x-y) = \mu(x) \cdot \chi^\alpha(x)\delta(x-y)$$

This states that $\mu(x)$ coincides with the local charge density at x. Clearly, this canonical action of Q on the decomposed spinor can be reproduced by the canonical action of

$$\bar{Q} = -\mu(x) \cdot [b^\dagger b - d^\dagger d]$$

This confirms that the holons b and d become (oppositely) charged under the standard coupling of a Weyl fermion to Maxwellian electromagnetism, while the spinon f^α is electrically neutral. Thus the spinless holons indeed carry the entire electric charge of the Weyl (Dirac) fermion while its entire spin is carried by the charge neutral spinon.

In the ultraviolet, individual fermions such as quarks and leptons behave like structureless point particles. Consequently in the ultraviolet region there must be a very strong confining interaction between their holon and spinon constituents. This is consistent with the verity, that the β-function of an abelian gauge theory such as the compact $U(1)$ interaction between holons and spinons should not display asymptotic freedom in the ultraviolet limit. Instead, it is natural to expect that the internal $U(1)$ interaction becomes strongly coupled and confining when we approach the ultraviolet limit. Thus the present slave-boson decomposition of a Dirac (Weyl) fermion is consistent with the experimental observation that at high energies and low densities elementary particles such as leptons and quarks behave asymptotically as structureless point particles.

But at low energy scales it is feasible that a compact $U(1)$ theory becomes weakly coupled. In an infrared environment where the constraint (4) is obeyed, a Weyl fermion may then become split into its independent holon and spinon constituents. It has been proposed [4], [3] that for an electron such a decomposition could take place in strongly correlated cuprate superconductors. The (d-wave) high-T_c superconductivity can then emerge in a phase where a spinon pairing becomes accompanied by a holon condensation,

$$<b> \neq 0$$

with a consequential spontaneous breaking of the internal $U(1)$ symmetry.

It is conceivable, that a slave-boson decomposition of a (relativistic) fermion could also occur in environments such as Early Universe when the density was very large, or in the interior of hadronic matter when energies are not very high. In these high density environments the number operators for the holons and spinons are presumably nonvanishing which implies that the ensuing Poisson brackets are nontrivial so that both spinons and holons can become dynamical physical degrees of freedom.

In order to test the relevance of the slave-boson decomposition in a given physical scenario, one needs in addition to substitute the decomposed fermion into the corresponding Hamiltonian. One can then verify whether or not the spinons and holons can indeed describe propagating degrees of freedom in the environment of interest, in a normal manner. In the case of the $t-J$ model, under conditions that are supposed to describe high-T_c superconductivity, the decomposed Hamiltonian does admit a natural intrepretation in terms of holons and spinons as particle-like excitations. This suggests, that a separation between spin and charge may take place. The theoretical and physical consequences of this scenario have been discussed widely in the literature and we refer to [3] for details.

GAUGE FIELDS

We are curious, whether a similar separation between spin and charge could also occur in the case of a Yang-Mills theory, and whether this could lead to an understanding of confinement [5], [6], [7]. For simplicity we shall only consider a pure $SU(2)$ Yang-Mills theory in a four dimensional space R^4 with Euclidean signature. But a generalization to more general gauge group $SU(N)$ in a Minkowskian signature space is straightforward.

We represent the gauge field as a linear combination

$$A_\mu = A_{\mu i}\sigma^i = C_\mu \sigma^3 + X_{\mu+}\sigma^+ + X_{\mu-}\sigma^- \tag{7}$$

where $\sigma^\pm = 1/2(\sigma^1 \pm i\sigma^2)$ and

$$X_{\mu\pm} = A_{\mu 1} \mp iA_{\mu 2}$$

Our slave-boson decomposition of A_μ entails a decomposition of $X_{\mu\pm}$ into its spin and charge constituents. For this, we introduce a complex vector field e_μ which we normalize according to

$$\begin{aligned} \vec{e}^2 &= 0 \\ \vec{e}\cdot\vec{e}^* &= 1 \end{aligned} \tag{8}$$

With ψ_1 and ψ_2 two complex scalars we can then write $X_{\mu\pm}$ as [5]

$$X_{\mu+} = X_{\mu-}^* = i\psi_1 e_\mu - i\psi_2^* e_\mu^* \tag{9}$$

Indeed, *any* four component complex vector can always be represented as a linear combination of the form (9). For this, it suffices to observe that an arbitrary, unconstrained four component complex vector describes eight independent real field degrees of freedom. On the other hand, the two complex fields ψ_1 and ψ_2 describe four, and the complex vector $\vec{e}$ when subject to the conditions (8) describes five independent field degrees of freedom. But one of these corresponds to the internal $U(1)$ rotation

$$\begin{aligned} \vec{e} &\longrightarrow e^{-i\xi}\vec{e} \\ \psi_1 &\longrightarrow e^{i\xi}\psi_1 \\ \psi_2 &\longrightarrow e^{i\xi}\psi_2 \end{aligned} \tag{10}$$

which leaves the r.h.s. of (9) intact. As a consequence, in the general case the r.h.s. of (9) also describes eight independent field degrees of freedom.

For simplicity, we may assume that the off-diagonal components $X_{\mu\pm}$ are subject to the maximal abelian gauge condition

$$D_\mu^{ij}[C]X_{\mu j} = (\partial_\mu \mp iC_\mu)X_{\mu\pm} \stackrel{def}{=} D_\mu^\pm X_{\mu\pm} \tag{11}$$

However, we shall not impose any condition on the diagonal component C_μ. As a consequence the gauge condition (11) removes two of the gauge degrees of freedom in A_μ. This leaves us with a $U(1) \in SU(2)$ gauge invariance, which corresponds to gauge transformations in the Cartan direction of $SU(2)$. Indeed, when we specify

$$g \to h = e^{i\omega\sigma^3} \tag{12}$$

we get

$$C_\mu\sigma^3 + X_{\mu+}\sigma^+ + X_{\mu-}\sigma^- \xrightarrow{h} (C_\mu + 2\partial_\mu\omega)\sigma^3 + e^{2i\omega}X_{\mu+}\sigma^+ + e^{-2i\omega}X_{\mu-}\sigma^- \tag{13}$$

while the condition (11) clearly remains intact.

When the $X_{\mu\pm}$ are subject to the condition (11), in the representation (9) there are *a priori* restrictions both on the scalars ψ_1 and ψ_2, and on the vector $\vec{e}$. But we now argue that (11) can be naturally interpreted as a restriction solely on the absolute values ρ_1 and ρ_2 of the complex fields ψ_1 and ψ_2. Indeed, consider the functional

$$\int d^4x\, X_{\mu+}X_{\mu-} = \int d^4x (|\psi_1|^2 + |\psi_2|^2) = \int d^4x (\rho_1^2 + \rho_2^2) \tag{14}$$

This is manifestly invariant under the abelian gauge transformation (13). But if we subject the *unconstrained* $X_{\mu\pm}$ to an arbitrary infinitesimal $SU(2)$ gauge transformation and demand that (14) remains stationary, the ensuing Euler-Lagrange equation coincides with the maximal abelian gauge condition [8]

$$\delta_g \int d^4x\, X_{\mu+}X_{\mu-} = 0 \Rightarrow (\partial_\mu \mp iC_\mu)X_{\mu\pm} \equiv D_\mu^\pm X_{\mu\pm} = 0$$

Notice that the functional (14) involves only the two absolute values ρ_1 and ρ_2. Since the Euler-Lagrange equation *i.e.* the maximal abelian gauge condition (11) gives two independent conditions, we can use it to solve for the two absolute values ρ_1 and ρ_2 in terms of the other variables. In the maximal abelian gauge (11) both of the ρ_1 and ρ_2 then acquire their (gauge invariant) extrema values along the $SU(2)$ gauge orbit.

We observe, that when we use the condition (11) and solve for ρ_1 and ρ_2, we introduce *no* restrictions on the complex vector $\vec{e}$. Nor do we introduce any restrictions on the phases of the complex fields ψ_1 and ψ_2. In particular, this means that the internal symmetry (10) remains intact when we evaluate the absolute values ρ_1 and ρ_2 at their gauge invariant extrema along the gauge orbit.

We note that in general there are Gribov ambiguities in the maximal abelian gauge condition. Consequently the extrema values of ρ_1 and ρ_2 on the orbit are not unique. Here we will not analyze the consequences that Gribov ambiguities might have.

The diagonal $U(1) \subset SU(2)$ gauge transformation (13) acts on the complex fields $\psi_{1,2}$ as follows,

$$\begin{aligned} \psi_1 &\to e^{2i\omega}\psi_1 \\ \psi_2 &\to e^{-2i\omega}\psi_2 \end{aligned} \qquad (15)$$

Here the phases differ from those in (10) by a relative sign. Since this $U(1)$ transformation leaves the vector $\vec{e}$ intact, only the complex fields ψ_1 and ψ_2 couple to the Cartan subgroup $U(1) \subset SU(2)$. On the other hand, the components e_μ transform as a vector under Lorenz transformations while the fields ψ_1 ad ψ_2 are scalars. This means that (9) entails a decomposition of $X_{\mu\pm}$ into two qualitatively very different sets of fields: The scalar fields ψ_1 and ψ_2 couple nontrivially to the abelian component of the $SU(2)$ gauge transformations *i.e.* carry a charge but have no spin. The complex vector $\vec{e}$ is neutral w.r.t. the abelian component of the gauge transformation but it carries the spin degrees of freedom of the $X_{\mu\pm}$.

As in the fermionic case, for consistency of the decomposition (9) we must assume that *both* condensates $\rho_{1,2}$ are nontrivial. This means, that for a spin-charge decomposition to occur in the quantum Yang-Mills theory we need *both* expectation values

$$<\rho_{1,2}> = \Delta_{1,2} \qquad (16)$$

to be nonvanishing. This condition then specifies the *material* environment where the separation between the spin and the charge of a gauge field can occur.

It is apparent that the present slave-boson decomposition of the gauge field is fully analogous to the slave-boson decomposition of the Dirac (Weyl) fermion: In both cases, the decomposition entails a separation between the carriers of spin, and the carriers of charge. Furthermore, in both cases the separation can only occur in a finite density environment. In the case of a fermion we need the μ in (5) to be non-vanishing and in the case of gauge field we need the condensates (16) to be non-vanishing. Furthermore, in both cases the decomposition introduces an internal, compact $U(1)$ that can be employed to argue that asymptotically in the short distance limit both the gauge field and the fermion become structureless point particles, with the spinon and holon confined to each other by the strong internal force. The internal spin-charge structures can then be visible only in the infrared region and in a finite density environment, when the internal $U(1)$ interaction becomes weak.

In analogy with high-temperature superconductivity, it becomes natural to propose that confinement in $SU(2)$ Yang-Mills theory is described by a a phase where spin-charge separation occurs and *both* condensates (16) are nonvanishing, with the ensuing vortices describing the confining strings. There are tentative numerical results [7], obtained by analysing the London limit of the Yang-Mills quantum theory, that indicate that confinement can indeed be related to the non-vanishing of *both* order parameters (16).

But until now, no serious lattice results have been presented to test this proposal. Such a serious lattice simulation would not only test whether the holon condensation could relate to confinement. It would also test the fundamental structure of Matter, whether the known elementary particles could indeed be composites of more fundamental constituents that describe their independent spin and charge degrees of freedom.

ACKNOWLEDGMENTS

I thank Ludvig Faddeev for various discussions and collaboration that led to the development of the ideas presented here. I also wish to thank the organizers of the *QCD@Work* 2005 for giving me an opportunity to present my results. I also wish to thank the Universities of Kyoto and Tokyo, and APCTP for hospitality during this work. This research has been supported by a VR Grant and by a STINT Thunberg scholarship.

REFERENCES

1. R.W. Haymaker, Phys. Rept. **315** (1999) 153; M. N. Chernodub and M. I. Polikarpov, in *Confinement, Duality and Non-Perturbative Aspects of QCD* (Plenum Press, New York, 1998)
2. N. Seiberg and E. Witten, Nucl. Phys. **B426** (1994) 19
3. P.A. Lee, N. Nagaosa and X.-G. Wen, cond-mat/0410445
4. P.W. Andersson, Science **235** (1987) 1196; L.D. Faddeev and L. Takhtajan, Phys. Lett. **A85** (1981) 375
5. L.D. Faddeev and A.J. Niemi, Phys. Lett. **B525** (2002) 195; Phys. Rev. Lett. **82** (1999) 1624
6. A.J. Niemi, JHEP 0408 (2004) 035
7. A.J. Niemi and N. Walet, Phys. Rev. **D72** (2005) 054007
8. F.V. Gubarev, L.Stodolsky and V.I. Zakharov, Phys. Rev. Lett. **86** (2001) 2220

Approximated Faddeev-Niemi Knotted Solitons

Andrzej Wereszczyński

Institute of Physics, Jagiellonian University, Reymonta 4, Kraków, Poland

Abstract. Application of the eikonal knots and their generalization in the context of the Skyrme-Faddeev-Niemi model is discussed. We show that eikonal knots appear in a very natural way in the SFN theory as solutions of the integrability condition. Moreover, they allow for analytical investigation of qualitative (Hopf index and geometry) as well as quantitative (energy) features of the SFN hopfions.

Keywords: Nonperturbative techniques; extended classical solutions
PACS: 11.15.Tk, 11.27.+d

FADDEEV-NIEMI MODEL AND THE EIKONAL EQUATION

The Skyrme-Faddeev-Niemi model [1], [2], [3] which widely believed to be a good candidate for the classical effective model for the nonperturbative low energy quantum gluodynamics, is given by the following action

$$S = \int d^4x \frac{1}{2} m^2 (\partial_\mu \vec{n})^2 - \frac{1}{4e^2} [\vec{n} \cdot (\partial_\mu \vec{n} \times \partial_\nu \vec{n})]^2, \qquad (1)$$

where the unit three-component vector field $\vec{n}$ represents the infrared relevant degrees of freedom of the underlying quantum theory. In the framework of this model particle-like excitations of the quantum color field i.e. glueballs are described as knotted solitons carrying topological Hopf index Q_H [4]. In spite of the fact that there is quite large numerical evidence for such objects [5], [6], the analytical results are scarcely known [7]. In the consequence, many important questions concerning geometrical features of the stable (lowest energy) and metastable states in a fixed topological sector are still unsolved.

This work is devoted for analytical investigation of the SFN model. In particular, we solve the integrability condition (and its generalization) of the SFN model and apply obtained knotted solutions to approximation of the SFN hopfions.

It has been recently shown [8] that one can define an integrable subsystem for the original (non-integrable) SFN model if the following condition is imposed

$$\mathcal{K}_\mu \partial^\mu u = 0, \qquad (2)$$

where

$$\mathcal{K}_\mu = m^2 \partial_\mu u - \frac{4}{e^2} \frac{K_\mu}{(1+|u|^2)^2} \quad \text{and} \quad K_\mu = (\partial^\nu u \partial_\nu u^*) \partial_\mu u - (\partial^\nu u)^2 \partial_\mu u^*. \qquad (3)$$

Here the complex scalar field is related with the vector field via the stereographic projection. The integrability is understood as existence of the infinite number (not

Noether) conserved currents

$$J_\mu = L_\mu \frac{\partial G}{\partial u} - L_\mu^* \frac{\partial G}{\partial u^*}, \qquad (4)$$

where G is any function of u and u^*.
The condition (2) can be satisfied in two ways. Namely, one can assume that the mass parameter is equal to zero or that the eikonal equation is obeyed

$$(\partial_\nu u)^2 = 0. \qquad (5)$$

The first possibility is trivial since it leads to solutions being unstable under the scale transformations. Thus, the eikonal equation is the unique integrability condition for the SFN theory. One should keep in mind that the full integrable subsystem consists of two parts: the non-dynamical constrain i.e. the eikonal equation (2) and the dynamical equation

$$\partial_\mu \left[m^2 \partial_\mu u - \frac{4}{e^2} \frac{\partial^\nu u \partial_\nu u^*}{(1+|u|^2)^2} \partial_\mu u \right] = 0. \qquad (6)$$

In this paper only the first equation will be analyzed. The simplest solutions of the integrability condition (2) take the form [9]

$$u = A \sinh^{\pm|n|} \eta \, \frac{\left(|m| \cosh \eta + \sqrt{n^2 + m^2 \sinh^2 \eta}\right)^{\pm|m|}}{\left(|n| \cosh \eta + \sqrt{n^2 + m^2 \sinh^2 \eta}\right)^{\pm|n|}} e^{i(m\xi + n\phi)}, \qquad (7)$$

where toroidal coordinates have been introduced: $x = \tilde{a} \sinh \eta \cos \phi / q$, $y = \tilde{a} \sinh \eta \sin \phi / q$, $z = \tilde{a} \sin \xi / q$, $q = \cosh \eta - \cos \xi$. Here, m, n are integer parameters and A is a complex constant. The point is that these solutions are topologically nontrivial and carry the Hopf index $Q_H = \pm mn$. The geometrical features of the solutions can be studied if we recall the definition of the position of the knot. It is given by a curve where the vector field takes antipodal value to the value at the spatial infinity. Thus, solutions (7) represent nothing else but unknots. It means that the core of the knot is a circle at $\eta = \infty$ and surfaces of constant n^3 component are toruses. It is dissimilar to SFN model where really knotted structures have been reported.
However, taking into account a huge symmetry of the eikonal equation [9], [10] we are able to construct more general (knotted, linked) and physically more relevant configurations [11]. In fact, the general topological solution is given as

$$u \to F(u), \qquad (8)$$

where F is any (anti)holomorphic function. Let us discus some examples in detail.

i) torus knots
Really knotted solutions is given as follow

$$u \to u + c_0, \qquad (9)$$

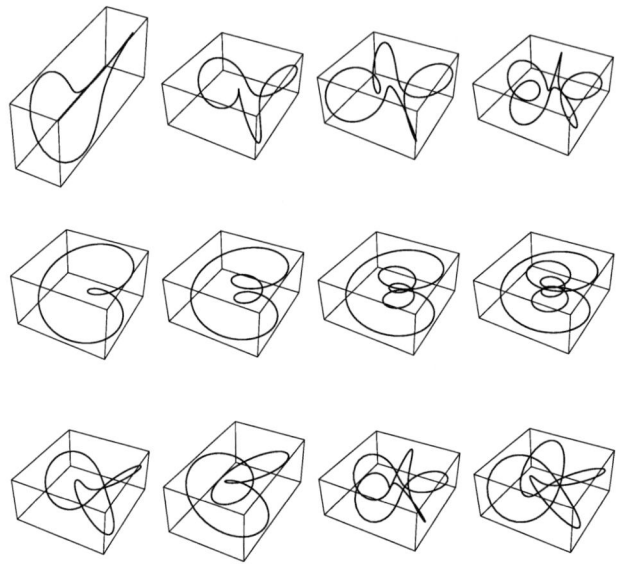

FIGURE 1. Position of the core of the eikonal knots for various (m,n) = (1,2),(1,3),(1,4),(1,5), (2,1),(3,1),(4,1),(5,1), (2,3),(3,2),(2,5),(3,4).

where c_0 is a nonzero complex constant and the parameters m,n in the basic solution u are assumed to be relative prime numbers. Then the core of the knot is a curve $m\xi + n\phi = \pi$ located on a torus with a constant radius η_0, $f(\eta_0) = |c_0|$ (see fig. 1).

ii) linked knots
Linked knots can be obtained taking the following solutions

$$u \to u^N + c_0, \tag{10}$$

where N is a positive number. Now, we have N elementary torus knots which are linked together (see fig. 2). Every knot is located on the same torus, with radius η_0, and carries the same topological charge $Q_H^{elem} = \pm mn$. To correctly calculate the total Hopf index one has to take into account the linking number L as well. Then we get this formula

$$Q_H = \pm(Nmn + L). \tag{11}$$

iii) other example
As an example of more sophisticated solution let us consider

$$u \to u^{N_1} + u^{N_2}, \tag{12}$$

where $N_1 > N_2$. In this case, in addition to $N_1 - N_2$ torus knots lying on the same torus and possessing $Q_H^{elem} = \pm mn$ topological charge, there is a unique central unknot located at $\eta = \infty$ which carries $Q_H^{centr} = \pm(N_1 - N_2 + 1)mn$ (see fig. 3).

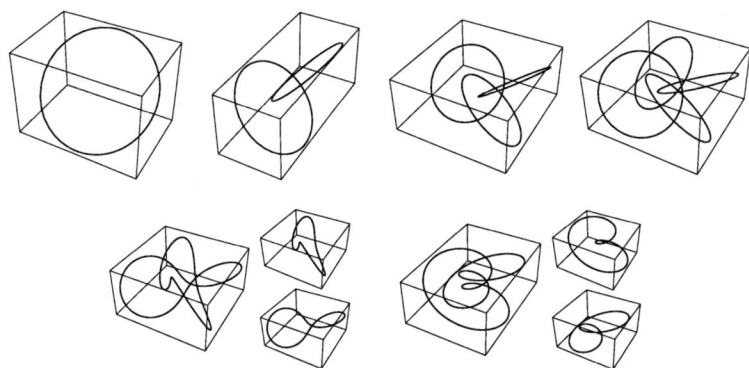

FIGURE 2. Position of the core of the linked eikonal knots for (m,n) = (1,1) and $N = 1,2,3,4$; (1,2) and $N = 2$; (2,1) and $N = 2$ with elementary knots.

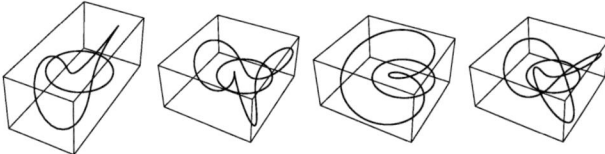

FIGURE 3. Position of the core of the linked eikonal knots for $N_1 = 2, N_2 = 1$ and various (m,n) = (1,2),(1,3),(2,1),(2,3).

As we see, the eikonal equation, which appears in a very natural way in the context of the SFN model as the integrability condition, allows for various knotted and linked configurations with arbitrary value of the Hopf charge. However, there is a restriction

TABLE 1. Energy of the eikonal knots and the SFN hopfions

Q_H	type of the knot (m,n)	E_{min}	E_{num}	accuracy
1	(1,1)	304.3	252.0	20.0%
2	(1,2)	467.9	417.5	12.0 %
	(2,1)	602.7		
3	(1,3)	658.1	578.5	13.8 %
	(3,1)	997.3		
4	(1,4)	855.5	743.0	15.0%
	(2,2)	914.3		
	(4,1)	1466.4		
5	(1,5)	1056.1	905.0	16.8 %
	(5,1)	2000.0		

for obtained knotted structures. Only torus knots (trefoil, solomon knot etc.) i.e. knots which can be plotted on a torus have been found. Apparently, there is a problem with construction of other, so-called hyperbolic knots (figure-8 etc.). Unfortunately, we do not know whether it is only a technical problem, and hyperbolic knots can be found using a better, more suitable Ansatz, or there is a profound reason which excludes such knots from the set of solutions of the eikonal equation.

Rather unexpectedly, the eikonal knots, if inserted to the total energy integral calculated for the SFN model, lead to finite energy configurations [11]. The lowest energy eikonal state E_{min} in a fixed topological sector approximates numerical hopfion E_{num} [5] with more less 15% accuracy (see tab. 1).

GENERALIZATION OF THE EIKONAL KNOTS

The eikonal equation can be generalized to the following form [12]

$$(\partial_\nu u)^2 (\partial_\nu u^*)^2 + \alpha (\partial_\nu u \partial^\nu u^*)^2 = 0, \qquad (13)$$

where α is a real constant. In fact, such a generalized eikonal equation appears in the context of integrability of nonlinear sigma models in higher dimensions, where allows us to define new integrable subsystems [12]. This equation possesses the following topological solutions

$$u(\eta,\xi,\phi) = A \sinh^{\pm a|k|} \eta \frac{\left(|m|\cosh\eta + \sqrt{k^2+m^2\sinh^2\eta}\right)^{\pm a|m|}}{\left(|k|\cosh\eta + \sqrt{k^2+m^2\sinh^2\eta}\right)^{\pm a|k|}} e^{i(m\xi+n\phi)}, \qquad (14)$$

where

$$a^2(\alpha) = \frac{(1-\alpha)}{(1+\alpha)} \pm \sqrt{\frac{(1-\alpha)^2}{(1+\alpha)^2} - 1}.$$

Obviously, such solutions carry the Hopf charge $Q_H = \pm mn$. Moreover, using the same symmetry as in the case of the eikonal equation we are able to find the general solution as $F(u)$. Thus, the geometrical as well as topological features of the generalized and standard eikonal knots are identical. They differ only via the profile function.

The generalized eikonal knots may be also applied to the SFN model. It has been checked that they give even better approximation to the energy. On an average, we have 2% improvement in the accuracy (see tab.2).

GENERAL TORUS KNOT APPROXIMATION

As we have shown the eikonal knots as well as their generalization provide good approximation to the SFN hopfions. This observation, combined with the fact that both types of the configurations differ only by the profile function, suggests that the general torus (un)knot Ansatz

$$u(\eta,\xi,\phi) = f(\eta) e^{i(m\xi+n\phi)} \qquad (15)$$

TABLE 2. Energy of the generalized eikonal knots and the SFN hopfions

Q_H	a	E_{min}	E_{num}	accuracy
1	1.170	296.0	252.0	17.5%
2	0.954	467.0	417.5	11.8 %
3	0.885	651.0	578.5	12.5 %
4	0.859	840.9	743.0	13.0%
5	0.859	1034.1	905.0	14.4 %

may father improve accuracy. Of course, in order to guarantee the proper topological behavior, the profile function should obey the boundary conditions: $f(\eta = 0) = \infty$ and $f(\eta = \infty) = 0$ (or inversely). In fact, such a general Ansatz has been previously applied to the SFN model [5], [6], [13]. Namely, it can be substituted into the total energy integral calculated for SFN action. Then, after minimizing it with respect to the profile function as well as the scale parameter, the best approximation in a fixed topological sector can be obtained (however this approximation is given only in a numerical form).

It is quite striking that the general torus knot Ansatz (15) allows us to achieve better approximation than in the eikonal case which defines the integrable sector for the model. It might indicate that there can be determined some new integrable sectors in the SFN model more adequate in the context of the knotted solitons.

This work is partially supported by Z. Wróblewski Stipends Founds and Foundation for Polish Science FNP. I am also indebted to organizers for financial support.

REFERENCES

1. L. Faddeev and A. Niemi, Nature **387**, 58 (1997); Phys. Rev. Lett. **82**, 1624 (1999).
2. Y. M. Cho, Phys. Rev.D **21**, 1080 (1980); Phys. Rev. D **23**, 2415 (1981); Phys. Rev. Lett. **46**, 302 (1981); Phys. Lett. B **603**, 88 (2004); Phys. Lett. B **616**, 101 (2005).
3. S. V. Shabanov, Phys. Lett. B **463**, 263 (1999); Phys. Lett. B **458**, 322 (1999); K.-I. Kondo, Phys. Lett. B **600**, 287 (2004); K.-I. Kondo, T. Murakami and T. Shinohara, hep-th/0504107.
4. L. Faddeev, A. Niemi and U. Wiedner, Phys. Rev. D **70**, 114033 (2004); A. Niemi and N. Walet, hep-ph/0504034.
5. R. A. Battye and P. M. Sutcliffe, Phys. Rev. Lett. **81**, 4798 (1998); Proc.Roy.Soc.Lond. A **455**, 4305 (1999).
6. J. Hietarinta and P. Salo, Phys. Lett. B **451**, 60 (1999); Phys. Rev. D **62**, 81701 (2000); J. Hietarinta, J. Jäykkä and P. Salo, Phys. Lett. A **321**, 324 (2004).
7. J. Sánchez-Guillén and L. A. Ferreira, in "Sao Paulo 2002, Integrable theories, solitons and duality" unesp2002/033, hep-th/0211277; F. Lin and Y. Yang, Commun. Math. Phys. **249**, 273 (2004); M. Hirayama and C.-G. Shi, Phys. Rev. D **69**, 045001 (2004).
8. H. Aratyn, L. A. Ferreira and A. H. Zimerman, Phys. Lett. B **456**, 162 (1999).
9. C. Adam, J. Math. Phys. **45**, 4017 (2004).
10. C. Adam and J. Sánchez-Guillén, JHEP 0501:004 (2005).
11. A. Wereszczyński, Eur. Phys. J. C **42**, 461 (2005); AIP Conf. Proc. 756, 293 (2005); Mod. Phys. Lett. A **20**, 1135 (2005).
12. A. Wereszczyński, Phys. Lett. B **621**, 201 (2005); C. Adam and J. Sánchez-Guillén, hep-th/0508011.
13. B. A. Fayzullaev, M. M. Musakhanov, D. G. Pak and M. Siddikov, Phys. Lett. B **609**, 442 (2005); S. Krusch and J. Speight, hep-th/0503067; N. Sawado, N. Shiiki and S. Tanaka, hep-ph/0507258.

Order, Disorder and Confinement

M. D'Elia*, A. Di Giacomo† and C. Pica†

*Dipartimento di Fisica dell'Università di Genova and INFN, Sezione di Genova, Via Dodecaneso 33, I-16146 Genova, Italy
† Dipartimento di Fisica dell'Università di Pisa and INFN, Sezione di Pisa, largo Pontecorvo 3, I-56127 Pisa, Italy

Abstract. Studying the order of the chiral transition for $N_f = 2$ is of fundamental importance to understand the mechanism of color confinement. We present results of a numerical investigation on the order of the transition by use of a novel strategy in finite size scaling analysis. The specific heat and a number of susceptibilities are compared with the possible critical behaviours. A second order transition in the $O(4)$ and $O(2)$ universality classes are excluded. Substantial evidence emerges for a first order transition. Results are in agreement with those found by studying the scaling properties of a disorder parameter related to the dual superconductivity mechanism of color confinement.

Keywords: QCD, Confinement, Deconfinement Transition
PACS: 11.15.Ha, 12.38.Aw, 14.80.Hv, 64.60.Cn

INTRODUCTION

Experiments indicate that confinement is an absolute property of matter. Indeed the upper limit on the number of free quarks per proton is $R = n_q/n_p \leq 10^{-27}$, while $R \sim 10^{-12}$ is expected from the Standard Cosmological Model. A reduction factor 10^{-15} is difficult to explain in natural ways unless $R = 0$, which means that confinement is related to some symmetry of the QCD vacuum. This also implies that the deconfining transition is associated to a change of symmetry of the vacuum, *i.e.* it is an order-disorder transition. This scenario must be confronted with direct studies of the deconfining phase transition: experimental studies going on with heavy ion collisions have not given definite answers yet and most of our knowledge relies on numerical simulations of QCD on the lattice. From this point of view, the case of two light degenerate dynamical flavours is of special interest. A schematic view of the phase diagram for $N_f = 2$ is shown in Fig. 1: m is the quark mass and μ is the baryon chemical potential.

In the $\mu = 0$ plane, as $m \to \infty$ quarks decouple and the system tends to the quenched limit. There the deconfining transition is an order-disorder first order phase transition, Z_3 is an associated symmetry and the Polyakov line $\langle L \rangle$ is an order parameter. Z_3 is explicitely broken by the inclusion of dynamical quarks and $\langle L \rangle$ is not a good order parameter, even if it works as such for quarks masses down to $m \simeq 2.5 - 3$ GeV.

At $m \simeq 0$ there is chiral phase transition at $T_c \simeq 170$ MeV, from a low temperature phase where chiral symmetry is spontaneously broken to an high temperature phase in which it is restored: the corresponding order parameter is the chiral condensate $\langle \bar{\psi}\psi \rangle$. At

---

[1] Talk presented by A. Di Giacomo

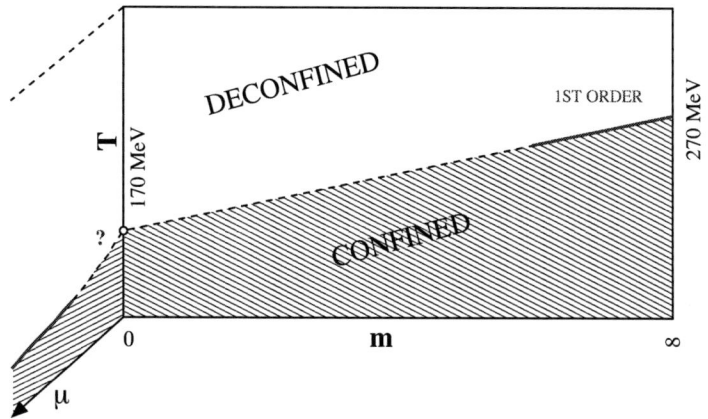

FIGURE 1. Schematic phase diagram of $N_f = 2$ QCD.

some temperature $T_A \geq T_c$ also the $U_A(1)$ symmetry, broken by the anomaly, is expected to be effectively restored. It is not clear which relation exists between the chiral transition and the deconfining transition: empirically the Polyakov line has a rapid increase at the transition temperature, indicating deconfinement. The transition line in Fig. 1 is defined by the maxima of a number of susceptibilities (C_V, χ_m, ...), all coinciding within errors, which indicate a rapid variation of the corresponding parameters across the line.

At $m \simeq 0$ a renormalization group analysis plus ε-expansion techniques can be made, assuming that the relevant degrees of freedom for the chiral transition are scalar and pseudoscalar fields [1, 2, 3]. If the $U_A(1)$ symmetry is effectively restored, *i.e.* if the η' mass vanishes at T_c, then there is no IR stable fixed point and the phase transition is first order; if not an IR fixed point exists, which can produce a second order phase transition in the $O(4)$ universality class.

In the first case the transition is first order also at $m \neq 0$ and most likely up to $m = \infty$. In the second case a phase transition is only present at $m = 0$, which goes into a continuous crossover as $m \neq 0$: that means that one can move continuously from confined to deconfined and that no true order parameter exists. This would be in contradiction with the deconfinement transition being associated to a change of symmetry: the issue is therefore fundamental.

The problem has been investigated on the lattice by several groups with staggered [4, 5, 6, 7, 8, 9, 10] or Wilson [11] fermions. The strategy used has either been to search for signs of discontinuity at the transition, or to study the scaling with m of different susceptibilities, or to study the magnetic equation of state. No clear discontinuities have been observed, but also no conclusive agreement of scaling with $O(4)$ critical indexes. We present the results of a big numerical effort aimed at clarifying the issue. We use non improved Kogut–Susskind action and lattices $4 \times L_s^3$ with $L_s = 16, 20, 24, 32$. A full account of our results can be found in [12].

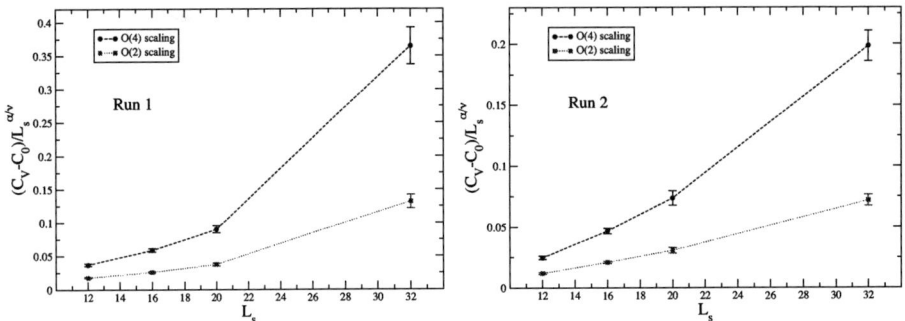

FIGURE 2. Specific heat peak value for Run1 (left) and for Run2 (right), divided by the appropriate powers of L_s to give a constant. Both the $O(4)$ and $O(2)$ critical behaviors are displayed.

RESULTS

The theoretical tool to investigate the order of a phase transition is finite size scaling. The extrapolation from finite size L_s to the thermodynamical limit is governed by critical indexes, which identify the order and the universality class of the transition. Around the chiral transition the system has two fundamental lengths: the correlation length ξ and the inverse quark mass $1/m_q$. ξ is usually traded with the reduced temperature $\tau = 1 - T/T_c$, $\xi \simeq \tau^{-\nu}$ as $\tau \to 0$. The effective action depends on the order parameter, as dictated by the symmetry, and as $\tau \to 0$ irrelevant terms can be neglected; the correlators of the order parameter describe the thermodynamics. The most important quantity is the specific heat, which shows the correct critical behaviour independently of the identification of the order parameter.

For the specific heat and the susceptibility of the order parameter the scaling laws are

$$C_V - C_0 \simeq L_s^{\alpha/\nu} \phi_c \left(\tau L_s^{1/\nu}, am_q L_s^{y_h} \right) ; \qquad (1)$$

$$\chi \simeq L_s^{\gamma/\nu} \phi_\chi \left(\tau L_s^{1/\nu}, am_q L_s^{y_h} \right) . \qquad (2)$$

C_0 stems from an additive renormalization. An alternative way to write them is

$$C_V - C_0 \simeq L_s^{\alpha/\nu} \tilde{\phi}_c \left(\tau (am_q)^{-1/(\nu y_h)}, am_q L_s^{y_h} \right) \qquad (3)$$

$$\chi \simeq L_s^{\gamma/\nu} \tilde{\phi}_\chi \left(\tau (am_q)^{-1/(\nu y_h)}, am_q L_s^{y_h} \right) . \qquad (4)$$

The values of the indexes characterize the transition: the values relevant to our analysis are listed in Table 1. $O(4)$ is the symmetry expected if the transition is second order, but it can break down to $O(2)$ by lattice discretization for Kogut–Susskind fermions [7] at non zero lattice spacing.

The scaling analysis is made difficult by the presence of two independent scales. The attitude taken in the previous literature has been to assume the volume large enough so to neglect the dependence on L_s: since at fixed am_q, β the susceptibilities must be analytic

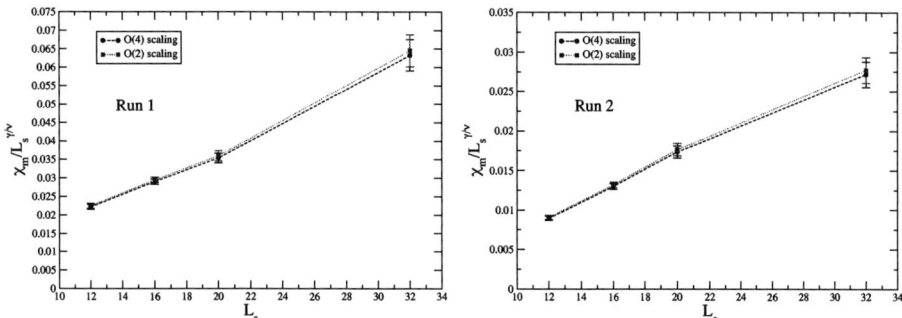

FIGURE 3. The same as figure 2 for the chiral susceptibility χ_m.

in the thermodynamical limit, at large L_s the dependence on $am_q L_s^{y_h}$ must cancel the dependence on L_s in front of the scaling functions in Eq.s (1) and (2). It follows that

$$C_V - C_0 \simeq (am_q)^{-\alpha/(\nu y_h)} f_c\left(\tau(am_q)^{-1/(\nu y_h)}\right) \quad (5)$$

$$\chi \simeq (am_q)^{-\gamma/(\nu y_h)} f_\chi\left(\tau(am_q)^{-1/(\nu y_h)}\right). \quad (6)$$

The peaks of $(C_V - C_0)$ and of χ should then scale as

$$(C_V - C_0)_{\max} \propto (am_q)^{-\alpha/(\nu y_h)}; \quad \chi_{\max} \propto (am_q)^{-\gamma/(\nu y_h)} \quad (7)$$

as $am_q \to 0$. The positions of the maxima scale as $\tau(am_q)^{-1/(\nu y_h)} = \text{const}$.

One can also consider to keep $\tau L_s^{1/\nu}$ fixed while taking $aL_s \gg 1/m_\pi$. This assumption should work better if L_s is still comparable to the correlation length, which may be the case close enough to the critical point. In this case the scaling laws are

$$C_V - C_0 \simeq (am_q)^{-\alpha/(\nu y_h)} f_c\left(\tau L_s^{1/\nu}\right); \quad \chi \simeq (am_q)^{-\gamma/(\nu y_h)} f_\chi\left(\tau L_s^{1/\nu}\right). \quad (8)$$

Eq.s (7) stay unchanged, the positions of the maxima scale as $\tau L_s^{1/\nu} = \text{const}$ and the width of the peaks are volume dependent.

We have instead followed a novel strategy which does not rely on any assumption: we have performed a number of simulations at different values of L_s and am_q keeping the variable $am_q L_s^{y_h}$ fixed and studying the scaling with the respect to the other one. In doing

TABLE 1. Critical exponents.

	y_h	ν	α	γ	δ
O(4)	2.487(3)	0.748(14)	-0.24(6)	1.479(94)	4.852(24)
O(2)	2.485(3)	0.668(9)	-0.005(7)	1.317(38)	4.826(12)
MF	9/4	2/3	0	1	3
1st Order	3	1/3	1	1	∞

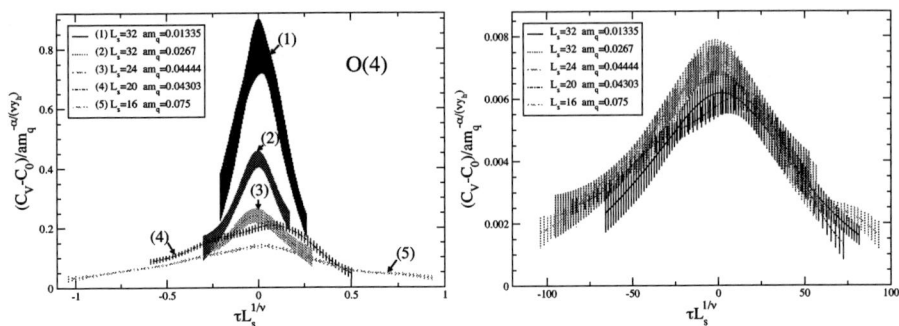

FIGURE 4. Comparison of specific heat scaling, Eq.8, for $O(4)$ (left) and first order (right).

so one has to assume a value for y_h: we have chosen that expected for $O(4)$, which is the same within errors as for $O(2)$. From Eq.s (1) and (2) it follows that, at fixed $am_q L_s^{y_h}$

$$(C_V - C_0)_{max} \propto L_s^{\alpha/\nu}; \quad \chi_{max} \propto L_s^{\gamma/\nu}. \tag{9}$$

as $L_s \to \infty$. If $O(4)$ or $O(2)$ is the correct symmetry, the values of α/ν and γ/ν should be consistent with the corresponding values listed in Table 1. We have run two such sets of Monte Carlo simulations, called in the following Run1 and Run2, with $am_q L_s^{y_h} = 74.7$ and $am_q L_s^{y_h} = 149.4$ respectively. The spatial lattice sizes L_s used for each of the two sets are $L_s = 12, 16, 20, 32$, the standard hybrid R algorithm [13] has been used to update configurations. In Figs. 2 and 3 we show the peak values of the specific heat and of the chiral susceptibility, divided by the appropriate power of L_s, as a function of L_s (see Eq. 9): scaling is clearly violated, $O(4)$ and $O(2)$ universality classes are excluded.

An alternative way to study the order of a transition is to look at scaling of pseudocritical couplings: one can try the two alternative scaling laws $\tau_c = k_\tau L_s^{1/\nu}$ or $\tau_c = k'_\tau (am_q)^{1/(\nu y_h)}$. The physical temperature $T = 1/(L_t a(\beta, m_q))$ is a function of both β and am_q, so that the reduced temperature τ can be expanded as a power series in $(\beta - \beta_0)$ and in am_q, where β_0 is the chiral critical coupling. Only the linear term in β was considered in the previous literature. We have found that the following terms are sufficient to fit the data

$$\tau \propto (\beta_0 - \beta) + k_m am_q + k_{m^2}(am_q)^2 + k_{m\beta} am_q (\beta_0 - \beta). \tag{10}$$

Our result is that it is not possible, within the present mass range, to discriminate among the possible critical behaviours by looking at pseudocritical couplings only. The inclusion of other terms in Eq. (10), besides the one linear in β solely considered in previous literature, is however crucial to obtain any scaling at all.

We can also use our data to perform the scaling analysis in the same way as done in previous literature, *i.e.* by making some assumption on the reaching of the thermodynamical limit: no universality class is chosen a priori in that case and one can test all the possible critical behaviours. We have found that assuming that $aL_s \gg 1/m_\pi$ but still $L_s \sim \xi$, *i.e.* Eqs. (8), works better: this is reasonable when ξ goes large. We have

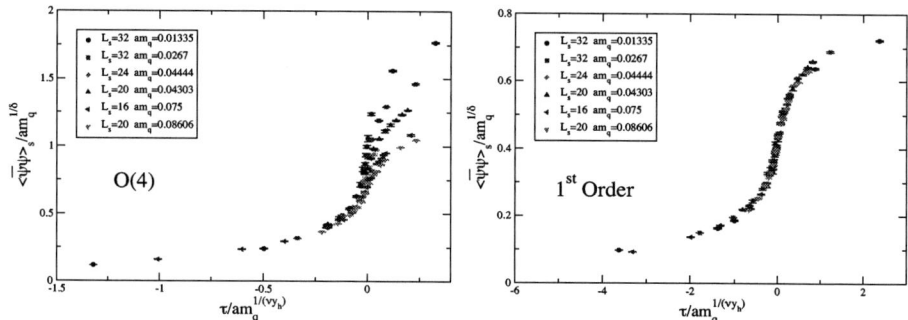

FIGURE 5. Scaling of the equation of state for $O(4)$ (left) and first order (right).

added to the data from Run1 and Run2, those from two other simulations performed at $L_s = 16$, $am_q = 0.01335$ and $L_s = 24$, $am_q = 0.04444$. In Fig.4 we show the scaling obtained for the specific heat: $O(4)$ is again clearly excluded, while a good agreement is found with a weak first order critical behaviour. A similar behaviour is observed for the chiral susceptibility.

We have also studied the magnetic equation of state, *i.e.* the scaling of $\langle \bar{\psi}\psi \rangle - \langle \bar{\psi}\psi \rangle_0 = am_q^{1/\delta} F(\tau m_q^{1/\nu y_h})$. Results are shown in Fig. 5: again the first order behaviour describes well the data while the second order is excluded.

Even if we have still not found clear signs of discontinuities in physical observables, our results give some evidence for a weak first order chiral transition which would persist also at $m \neq 0$. That would be in agreement with the deconfining transition being order-disorder and with confinement being associated to some symmetry of the QCD vacuum. This naturally leads to look for such a symmetry and for an order parameter which describes the deconfining transition both with and without dynamical quarks. One candidate is dual superconductivity of the vacuum and the related parameter is the vacuum expectation value of a magnetically charged operator $\langle \mu \rangle$. Indeed it has been shown that it is good order parameter also in full QCD [14] and its critical behaviour across the $N_f = 2$ transition has been studied in Ref. [15]. In Fig. 6 we show the scaling of its susceptibility $\rho = d/d\beta \ln\langle \mu \rangle$: again results seem to indicate a first order critical behaviour.

CONCLUSIONS

We have argued that the study of the order of the chiral phase transition for $N_f = 2$ is of fundamental importance to understand confinement. We have shown that the analysis of pseudocritical coupling alone cannot discern between the possible critical behaviours. By adopting a novel strategy which reduces the finite size scaling analysis to a one scale problem, we have been able to exclude a $O(4)$ ($O(2)$) second order critical behaviour, while we have found consistency with a weak first order critical behaviour both in the scaling of susceptibilities and in the equation of state. This would be in agreement with confinement being an absolute property of matter related to some symmetry and with the

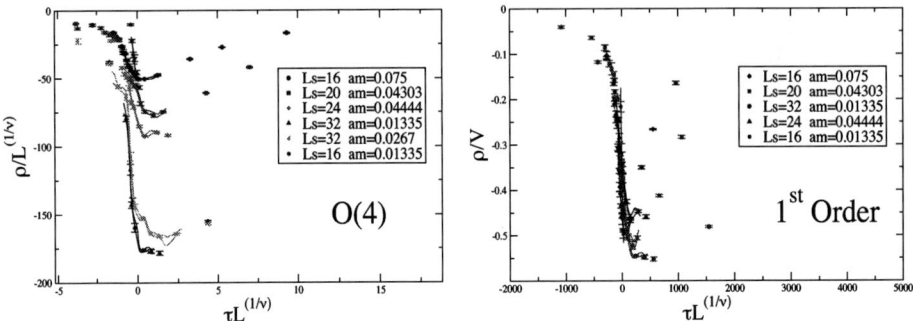

FIGURE 6. Scaling of the susceptibility ρ of the disorder parameter $\langle \mu \rangle$ for $O(4)$ (left) and first order (right).

deconfining transition being order-disorder: indeed the analysis of the scaling properties of the susceptibility of an order parameter associated with the dual superconductivity mechanism of color confinement leads to analogous results. However we have still not found any clear evidence for discontinuities in physical observables. The issue is still open and we plan in the future to investigate it more deeply by making simulations with improved actions and algorithms and with $am_q L_s^{y_h}$ fixed according to first order.

ACKNOWLEDGMENTS

This work has been partially supported by MIUR, Program "Frontier problems in the Theory of Fundamental Interactions".

REFERENCES

1. R. D. Pisarski and F. Wilczek, *Phys. Rev. D* **29**, 338 (1984).
2. F. Wilczek, *Int. J. Mod. Phys. A* **7**, 3911 (1992).
3. K. Rajagopal and F. Wilczek, *Nucl. Phys. B* **399**, 395 (1993).
4. M. Fukugita, H. Mino, M. Okawa and A. Ukawa, *Phys. Rev. Lett.* **65**, 816 (1990).
5. M. Fukugita, H. Mino, M. Okawa and A. Ukawa, *Phys. Rev. D* **42**, 2936 (1990).
6. F. R. Brown, F. P. Butler, H. Chen, N. H. Christ, Z. Dong, W. Schaffer, L. I. Unger and A. Vaccarino, *Phys. Rev. Lett.* **65**, 2491 (1990).
7. F. Karsch, *Phys. Rev. D* **49**, 3791 (1994).
8. F. Karsch and E. Laermann, *Phys. Rev. D* **50**, 6954 (1994).
9. S. Aoki et al. (JLQCD collaboration), *Phys. Rev. D* **57**, 3910 (1998).
10. C. Bernard, C. DeTar, S. Gottlieb, U. M. Heller, J. Hetrick, K. Rummukainen, R.L. Sugar and D. Toussaint, *Phys. Rev. D* **61**, 054503 (2000).
11. A. A. Khan et al. (CP-PACS collaboration), *Phys. Rev. D* **63**, 034502 (2001).
12. M. D'Elia, A. Di Giacomo and C. Pica, *arXiv*:hep-lat/0503030.
13. S. A. Gottlieb, W. Liu, D. Toussaint, R. L. Renken and R. L. Sugar, *Phys. Rev. D* **35**, 2531 (1987).
14. J. M. Carmona, M. D'Elia, L. Del Debbio, A. Di Giacomo, B. Lucini and G. Paffuti, *Phys. Rev. D* **66**, 011503 (2002).
15. M. D'Elia, A. Di Giacomo, B. Lucini, G. Paffuti and C. Pica, *Phys. Rev. D* **71**, 114502 (2005).

Topological susceptibility in the SU(3) gauge theory

Luigi Del Debbio

CERN, Department of Physics, TH Division, CH-1211 Geneva 23, Switzerland

Abstract. We compute the topological susceptibility for the SU(3) Yang–Mills theory by employing the expression of the topological charge density operator suggested by Neuberger's fermions. In the continuum limit we find $r_0^4 \chi = 0.059(3)$, which corresponds to $\chi = (191 \pm 5 \,\text{MeV})^4$ if F_K is used to set the scale. Our result supports the Witten–Veneziano explanation for the large mass of the η'. Comments on the large-volume distribution of the topological charge are presented.

Keywords: Witten-Veneziano mechanism, Lattice QCD
PACS: 11.15.Ha, 11.30.Rd, 11.10.Gh, 12.38.Gc

INTRODUCTION

The topological susceptibility in the pure Yang–Mills (YM) gauge theory can be defined in Euclidean space-time as

$$\chi = \int d^4x \, \langle q(x)q(0)\rangle, \tag{1}$$

where the topological charge density $q(x)$ is given by

$$q(x) = -\frac{1}{32\pi^2} \varepsilon_{\mu\nu\rho\sigma} \text{Tr}\left[F_{\mu\nu}(x)F_{\rho\sigma}(x)\right]. \tag{2}$$

The susceptibility χ characterizes the SU(3) pure gauge theory, and plays a crucial rôle in the QCD-based explanation of the large mass of the η' meson proposed by Witten and Veneziano (WV) [1, 2]. The WV mechanism predicts that at the leading order in N_f/N_c, where N_f and N_c are the number of flavors and colors respectively, the contribution due to the anomaly to the mass of the $U_\text{A}(1)$ particle is given by [1, 2, 3, 4, 5]

$$\frac{F_\pi^2 m_{\eta'}^2}{2N_\text{f}} = \chi, \tag{3}$$

where F_π is the pion decay constant.

The lattice formulation of gauge theories is at present the only approach where nonperturbative computations can be performed with controlled systematic errors. A crucial progress in the study of topology on the lattice came with the discovery of an explicit realization of Ginsparg–Wilson (GW) fermions [6, 7, 8, 9], preserving an exact chiral symmetry at finite lattice spacing [10]. For a review of recent results and references to the original works see e.g. [11]. In this framework, the chiral anomaly is recovered à la Fujikawa [12] with the topological charge density operator defined as [13]:

$$q(x) = -\frac{\bar{a}}{2} \text{Tr}\left[\gamma_5 D(x,x)\right], \tag{4}$$

where the trace runs over spin and color indices. These developments made it possible for the first time to find an unambiguous definition of the topological susceptibility with a finite continuum limit [4, 14, 15], which is independent of the details of the lattice definition [15]. Using the charge density suggested by GW fermions: $Q \equiv \sum_x q(x) = n_+ - n_-$, with n_+ (n_-) the number of zero modes of D with positive (negative) chirality in a given background, the suggestive formula

$$\chi = \lim_{\substack{a \to 0 \\ V \to \infty}} \frac{\langle Q^2 \rangle}{V} \qquad (5)$$

is recovered, where V is the volume.

The results of the investigation reported here, and first published in Ref. [16], show the first extrapolation of the topological susceptibility to the continuum limit with controlled systematic errors. The result for the adimensional scaling quantity computed on the lattice is $r_0^4 \chi = 0.059(3)$, r_0 being a low-energy reference scale [17]. In physical units, it corresponds to $\chi = (191 \pm 5\,\mathrm{MeV})^4$, if F_K is used to set the scale. Our result supports the WV explanation for the large mass of the η' meson within QCD. Some additional comments on the large-volume limit of the topological charge distribution are also presented.

LATTICE COMPUTATION

Details on the generation of the gauge configurations can be found in Refs. [18, 19]. Table 1 shows the list of simulated lattices, where the bare coupling constant $\beta = 6/g_0^2$, the linear size L/a in each direction and the number of independent configurations are reported for each lattice.

The topological charge density is defined as in Eq. (4), with D being the massless Neuberger–Dirac operator:

$$D = \frac{1}{\bar{a}}\left[1 + \gamma_5 \mathrm{sign}(H)\right] \qquad (6)$$

$$H = \gamma_5(aD_\mathrm{w} - 1 - s), \qquad \bar{a} = \frac{a}{1+s}. \qquad (7)$$

Here s is an adjustable parameter in the range $|s| < 1$, and D_w denotes the standard Wilson–Dirac operator (the notational conventions not explained here are as in Ref. [20]). For a given gauge configuration, the topological charge is computed by counting the number of zero modes of D with the algorithm proposed in Ref. [20]. Let us emphasize that, as s is varied, D defines a one-parameter family of fermion discretizations, which correspond to the same continuum theory but with different discretization errors at finite lattice spacing. Our analysis includes data sets computed for $s = 0.4$ and $s = 0.0$. Most of the data were taken from Refs. [19] and [18] for $s = 0.4$ and $s = 0.0$ respectively. The number of configurations were increased, where necessary, in order to achieve homogeneous statistical errors of the order of 5% for each data point. Some new lattices were added so as to perform careful studies of the systematic uncertainties which we describe below, before presenting the physical results.

TABLE 1. Simulation parameters and results. For lattices A_1–D and E–J, $s = 0.4$ and $s = 0.0$ respectively.

lat	β	L/a	r_0/a	L[fm]	N_{conf}	$\langle Q^2 \rangle$	$r_0^4 \chi$
A_1	6.0	12	5.368	1.12	2452	1.633(48)	0.0654(22)
A_2	6.1791	16	7.136	1.12	1138	1.589(76)	0.0629(32)
A_3	5.8989	10	4.474	1.12	1460	1.737(72)	0.0696(30)
A_4	6.0938	14	6.263	1.12	1405	1.535(63)	0.0615(27)
B_0	5.8458	12	4.032	1.49	2918	5.61(16)	0.0715(22)
B_1	6.0	16	5.368	1.49	1001	5.58(28)	0.0707(37)
B_2	6.1366	20	6.693	1.49	963	4.81(24)	0.0604(32)
B_3	5.9249	14	4.697	1.49	1284	5.59(24)	0.0708(33)
C_0	5.8784	16	4.301	1.86	1109	15.02(72)	0.0784(39)
C_1	6.0	20	5.368	1.86	931	12.76(95)	0.0662(50)
D	6.0	14	5.368	1.30	1577	3.01(12)	0.0651(27)
E	5.9	12	4.483	1.34	1349	2.79(12)	0.0543(24)
F	5.95	12	4.917	1.22	1291	1.955(79)	0.0551(24)
G	6.0	12	5.368	1.12	3586	1.489(37)	0.0596(18)
H	6.1	16	6.324	1.26	962	2.45(13)	0.0599(33)
J	6.2	18	7.360	1.22	1721	2.114(76)	0.0591(24)

Statistical errors are computed assuming that the measurements are statistically independent; such an assumption is supported by previous experience about the autocorrelation times of the topological charge [16, 21, 22]. Besides the statistical errors, the systematic uncertainties stem from finite-volume effects and from the extrapolation needed to reach the continuum limit.

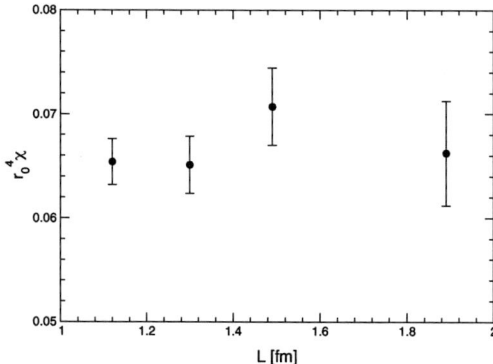

FIGURE 1. The topological susceptibility, in units of r_0^{-4}, as a function of the linear lattice size, in fm, at $\beta = 6.0$.

Let us begin by discussing the finite-volume effects. The pure gauge theory has a mass gap, and therefore the topological susceptibility approaches the infinite-volume limit exponentially fast with L. Since the mass of the lightest glueball is around 1.5 GeV, finite-volume effects are expected to be far below our statistical errors as soon as $L \geq 1$ fm. In order to further verify that no sizeable finite-volume effects are present in our data, we simulated four lattices at $\beta = 6.0$ but with different linear sizes $L = 1.12, 1.30, 1.49, 1.86$ fm. The results obtained for χ are shown in Fig. 1, where no dependence on L is visible, hence confirming that finite-volume effects are below our statistical errors.

In order to gain some insight on the large-volume limit of the topological charge distribution, let us consider the partition function for QCD with a θ-term, $Z(\theta) = e^{-VF(\theta)}$, and let us assume that the free energy density $F(\theta)$ can be expanded in a Taylor series:

$$F(\theta) = \sum_n (-)^{n+1} C_{2n} \frac{1}{(2n)!} \theta^{2n}, \tag{8}$$

where the C_n are the connected moments of the topological charge probability distribution:

$$C_n = (-)^{n/2+1} \left.\frac{d^n F}{d\theta^n}\right|_{\theta=0} = \frac{1}{V}\langle v^n \rangle_c, \tag{9}$$

All moments are of order V^0, the second moment $C_2 = \langle Q^2 \rangle / V$ is the topological susceptibility of the pure gauge theory, the fourth moment is $C_4 = \left(\langle Q^4 \rangle - 3\langle Q^2\rangle^2\right)/V$, and so on. The probability distribution itself is given by:

$$P_v = \int \frac{d\theta}{2\pi} e^{-i\theta v} e^{-VF(\theta)} \tag{10}$$

A change of the integration variable, $y = \sqrt{V}\theta$, yields a form for the probability distribution that is better suited for discussing its large-volume limit:

$$P_v = \int \frac{d\theta}{2\pi\sqrt{V}} e^{-iyv/\sqrt{V}} \exp\left[-\frac{C_2}{2} y^2 + \frac{C_4}{4!V} y^4 - \frac{C_6}{6!V^2} y^6 + \ldots\right]. \tag{11}$$

Separating the Gaussian part of the exponential and expanding the rest of the exponential, the integral can be performed exactly:

$$P_v = \frac{e^{-v^2/2C_2 V}}{\sqrt{2\pi C_2 V}} \left\{1 + \frac{1}{V}\mathscr{P}_2(v^2/V) + \frac{1}{V^2}\mathscr{P}_4(v^2/V) + \ldots\right\}, \tag{12}$$

where each $\mathscr{P}_n$ indicates a polynomial function of order n, that we do not need to specify. Some generic features of the charge distribution can be read directly from Eq. (12).

- a Gaussian distribution is obtained *if and only if* the free energy is exactly a quadratic function of θ, i.e. if $C_n = 0$ for $n > 2$. Evidence for deviations from a Gaussian distribution were reported in Refs. [21, 23].

- if $\frac{v^2}{V} \ll 1$ then all polynomials $\mathscr{P}$ in Eq. (12) are parametrically of order V^0, and therefore are suppressed by powers of V. Hence, an approximately Gaussian distribution is expected for small values of $v/\sqrt{V}$. This limit corresponds to the Gaussian distribution that is obtained by saddle-point approximation [24].
- for a given value of V, corrections to the Gaussian distribution appear as soon as $v^2/V \sim \sqrt{V}$. In order to obtain the *exact* probability distribution, one should then sum up all the terms in the expansion above.
- note that, when $v^2/V \sim \sqrt{V}$, the exponential suppression factor in Eq. (12) is already of the order of $e^{-\sqrt{V}}$.

The corrections to the Gaussian behavior do not modify the value of C_2; but they have to be taken into account when considering higher moments of the distribution. Much higher statistics are required in order to highlight the deviations from a Gaussian distribution [21]; higher momenta of the topological charge distribution measured on our data are all compatible with zero within large statistical errors.

Let us now consider the continuum limit. As pointed out in the introduction, the topological susceptibility defined from the index of the Neuberger operator is not plagued by power divergences and does not require multiplicative renormalization. This is a direct consequence of the lattice chiral symmetry, and is at variance with what happens for other definitions used in the past to compute χ. At finite lattice spacing, χ is affected by discretization effects starting at $O(a^2)$, which are not universal, and, in our case, depend on the value of s chosen to define the Neuberger operator. In order to compare results at different lattice spacings, and to extrapolate them to the continuum limit, we adopt r_0 as the reference scale; this choice is motivated by its precise determination in the range of β explored in this work [17]. The values of the adimensional quantity $r_0^4 \chi$ that we obtain are reported in Table 1. Data, displayed in Fig. 2 as a function of a^2/r_0^2, show sizeable $O(a^2)$ effects for both the $s = 0.4$ and $s = 0.0$ samples. For $\beta \leq 6.0$, the difference between the two discretizations is statistically significant. Within our statistical errors, and in the range where our simulations are performed, our results suggest a linear dependence in a^2. For the $s = 0.4$ sample, the value of χ^2 per degree of freedom, χ^2_{dof}, clearly disfavors a constant behavior, while a linear fit of the form

$$r_0^4 \chi(s) = c_0 + c_1(s) \left(\frac{a}{r_0}\right)^2 \tag{13}$$

yields a value of $c_0 = 0.056(3)$ with $\chi^2_{\text{dof}} \approx 0.79$. The quadratic fit in a^2/r_0^2 yields an extrapolated value compatible with that of the linear one, but with an error three times larger, and the coefficient of the quadratic term compatible with zero. For the $s = 0.0$ sample, all three fits give good values of χ^2_{dof}, and for the linear one we obtain $c_0 = 0.064(4)$ with $\chi^2_{\text{dof}} \approx 0.68$, which is compatible with the outcome of the same fit for $s = 0.4$. The agreement between the two extrapolations indicates that we reached the scaling regime. This is confirmed by the compatibility of the results in the two data sets for $\beta > 6.0$. A robust estimate of χ in the continuum limit can thus be obtained by performing a combined linear fit of the data. This fit gives a very good value of χ^2_{dof} when all sets are included, and is very stable if some points at larger values of a^2/r_0^2 are

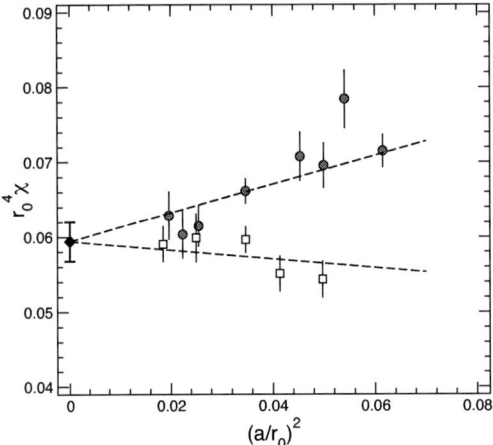

FIGURE 2. Continuum extrapolation of the adimensional product $r_0^4\chi$. The $s = 0.0$ and $s = 0.4$ data sets are represented by black circles and white squares respectively. The dashed lines represent the results of the combined fit described in the text. The filled diamond at $a = 0$ is the extrapolated value in the continuum limit.

removed. In particular a combined fit of all points with $a^2/r_0^2 < 0.05$ gives $c_0 = 0.059(3)$ with $\chi^2_{\text{dof}} \approx 0.73$, and the error is expected to be Gaussian.

PHYSICAL RESULTS

From the previous analysis, our best result for the topological susceptibility is the one obtained from a combined fit of the two sets of data with $a^2/r_0^2 < 0.05$:

$$r_0^4 \chi = 0.059 \pm 0.003, \tag{14}$$

which is the main result of this work. Since r_0 is not directly accessible to experiments, we express our result in physical units by using the lattice determination of $r_0 F_K = 0.4146(94)$ in the pure gauge theory with valence quarks [25] and, taking $F_K = 160(2)$ MeV as an experimental input, we obtain

$$\chi = (191 \pm 5\,\text{MeV})^4, \tag{15}$$

which has to be compared with [2]

$$\left. \frac{F_\pi^2}{6}\left(m_\eta^2 + m_{\eta'}^2 - 2m_K^2\right) \right|_{\text{exp}} \simeq (180\,\text{MeV})^4. \tag{16}$$

Notice that, since Eq. (3) is valid only at the leading order in a N_f/N_c expansion, the ambiguity in the conversion to physical units in the pure gauge theory is of the same order as the neglected terms.

Our result supports the fact that the bulk of the mass of the pseudoscalar singlet meson is generated by the anomaly through the Witten–Veneziano mechanism.

ACKNOWLEDGMENTS

I would like to thank the Organizers of the Workshop for providing a very nice setting to this meeting. The work presented here was obtained in collaboration with L. Giusti and C. Pica: working with them has proved a very pleasant experience. I would also like to thank E. Vicari for many discussions on this subject, and P. Faccioli for interesting comments.

REFERENCES

1. E. Witten, *Nucl. Phys.* **B156**, 269 (1979).
2. G. Veneziano, *Nucl. Phys.* **B159**, 213–224 (1979).
3. E. Seiler, and I. O. Stamatescu (1987), mPI-PAE/PTh 10/87.
4. L. Giusti, G. C. Rossi, M. Testa, and G. Veneziano, *Nucl. Phys.* **B628**, 234–252 (2002), `hep-lat/0108009`.
5. E. Seiler, *Phys. Lett.* **B525**, 355–359 (2002), `hep-th/0111125`.
6. P. H. Ginsparg, and K. G. Wilson, *Phys. Rev.* **D25**, 2649 (1982).
7. H. Neuberger, *Phys. Lett.* **B417**, 141–144 (1998), `hep-lat/9707022`.
8. H. Neuberger, *Phys. Rev.* **D57**, 5417–5433 (1998), `hep-lat/9710089`.
9. H. Neuberger, *Phys. Lett.* **B427**, 353–355 (1998), `hep-lat/9801031`.
10. M. Lüscher, *Phys. Lett.* **B428**, 342–345 (1998), `hep-lat/9802011`.
11. L. Giusti, *Nucl. Phys. Proc. Suppl.* **119**, 149–160 (2003), `hep-lat/0211009`.
12. K. Fujikawa, *Phys. Rev. Lett.* **42**, 1195 (1979).
13. P. Hasenfratz, V. Laliena, and F. Niedermayer, *Phys. Lett.* **B427**, 125–131 (1998), `hep-lat/9801021`.
14. L. Giusti, G. C. Rossi, and M. Testa, *Phys. Lett.* **B587**, 157–166 (2004), `hep-lat/0402027`.
15. M. Luscher, *Phys. Lett.* **B593**, 296–301 (2004), `hep-th/0404034`.
16. L. Del Debbio, L. Giusti, and C. Pica, *Phys. Rev. Lett.* **94**, 032003 (2005), `hep-th/0407052`.
17. M. Guagnelli, R. Sommer, and H. Wittig, *Nucl. Phys.* **B535**, 389–402 (1998), `hep-lat/9806005`.
18. L. Del Debbio, and C. Pica, *JHEP* **02**, 003 (2004), `hep-lat/0309145`.
19. L. Giusti, M. Luscher, P. Weisz, and H. Wittig, *JHEP* **11**, 023 (2003), `hep-lat/0309189`.
20. L. Giusti, C. Hoelbling, M. Luscher, and H. Wittig, *Comput. Phys. Commun.* **153**, 31–51 (2003), `hep-lat/0212012`.
21. L. Del Debbio, H. Panagopoulos, and E. Vicari, *JHEP* **08**, 044 (2002), `hep-th/0204125`.
22. L. Del Debbio, G. M. Manca, and E. Vicari, *Phys. Lett.* **B594**, 315–323 (2004), `hep-lat/0403001`.
23. M. D'Elia, *Nucl. Phys.* **B661**, 139–152 (2003), `hep-lat/0302007`.
24. R. Brower, S. Chandrasekharan, J. W. Negele, and U. J. Wiese, *Phys. Lett.* **B560**, 64–74 (2003), `hep-lat/0302005`.
25. J. Garden, J. Heitger, R. Sommer, and H. Wittig, *Nucl. Phys.* **B571**, 237–256 (2000), `hep-lat/9906013`.

Finite temperature lattice gauge theories in external fields

Leonardo Cosmai

INFN - Sezione di Bari, I-70126 Bari, Italy

Abstract. We study finite temperature lattice gauge theories in external background fields. We found that for non abelian gauge theories the deconfinement temperature depends on the strength of a constant chromomagnetic background field.

Keywords: Lattice Gauge Theories, Color Confinement, Finite Temperature Quantum Field Theory

PACS: 11.15.Ha, 11.10.Wx

INTRODUCTION

Up to now there is no totally convincing explanation of the confinement phenomenon (for recent reviews on confinement see [1–3]) and a full understanding of the QCD vacuum dynamics is still lacking. Indeed, as recently observed [4] in connection with dual superconductivity picture, even if magnetic monopoles do condense in the confinement mode, the actual mechanism of confinement could depend on additional dynamical forces. Therefore we feel that it is important to explore any new paths that possibly may suggest new hints for understanding the QCD vacuum.

An external field can be useful to probe the vacuum structure of a quantum field theory. To investigate the vacuum structure of lattice gauge theories we introduced [5] a lattice effective action $\Gamma[\vec{A}^{\text{ext}}]$ for the external static background field $\vec{A}^{\text{ext}}$

$$\Gamma[\vec{A}^{\text{ext}}] = -\frac{1}{L_t} \ln \left\{ \frac{\mathscr{Z}[\vec{A}^{\text{ext}}]}{\mathscr{Z}[0]} \right\} \tag{1}$$

where L_t is the lattice size in time direction and $\vec{A}^{\text{ext}}(\vec{x})$ is the continuum gauge potential of the external static background field. $\mathscr{Z}[\vec{A}^{\text{ext}}]$ is the lattice partition functional

$$\mathscr{Z}[\vec{A}^{\text{ext}}] = \int_{U_k(\vec{x}, x_t=0) = U_k^{\text{ext}}(\vec{x})} \mathscr{D}U \, e^{-S_W}, \tag{2}$$

with S_W the standard pure gauge Wilson action.

The functional integration is performed over the lattice links, but constraining the spatial links belonging to a given time slice (say $x_t = 0$) to be

$$U_k(\vec{x}, x_t = 0) = U_k^{\text{ext}}(\vec{x}), \quad (k = 1, 2, 3), \tag{3}$$

$U_k^{\text{ext}}(\vec{x})$ being the lattice version of the external continuum gauge potential $\vec{A}^{\text{ext}}(x) = \vec{A}_a^{\text{ext}}(x) \lambda_a / 2$. Note that the temporal links are not constrained.

In the case of a static background field which does not vanish at infinity we must also impose that, for each time slice $x_t \neq 0$, spatial links exiting from sites belonging to the spatial boundaries are fixed according to eq. (3). In the continuum this last condition amounts to the requirement that fluctuations over the background field vanish at infinity.

If we now consider the gauge theory at finite temperature $T = 1/(aL_t)$ in presence of an external background field, the relevant quantity turns out to be the free energy functional defined as

$$\mathscr{F}[\vec{A}^{\text{ext}}] = -\frac{1}{L_t} \ln \left\{ \frac{\mathscr{Z}_\mathscr{T}[\vec{A}^{\text{ext}}]}{\mathscr{Z}_\mathscr{T}[0]} \right\}. \quad (4)$$

$\mathscr{Z}_\mathscr{T}[\vec{A}^{\text{ext}}]$ is the thermal partition functional [6] in presence of the background field $\vec{A}^{\text{ext}}$, and is defined as

$$\mathscr{Z}_T\left[\vec{A}^{\text{ext}}\right] = \int_{U_k(\vec{x},L_t)=U_k(\vec{x},0)=U_k^{\text{ext}}(\vec{x})} \mathscr{D}U \, e^{-S_W}. \quad (5)$$

In eq. (5), as in eq. (2), the spatial links belonging to the time slice $x_t = 0$ are constrained to the value of the external background field, the temporal links are not constrained. On a lattice with finite spatial extension we also usually impose that the links at the spatial boundaries are fixed according to boundary conditions eq. (3), apart from the case in which the external background field vanishes at spatial infinity (as happens for the monopole field), where the choice of periodic boundary conditions in the spatial direction is equivalent to eq. (3) in the thermodynamical limit. If the physical temperature is sent to zero, the thermal functional eq. (5) reduces to the zero-temperature Schrödinger functional eq. (2). The free energy functional eq. (4) corresponds to the free energy, $F[\vec{A}^{\text{ext}}]$, in presence of the external background field evaluated with respect to the free energy, $F[0]$, with $\vec{A}^{\text{ext}} = 0$. When the physical temperature is sent to zero the free energy functional reduces to the vacuum energy functional eq. (1).

Since our lattice has the topology of a torus, the magnetic field turns out to be quantized $a^2 \frac{gH}{2} = \frac{2\pi}{L_1} n_{\text{ext}}$, n_{ext} integer. Moreover, since the free energy functional $\mathscr{F}[\vec{A}^{\text{ext}}]$ is invariant for time independent gauge transformations of the background field $\vec{A}^{\text{ext}}$, it follows that for a constant background field the free energy $F[\vec{A}^{\text{ext}}]$ is proportional to the spatial volume $V = L_s^3$, and the relevant quantity is the density of free energy $f[\vec{A}^{\text{ext}}] = \frac{1}{V} F[\vec{A}^{\text{ext}}]$. We evaluate by numerical simulations $f'[\vec{A}^{\text{ext}}]$ the derivative with respect to the coupling β of the free energy density $f[\vec{A}^{\text{ext}}]$ at fixed external field strength gH.

We present here results obtained in studying vacuum dynamics of U(1), SU(2), and SU(3) lattice gauge theories under the influence of an abelian chromomagnetic background field. In particular we give numerical evidence that, for non abelian theories, the deconfinement temperature depends on the strength of the external abelian chromomagnetic background field. This is at variance with the case of an abelian monopole background field that does not modify the deconfinement temperature. The main aim is to ascertain if the dependence of the deconfinement temperature on the strength of an applied external constant abelian chromomagnetic field is a peculiar feature of non abelian gauge theories.

SU(3) IN (3+1) DIMENSIONS

It is known that the pure SU(3) gauge system undergoes a deconfinement phase transition at a given critical temperature. We studied the possible dependence of the critical temperature from the strength of a constant abelian chromomagnetic background field. The critical coupling β_c has been evaluated by measuring $f'[\vec{A}^{\text{ext}}]$, the derivative of the free energy density with respect to β, as a function of β. Indeed we found that $f'[\vec{A}^{\text{ext}}]$ displays a peak in the critical region where it can be parameterized as

$$\frac{f'(\beta, L_t)}{\varepsilon'_{\text{ext}}} = \frac{a_1(L_t)}{a_2(L_t)[\beta - \beta^*(L_t)]^2 + 1}. \tag{6}$$

In eq. (6) we normalize f' to $\varepsilon'_{\text{ext}}$, the derivative of the classical energy due to the external applied field

$$\varepsilon'_{\text{ext}} = \frac{2}{3}[1 - \cos(\frac{gH}{2})] = \frac{2}{3}[1 - \cos(\frac{2\pi}{L_1}n_{\text{ext}})]. \tag{7}$$

We have checked that the evaluation of the critical coupling $\beta^*(L_t)$ by means of $f'[\vec{A}^{\text{ext}}]$ is consistent with the usual determination obtained through the temporal Polyakov loop susceptibility.

We varied the strength of the applied external abelian chromomagnetic background field to study quantitatively the dependence of T_c on gH.

The lattice data for the SU(3) critical temperature in units of the string tension (see Fig. 1) can be reproduced by the linear fit $\frac{T_c}{\sqrt{\sigma}} = \alpha \frac{\sqrt{gH}}{\sqrt{\sigma}} + \frac{T_c(0)}{\sqrt{\sigma}}$ with $\frac{T_c(0)}{\sqrt{\sigma}} = 0.643(15)$, $\alpha = -0.245(9)$. The critical field can now be expressed in units of the string tension $\frac{\sqrt{gH_c}}{\sqrt{\sigma}} = 2.63 \pm 0.15$. Assuming $\sqrt{\sigma} = 420$ MeV, we find for the critical field $\sqrt{gH_c} = (1.104 \pm 0.063)$GeV corresponding to $gH_c = 6.26(2) \times 10^{19}$ Gauss.

SU(2) IN (3+1) DIMENSIONS

We studied the SU(2) lattice gauge theory in a constant abelian chromomagnetic field. As for SU(3), the deconfinement temperature turns out to depend on the strength of the applied chromomagnetic field.

We evaluated the critical coupling $\beta^*(L_t, n_{\text{ext}})$ on a $64^3 \times 8$ lattice versus the strength of the external chromomagnetic field. As in previous section the critical coupling has been found by locating the peak of the derivative of the free energy density with respect to the gauge coupling β. The critical temperature can be expressed in units of the string tension and can be fitted by means of a linear law. As in the SU(3) case discussed in previous section, we found that the linear fit works quite well and we get $\frac{T_c(0)}{\sqrt{\sigma}} = 0.710(13)$, $\alpha = -0.126(5)$. The value obtained for $T_c(0)/\sqrt{\sigma}$ is in good agreement with the value $T_c/\sqrt{\sigma} = 0.694(18)$, without external field, obtained in the literature. The critical field in string tension units that turns out to be $\frac{\sqrt{gH_c}}{\sqrt{\sigma}} = 5.33 \pm 0.33$. Note that the critical field $\sqrt{gH_c}/\sqrt{\sigma}$ is about a factor 2 greater than the SU(3) critical value.

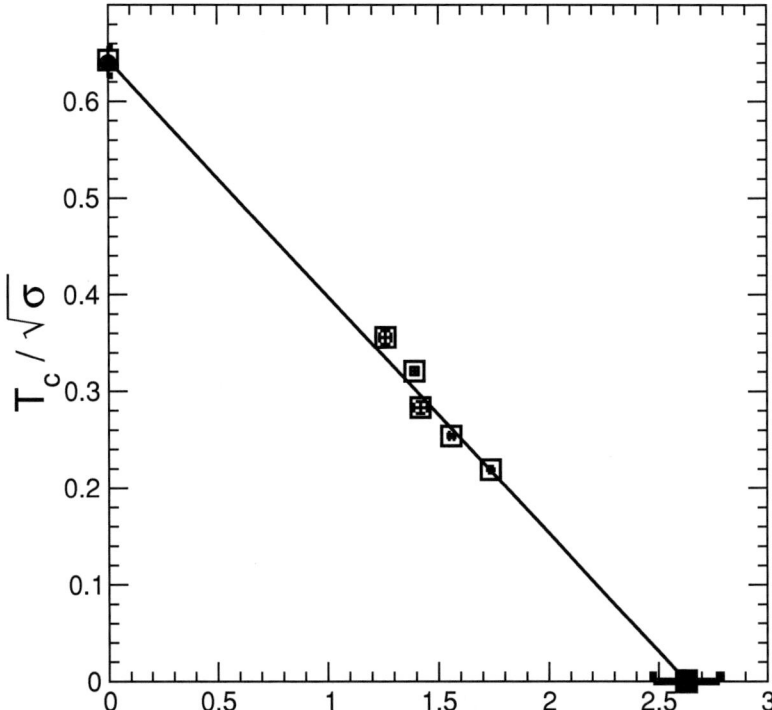

FIGURE 1. SU(3) in (3+1) dimensions. The critical temperature T_c estimated on a $64^3 \times 8$ lattice in units of the string tension versus the square root of the field strength $\sqrt{gH}$ in units of the string tension. Solid line is the linear fit. In correspondence of zero vertical axis: open circle is $T_c/\sqrt{\sigma}$ at zero external field; full circle is the determination of $T_c/\sqrt{\sigma}$ obtained in the literature [7] The full square is the critical field in units of the string tension.

Our results indicate a dependence of the deconfinement temperature on the strength of a constant abelian chromomagnetic background field. On the other hand, we found that such an effect is absent for four dimensional U(1) lattice gauge theory, so that we may conclude that it is peculiar of non abelian gauge theories. In Fig. 2 we display the phase diagram for four dimensional SU(2) and SU(3) gauge theories.

SU(3) IN (2+1) DIMENSIONS

In (3+1) dimensions we found that the deconfinement temperature for non abelian gauge theories SU(2) and SU(3) in presence of an abelian constant chromomagnetic background field depends on the strength of the applied field, and eventually becomes

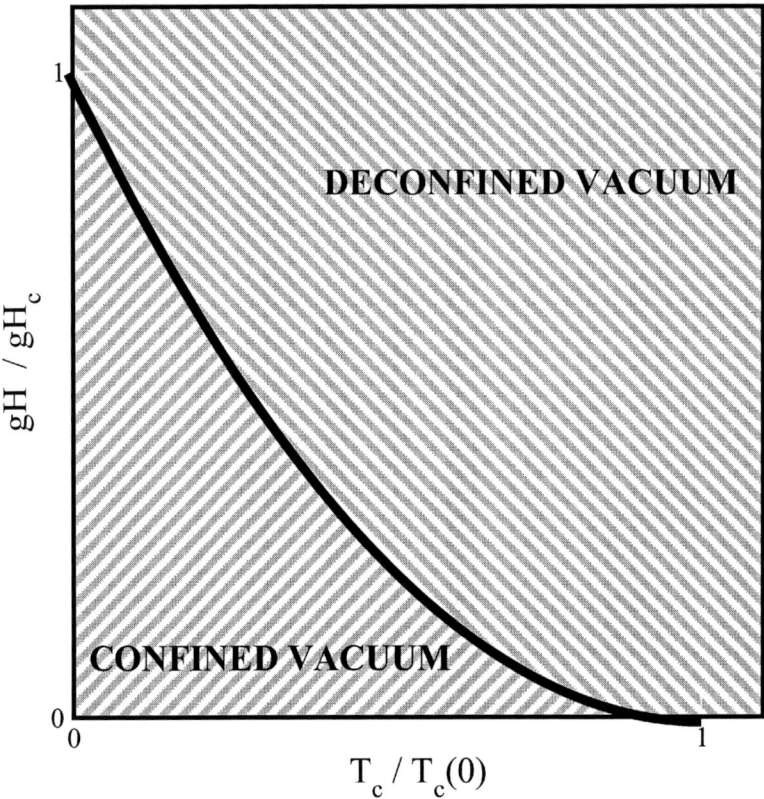

FIGURE 2. Phase diagram of four dimensional SU(2) and SU(3) gauge theories.

zero for a critical value of the field strength. One may wonder if this phenomenon, which we found to be peculiar of non abelian gauge theories, continues to hold in (2+1) dimensions. To this purpose we consider here the non abelian SU(3) lattice gauge theory to be contrasted with the abelian U(1) lattice gauge theory at finite temperature.

As is well known gauge theories in (2+1) dimensions possess a dimensionful coupling constant, namely g^2 has dimension of mass and so provides a physical scale. In (2+1) dimensions the chromomagnetic field H^a is a (pseudo)scalar $H^a = \frac{1}{2}\varepsilon_{ij}F_{ij}^a = F_{12}^a$. As in the four dimensional case since we assume to have a lattice with toroidal geometry the field strength is quantized. We computed the derivative of the free energy density eq. (4) on a $L \times 256 \times 4$ lattice, with $L = 256, 512$ and several values of the external field strength parameterized by n_{ext}. We locate the critical coupling β_c as the position of the maximum of the derivative of the free energy density at given external field strength. As for SU(3) in (3+1) dimensions, the value of β_c depends on the field strength. Using the parameterization for the string tension given in eq. (C9) of ref. [8], we find that, as in (3+1) dimensions, $T_c/\sqrt{\sigma}$ depends linearly on the applied field strength (see Fig. 3). The linear fit gives $\frac{T_c(0)}{\sqrt{\sigma}} = 1.073(87)$ $\alpha = -0.193(76)$ that implies a critical field in string tension units $\sqrt{gH_c}/\sqrt{\sigma} = 5.5 \pm 3.7$. Note that value for $T_c(0)/\sqrt{\sigma}$ in the present work is in fair agreement with $T_c/\sqrt{\sigma} = 0.972(10)$ without external field obtained in ref. [9]. To check possible finite volume effects, we performed a lattice simulation with $L_t = 8$. The result, displayed in Fig. 3, shows that within statistical uncertainties our estimate of the critical temperature from the simulation with $L_t = 8$ is in agreement with result at $L_t = 4$.

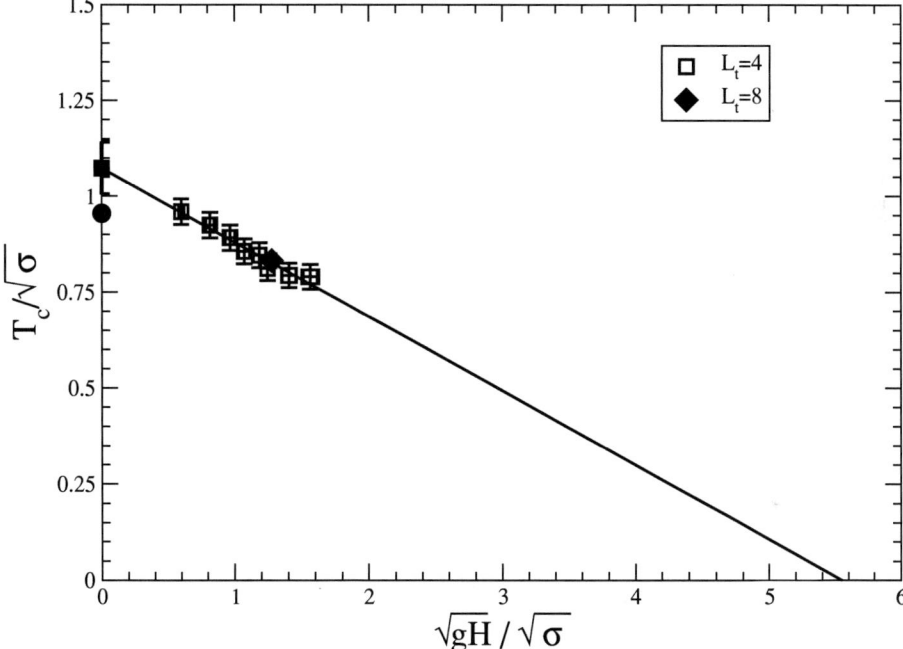

FIGURE 3. SU(3) in (2+1) dimensions. The critical temperature T_c estimated on $256^2 \times 4$, $512^2 \times 4$ and $512 \times 256 \times 8$ lattices in units of the string tension versus the square root of the field strength $\sqrt{gH}$ in units of the string tension. Open squares refer to $L_t = 4$, diamond to $L_t = 8$. Solid line is the linear fit. In correspondence of $\sqrt{gH}/\sqrt{\sigma} = 0$: full square represents $T_c/\sqrt{\sigma}$ at zero external field obtained by a linear extrapolation, full circle is the value given in ref. [8].

U(1) IN (2+1) DIMENSIONS

It is known that [10] compact quantum electrodynamics in (2+1) dimensions at zero temperature confines external charges for all values of the coupling and that the confining mechanism is the condensation of magnetic monopoles which gives rise to a linear confining potential and a non-zero string tension. On the other hand, at finite temperature the gauge system undergoes a deconfinement transition which appears to be of the Kosterlitz-Thouless type [11].

We studied lattice U(1) gauge theory in an uniform external magnetic field. We performed numerical simulations on $512 \times 256 \times 4$ and $512 \times 64 \times 8$ lattices. To determine the critical coupling β_c, we measured the derivative of the free energy density. Contrary to the case of (2+1) and (3+1) non abelian lattice gauge theories, we do not find a dependence of the critical value of the coupling β_c on the magnetic field strength. By increasing the temporal size to $L_t = 8$ the critical coupling increases and is still independent of the external magnetic field strength. Therefore we can conclude that even in (2+1) dimensional case the critical coupling does not depend on the strength of the external magnetic field as for U(1) lattice gauge theories in (3+1) dimensions.

CONCLUSIONS

We have investigated U(1), SU(2), and SU(3) pure gauge theories both in (3+1) and (2+1) dimensions in presence of an uniform abelian (chromo)magnetic field. For non abelian gauge theories we found that, both in (3+1) and (2+1) dimensions, there is a critical field gH_c such that for $gH > gH_c$ the gauge systems are in the deconfined phase. This effect resembles what happens in ordinary superconductors and therefore we named it as "reversible vacuum color Meissner effect". On the other hand our numerical results for abelian gauge theories do not show any dependence of the critical coupling from the strength of an external magnetic field.

The dependence of the deconfinement temperature on the strength of the abelian chromomagnetic field in non abelian gauge theories could be intuitively understood by considering that strong enough chromomagnetic fields would force long range color correlations such that the gauge system gets deconfined.

On a more speculative side, one may thus imagine the confining vacuum of non abelian gauge theories as a disordered chromomagnetic condensate which confines color charges due to both the presence of a mass gap and the absence of long range color correlations, as argued by R. P. Feynman for QCD in (2+1) dimensions [12].

REFERENCES

1. G. Ripka, *Dual superconductor models of color confinement*, vol. 639 of *Lecture Notes in Phys.*, Springer-Verlag, 2004.
2. J. Greensite, *Prog. Part. Nucl. Phys.* **51**, 1 (2003), hep-lat/0301023.
3. R. W. Haymaker, *Phys. Rept.* **315**, 153–173 (1999), hep-lat/9809094.
4. G. 't Hooft (2004), hep-th/0408183.
5. P. Cea, and L. Cosmai, *Phys. Rev.* **D60**, 094506 (1999), hep-lat/9903005.
6. D. J. Gross, R. D. Pisarski, and L. G. Yaffe, *Rev. Mod. Phys.* **53**, 43 (1981).
7. M. J. Teper (1998), hep-th/9812187.
8. M. J. Teper, *Phys. Rev.* **D59**, 014512 (1999), hep-lat/9804008.
9. J. Engels, et al., *Nucl. Phys. Proc. Suppl.* **53**, 420–422 (1997), hep-lat/9608099.
10. A. M. Polyakov, *Nucl. Phys.* **B120**, 429–458 (1977).
11. P. D. Coddington, A. J. G. Hey, A. A. Middleton, and J. S. Townsend, *Phys. Lett.* **B175**, 64 (1986).
12. R. P. Feynman, *Nucl. Phys.* **B188**, 479 (1981).

CP Violation in B-Physics

Fernando Ferroni

Universita` di Roma 'La Sapienza' & INFN Roma

Abstract. The two competing B-factories (KEKB and PEPII) are performing spectacularly well and the experiments they host (Belle and BaBar) have harvested an impressive amount of data. CP violation in B-physics has been established and it has become a precision measurement.

Keywords: CP. B-Physics.
PACS: 13.25.Hw

INTRODUCTION

The mystery of our existence is based on the premise that although our theory of fundamental sub-atomic particle interactions (the Standard Model) places matter and anti-matter at nearly equal footing, our Universe appears to be composed of matter only. Indeed the Standard Model accounts for small differences in the interactions of matter and anti-matter through a phenomena known as CP violation. This difference however falls short, by orders of magnitude, in accounting for the observed matter asymmetry in the Universe.

Therefore investigations of the matter/anti-matter asymmetries it is of vital importance for the possible understanding of what could exist beyond the Standard Model.

Two powerful particle accelerators, KEKB [1] and PEPII [2] were built to study CP violation in the B meson decays.

B-physics has some distinct advantage over K-physics for studying CP violation. Indeed the asymmetries expected in some decay channel of the B-mesons can be very large. The amount of CP violation in the theoretical framework of the Standard Model is fixed by the area of the unitary triangle ($O(10^{-6})$) and therefore, with the caveat of choosing rather rare decays, the ratio of the difference of the decay width with their sum can easily be of ($O(1)$).

Better that any word the following figure (Fig. 1), showing the direct CP asimmetry observed in the decay of $B^0 \to K^+\pi^-$ makes the case.

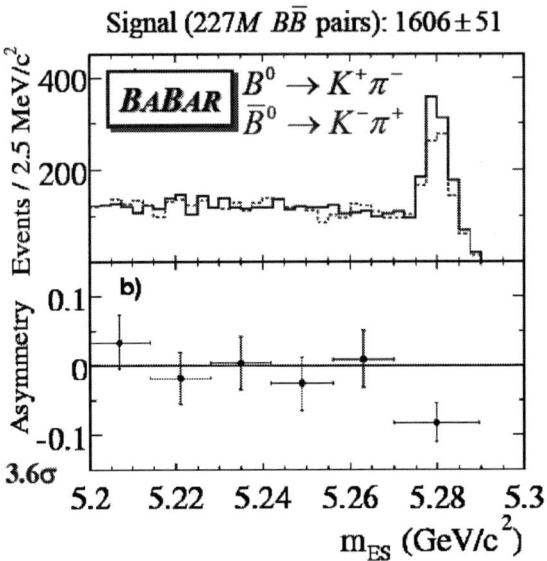

FIGURE 1. Difference (above) and asymmetry (below) in the decay of B^0 and its antiparticle in $K\tilde{\pi}$

The requirement of studying rare decays calls for accelerators producing a lot of B-mesons. The ideal environment is found at an e+e- collider running at the center of mass energy of the Y(4S) resonance with a luminosity much in excess of 10^{33} cm^{-2}s^{-1}. This means in practice producing B-meson pairs at a rate in excess of 1 Hz and therefore at the end of a conventional year of running (10^7 s) accumulating up to 100 millions of them allowing the exploration of decay channels with an effective branching fractions (including the specific decay chain and detection efficiency) of less than 10^{-6}. One peculiarity of the game is that the CP asymmetry cannot be integrated over the decay time due to the coherent nature of the Y(4S) state. This means in practice that one shall measure the time dependence of the decay asymmetry or in more experimental language the knowledge of the position of the decay vertices of the two B-meson is needed. The two vertices are in average separated by one lifetime and if the collision were to happen in a conventional collider the value would be of some tens of micron, too little to be measured precisely enough. The solution is in building an energy asymmetric collider where one can create a boost (O(0.5)) that brings the separation to some hundreds of micron, a measure easily accessible to silicon vertex detectors performances.

CP VIOLATION IN STANDARD MODEL

In the three generation Standard Model the coupling between the quarks is described by the CKM matrix [4]. The unitarity of the CKM matrix implies definite relations between its elements.

Three of these relations requires the sum of three complex quantities to vanish so that they can be geometrically represented in the complex plane as a triangle. These are the unitary triangles.

In two of these triangles one side is however much too shorter than the other two in such a way that they almost collapse to a line making unattractive their experimental test.

The most exciting physics lies in the third of them, the so-called B_d-physics triangle (Fig. 2).

$V_{ud}V_{ub} + V_{cd}V_{cb} + V_{td}V_{tb} = 0$

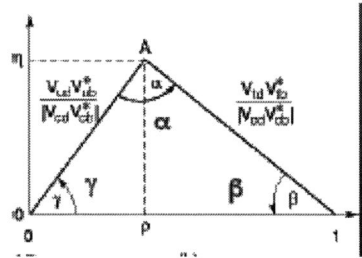

FIGURE 2. The unitary triangle for the B_d physics.

The openness of this triangle points to possible large CP asymmetries in B decays.

The name of the game is to measure all the three angles α, , β, γ. also known as ϕ_1, ϕ_2, ϕ_3, and show that their sum makes up to π. As an intermediate goal it is possible to measure the easiest and cleanest of them (β or ϕ_1) and compare the experimental result with the predictions obtained by using the experimental information on triangle sides and some result from lattice QCD calculations.

The challenge is in finding the processes that are interpretable in terms of theory prediction.

The possible manifestations of CP violation can be classified in a model-independent way:

- CP violation in decay, which occurs in both charged and neutral decays, when the amplitude for a decay and its CP conjugate process have different magnitudes;
- CP violation in mixing, which occurs when the two neutral mass eigenstates cannot be chosen to be CP eigenstates;
- CP violation in the interference between decays with and without mixing, which occurs in decays into final states that are common to B^0 and anti B^0. It often occurs in combination with the other two types but, important for B-factories physics, there are cases when, to an excellent approximation, it is the only effect.

The best opportunities are found in the processes of the third type. Indeed the CP violation in decay (direct CP) although experimentally accessible has to deal with to date non-calculable strong phases, while the one in mixing has to face the almost impossible experimental problem of telling a decay rate difference at a part in 10^6 in decays that are pretty frequent, like the semileptonic ones that occurs about 10% of the time.

The reference mode for measuring CP violation in B-physics has been identified since long time It is properly called the 'golden channel' [5]. It is the decay $B^0 \rightarrow J/\Psi K^0$. The reason is that in this decay the mixing, mediated by V_{td}, carries the phase $\beta(\phi_1)$ while the decay mediated by V_{cb} is real, ending up in an effect proportional to $\sin 2\beta$. The non-tree amplitudes are tiny and however they carry the same weak phase.

This channel is not only theoretically golden but also experimentally accessible in a rather easy way.

COLLIDERS, EXPERIMENTS AND TECHNIQUE OF THE MEASUREMENT

The machines that have been built to accept the challenge posed by this fundamental measurement are KEKB in Tsukuba (Japan) and PEPII at SLAC (USA).

Their asymmetric energy configuration matches the request and their performance in term of luminosity has been just extraordinary. KEKB has reached in terms of instantaneous luminosity the value of $1.5 \; 10^{34}$ cm^{-2}s^{-1} and has delivered to its experiment 450 fb^{-1} of integrated luminosity corresponding to roughly a billion B decays (250 fb^{-1} for PEPII).

BaBar [6] and Belle [7] are two classic style e+e-detectors. Their structure is pretty similar. From inside out they exhibit a Silicon Vertex Detector followed by a low mass Drift Chamber for tracking charged particles, a Cerenkov system for telling kaons from pions, and a CsI crystal electro-magnetic calorimeter. These detectors are immersed in a solenoidal field (1.5 T) provided by a superconducting magnet. Beyond the solenoid and interspersed in the iron of the return yoke resistive proportional chambers help in discriminating muons from hadrons and provide a determination of the impact point of neutral hadrons.

A measurement of A_{CP} requires:
- the reconstruction of a B_{CP} decay like $B^0 \rightarrow J/\Psi K^0$.
- the determination of the experimental $\Delta t(z)$ resolution. Typical Δz resolution from data is 180 μ dominated by the tag B.
- the assignment of the B flavor to the CP decay and the fraction of events in which the tag assignment is incorrect. For flavor tagging the information from the other B decay in the event is exploited by looking for the presence of leptons, kaons and/or soft pions. The effective efficiency of the tagging algorithm is of the order of 30%.

The steps required in the analysis and the typical figure of merit are shown in Fig. 3.

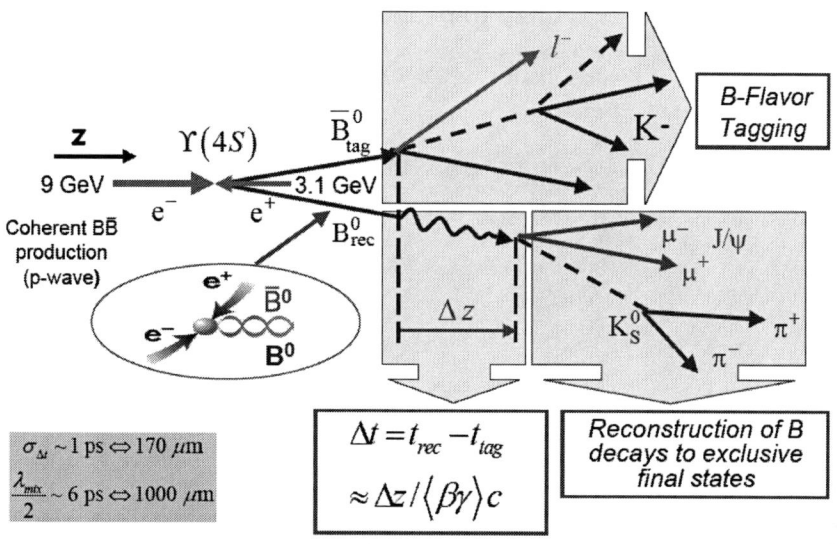

FIGURE 3. The steps needed for the CP time dependent measurement.

THE PHYSICS PROGRAMME AND RESULTS

The physics programme as evolved with the experience of the couple of years of run and the billion of B-decays analyzed can be summarized as follows:

- precision measurement of the angle β in the charmonium modes
- alternative measurement of the same angle in the b> sssbar decays where the tree diagram is more or less absent, the amplitude is dominated by penguin diagrams that carry the phase βnd where the possible presence of loops of new virtual particles (i.e. SUSY) would change the weak phase revealing evidence of New Physics beyond the Standard Model
- measurement of the angle α in the charmless two body decays
- measurement of the angle γ in a variety of modes, the most promising of all relying in a Dalitz analysis of the $B^+ \rightarrow D^0 K^+$ ($D^0 \rightarrow K_S \pi^+ \pi^-$) decay

sin2β in charmonium modes

The measurement has become very precise, the world average is at the time of this workshop: $\sin2\beta_{exp}$ = 0.726 +/ The prediction from the analysis of the unitary triangle is: $\sin2\beta_{pre}$ = 0.734 +/- 0.043.
Excellent agreement that confirms the invincibility of the Standard Model.

sin2β in penguin modes

The prototype of this kind of modes is B→ ΦK^0. Its decay diagram is shown in the following figure (Fig. 4).

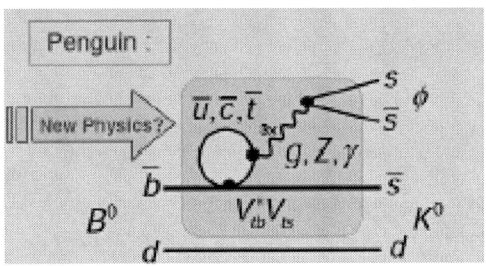

FIGURE 4. B→ ΦK^0 decay diagram.

It is a penguin diagram with the dominant contribution coming from the top loop. The dominant phase of the decay is V_{ts} (real) so that the expected CP violation phase is the one from the mixing leading to β. It is clear though that in case SUSY particles would mind to circulate in the loop they could bring a different phase. So the comparison between the value of sin2β measured in this mode with the one obtained in charmonium modes could show the presence of new particles not yet discovered. There are many modes that can be used for this test with some caveat. They are not all theoretically clean at the same level. It is rather arbitrary therefore to take their average for eventually enhancing the significance of an effect. It is however suggestive that the ones so far measured, systematically indicate a value of sin2β somewhat lower than expected. The following figure (Fig. 5) shows the compilation [8] of the results for several modes.

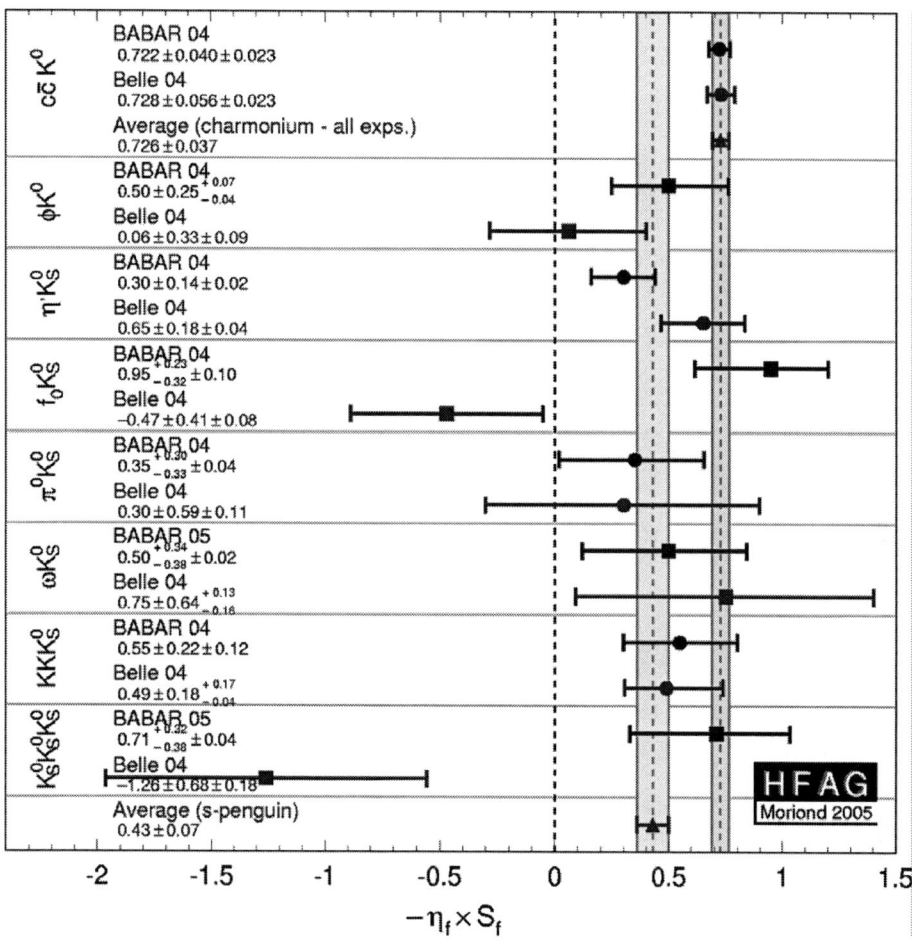

FIGURE 5. sin2β as determined in various b-→s penguin dominated channels with respect to to the value obtained in charmonium modes.

Not much can be said at this moment except that a significant increase of statistics would be desirable. An encouragement to the B-factories to keep going at full steam.

sin2β in charmless two body decays

The decay of B-mesons in charmless two body decays is the key for the measurement of the angle α. Indeed the mixing provide the phase β and the V_{ub} transition add the phase γ bringing to α= π − (β+γ).
This would be absolutely true only if the decay were dominated by the tree diagram. Unfortunately at least in the simpler of all the charmless two body decays (B→ ππ) it is observed an important, if not dominant, contribution of a penguin decay that brings a different weak phase. This is diagrammatically seen in Fig. 6.

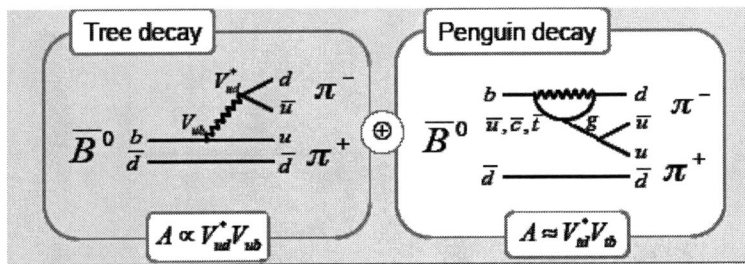

FIGURE 6. Tree and penguin diagram contributing to B→ ππ decay.

The solution in case of large statistics is known. It consists in a isospin analysis [9] making use of all the three B→ ππ decay amplitudes as shown in Fig. 7.

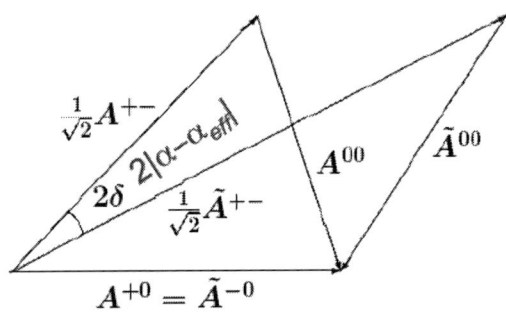

FIGURE 7. Isospin analysis triangle.

This analysis however requires one of the following conditions to be verified to be carried out safely:

- the three sides are of the same order of magnitude and the statistics is large
- the 00 amplitude is very small so that the triangle effectively collapses.

In the case of B→ ππ neither of the conditions is verified since statistics is limited and the 00 amplitude is sizeable.

The way out is found in the B→ ρ ρ decay. Here the 00 amplitude, for which only un upper limit has been determined, results to be at least 30 times smaller than the +- one. The triangle is therefore very squashed and the difference between the value of α determined (effective) and the true one is small.

The value [10] obtained by BaBar is α=(96+/-10+/-4+/-11)⁰ where the last error reflects the conservative estimate of the yet unknown contribution of the penguin amplitude.

A combination [11] of all the available information is presented in Fig.8.

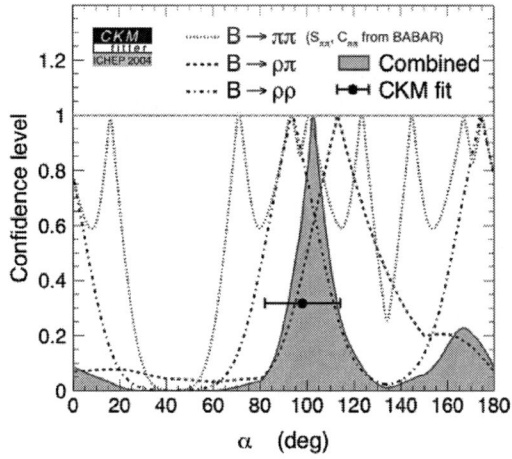

FIGURE 8. Confidence level as a function of the angle α for the three modes (B→ ππ, B→ ρπ, B→ ρρ) used. Individual contributions are shown as well as their combination (shaded).

The result is α=(103+/-11)⁰.

As can be seen the most precise result is due to B→ ρρ. However the B→ ρπ channel is crucial in resolving the ambiguity. Very modest indeed is the contribution of B→ ππ.

γ in the B⁺→ D⁰ K⁺ decays

The difficulty in measuring this angle at the B-factories comes from the problem that when using the B⁰ decays you cannot get rid of the mixing phase β. Indeed this means that one shall move to the charged B decays occurring via the b → u transitions whose phase is indeed γ. Here however, at first glance the necessary interference between two diagrams leading to the same final state cannot take place. There are several possibilities that have been pointed out [12] for circumventing this problem. Here we discuss the results obtained by exploiting the most promising [13] of all methods discovered so far.

The basic idea is of making use of the interference between the two diagrams (one Cabibbo favoured and the other Cabibbo suppressed) that lead to the same final state $B^+ \to (K_S\pi^+\pi^-)K^+$ shown in Fig. 9)

FIGURE 9. Cabibbo-favored and Cabibbo-soppressed diagrams for the $B^+ \to (K_S\pi^+\pi^-)K^+$ final state.

In the more than reasonable assumption that the $D^0 \to K_S\pi^+\pi^-$ is dominated by quasi-two body amplitudes a simultaneous fit to $D^0 \to K_S\pi^+\pi^-$ Dalitz plots of B^+ and B^- data yields three parameters (r, γ, δ), respectively the amount of Cabibbo suppression, the angle γ of the unitary triangle and the strong phase. The results [14] obtained by Belle for three different modes are shown below,

	r	ϕ_3 (°)	δ (°)
$DK^\pm$	$0.21 \pm 0.08 \pm 0.03 \pm 0.04$	$64 \pm 19 \pm 13 \pm 11$	$157 \pm 19 \pm 11 \pm 21$
$D^*K^\pm$	$0.12^{+0.16}_{-0.11} \pm 0.02 \pm 0.04$	$75 \pm 57 \pm 11 \pm 11$	$321 \pm 57 \pm 11 \pm 21$
$DK^{*\pm}$	$0.25^{+0.17}_{-0.18} \pm 0.09 \pm 0.04 \pm 0.08$	$112 \pm 35 \pm 9 \pm 11 \pm 8$	$353 \pm 35 \pm 8 \pm 21 \pm 49$

and the combination of the first two lines gives:

$\phi_3(\gamma) = (68+14-15+/-13+/-11)^0$

where the last error has to do with the systematic related to the Dalitz fit.
One aspect of this analysis is the strong dependence on the value of r.
In spite of all the difficulties the combination of all the available results (any method) from the two experiments gives an impressive constrain [15] to the value of the angle γ on the unitary plane (see Fig. 10).

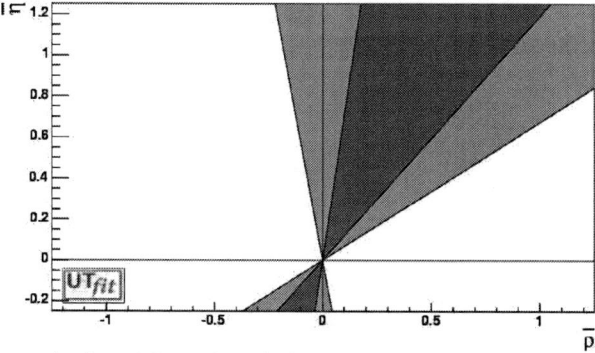

FIGURE 10. Determination of the angle γ obtained combining the experimental information from all methods and both experiments.

Status of the unitary triangle

The fit technology to the unitary triangle is a consolidated art by now. Two different approaches [11,15] exist. They differ in the statistical methodology (frequentist vs. Bayesian) and in the choice of some either experimental or theoretical (lattice computation) result. The final value of the coordinates of the triangle apex (i.e. the amount of CP violation in the Standard Model) are however very close by. One impressive feature of the triangle analysis is that the triangle apex can be determined with precision even by using the sole results of CP violating measurement in B-physics, namely the angles (see Fig. 11).

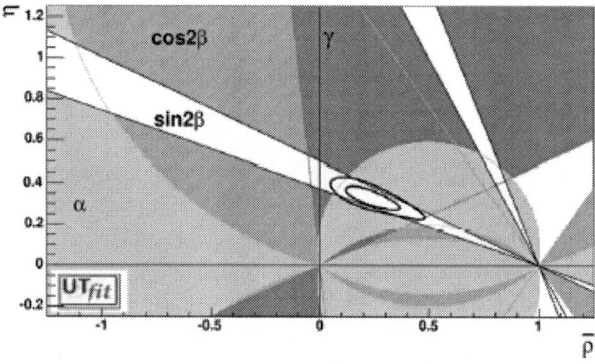

FIGURE 11. Constrain on the apex of the unitary triangle as obtained by fitting all together the results of BaBar and Belle on the angles α, β and γ.

As far as the complete analysis the graphical result is shown in Fig 12. The values of the apex coordinates (ρ,η) are:
$\rho = 0.207 +/- 0.038$
$\eta = 0.341 +/- 0.023$

CP-violation in B-physics exists and it is compatible with a Standard Model interpretation.

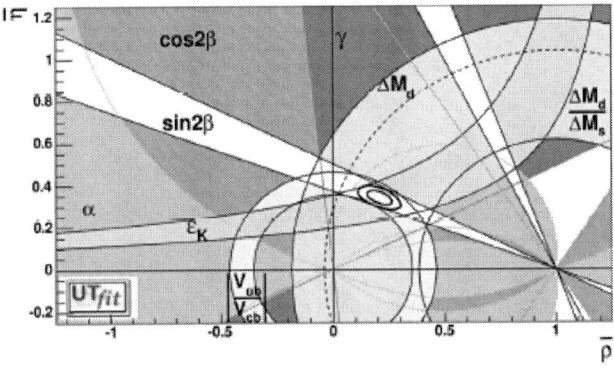

FIGURE 12. Constrain on the apex of the unitary triangle as obtained by fitting all the experimental available results as well as the lattice simulations.

CONCLUSIONS

The primary goal of the B-factories has been achieved. The violation of the CP symmetry has been observed in the B system. The measurement of the angle $\beta(\phi_1)$ of the unitary triangle has reached a precision level. This allows to say that the CP violation observed both in K-physics and B-physics belongs to the all mighty Standard Model of the Electroweak Interactions.

The long march to the measurement of the other two angles has started. The most promising methodologies have been already identified, statistics is the sole issue. The hunt to α ($_2\phi$) has already yielded a significant result. The angle $\gamma(\phi_3)$ is still opposing a fierce resistance, however one should remember that the possibility of measuring it was not even considered in the initial program of the experiments.

The final hope is still to find a deviation between the observed and predicted quantities that points to some exciting New Physics beyond the Standard Model. This might be achievable during the life span of the existing B-factories that will eventually integrate of the order of 1-2 ab^{-1} each till the completion of their cycle foreseen for the end of the decade.

The relay will then be hopefully taken both by a Super B-factory with a luminosity improved by a factor 20-50 with respect to now and an hadronic experiment (LHCb [16]) that will disclose the doors of the B_s physics.

REFERENCES

1. S. Kurokawa and E. Kikutani, *Nuclear Instrument and Methods in Physics Research*, **A499**, 1 (2003).
2. PEP-II- An asymmetric B Factory, Conceptual Design Report, SLAC-418, LBL-5379, 1993.
3. A. Garren et al., *IEEE Part. Accel.*, 1847 (1989).
4. N. Cabibbo, *Phys. Rev. Letters* **10**, 531 (1963). M.Kobayashi and T. Maskawa, *Progress Theoretical Physics* **49**, 652 (1973).
5. A.B. Carter and A.I. Sanda, *Phys. Rev.* **D23**, 1567 (1981). I.I. Bigi and A.I. Sanda, *Nucl. Phys.* **B193**, 85 (1981).
6. B. Aubert et al. *Nuclear Instrument and Methods in Physics Research* **A479**, 1 (2002).
7. A. Abashian et al., *Nuclear Instrument and Methods in Physics Research*, **A479**, 117 (2002).
8. http://www.slac.stanford.edu/xorg/hfag/
9. M. Gronau and D. London, *Phys. Rev. Letters* **65**, 3381 (1990).
10. B. Aubert et al., *Phys. Rev. Letters* **95**, 041805 (2005).
11. http://ckmfitter.in2p3.fr/
12. M. Gronau and D. London, *Phys. Lett.* **B253**, 483 (1991). M. Gronau and D. Wyler, *Phys. Lett.* **B265**, 172 (1991).
13. A. Giri et al., *Phys. Rev.* **D68**, 054018 (2003).
14. http://belle.kek.jp/belle/talks/LP05/Abe.pdf
15. http://utfit.roma1.infn.it/
16. http://lhcb-new.web.cern.ch/LHCb-new/

New Physics Search in B and kaon physics

Tobias Hurth

CERN, Dept. of Physics, Theory Unit, CH-1211 Geneva 23, Switzerland
SLAC, Stanford University, Stanford, CA 94309, USA

Abstract. With the running B, kaon and neutrino physics experiments, flavour physics takes centre stage within today's particle physics. We discuss the opportunities offered by these experiments in our search for new physics beyond the SM. We focus on rare B and kaon decays, highlighting specific observables in an examplary mode. Moreover, we briefly discuss the restrictive role of long-distance strong interactions and some new tools such as QCD factorization and SCET to handle them.

Keywords: B decays, New Physics
PACS: PACS: 12.60.Cn, 14.40.Nd

INTRODUCTION

There are three main issues within the present B and kaon physics programme: (i) the search for new degrees of freedom beyond the standard model (SM) in flavour- or CP-violating processes; (ii) the question of the precise mechanism of CP violation and (iii) the search for a quantitative understanding of the strong interactions within flavour observables.

Rare B and kaon decays representing loop-induced processes are highly sensitive probes for new degrees of freedom beyond the SM. Through virtual (loop) contributions of new particles to such observables, one can investigate high-energy scales even before such energies are accessible at collider experiments.

Such flavour information is complementary to the collider data of the Large Hadron Collider (LHC); for example the present flavour data from the B factories in the $b \to s$ sector is not very restrictive (yet). Within a supersymmetric new physics scenario, present bounds on squark mixing still allow for large contributions to flavour-violating squark decays at tree level, which can be measured at the LHC. In these cases, additional information from future flavour experiments will be necessary to interpret those LHC data properly.

The CKM prescription of CP violation with one single phase is very predictive. Before the start of the B factories, the neutral kaon system was the only environment where CP violation had been observed. It was difficult to decide if the CKM description of the SM really accounted quantitatively for the CP violation observed in the kaon system, because of the large theoretical uncertainties due to long-range strong interactions. The rich data sets from the present and planned B experiments now allow for an independent and really quantitative test of the CKM-induced CP-violating effects in several independent channels.

Quark-flavour physics is governed by the interplay of strong and weak interactions.

One of the main difficulties in examining the observables in flavour physics is the influence of the long-distance strong interaction. The resulting hadronic uncertainties restrict the opportunities in flavour physics significantly. If new physics does not show up in B physics through large deviations, as recent experimental data indicate, the focus on theoretically clean variables within the indirect search for new physics is mandatory.

Nevertheless there are new tools, such as QCD factorization and the soft-collinear effective theory (SCET), to tackle the strong interaction within B decays. The large data sets from the B experiments should be used to sharpen these new tools and improve our present understanding of the strong interaction.

EXPERIMENTAL ROADMAP

The present experimental roadmap of flavour physics offers great opportunities. Several B-physics experiments are successfully running at the moment and, in the upcoming years, new facilities will start to explore B physics with increasing sensitivity and within various experimental settings. There are two B factories, operating at the $\Upsilon(4S)$ resonance in an asymmetric mode, successfully obtaining data, namely the BABAR experiment at SLAC (Stanford, USA) [1] and the BELLE experiment at KEK (Tsukuba, Japan) [2]. An upgrade of the BELLE machine is planned; and there is also a vision of Super-BELLE [3]. After the present hadronic B-physics programme at FERMILAB (Batavia, USA) [4], there are strong B-physics programmes planned at three LHC experiments at CERN in Geneva, especially at LHCb [5]. There is also a future option of a B-physics programme at a future linear collider via a Giga-Z factory (see [6]).

The main motivation for a B physics programme at hadron colliders is the huge b-quark production cross section with respect to the one at e^+e^- machines, and the opportunity to analyse also the B_s system. Nevertheless, a future Super-B factory would be fully competitive but also complementary to the planned hadronic B physics programme (see for details [8, 7]).

There are many further sectors of flavour physics that offer important experimental opportunities. K decays such as $K \to \pi \nu \bar{\nu}$ and $K_L \to \pi^0 \ell^+ \ell^-$ are extremely sensitive to possible new degrees of freedom and are largely unexplored. In fact, at present we have fewer constraints on short-distance-dominated $s \to d$ quark transitions than on $b \to s$ ones. In the presence of new physics, charm physics could provide important inputs by future e^+e^- and fixed-target experiments. Searches for electric dipole moments of various particles are a very important source of information on the flavour and CP structure. Open questions in neutrino physics, regarding their masses, their mixing and their particle nature, are actively being attacked in the present and future experimental programme. The study of the correlation of neutrino properties with flavour phenomena in the charged-lepton and in the quark sector, e.g. charged-lepton flavour violation, is also an important target. Pushing the present limits on $\mu \leftrightarrow e$ and $\mu \leftrightarrow \tau$ transitions might lead to important insight.

HADRONIC UNCERTAINTIES

The crucial problem in the new physics search within flavour physics is the optimal separation of new physics effects and hadronic uncertainties. This can be successfully solved only for a selected number of *golden* observables in flavour physics, where

hadronic physics can be disentangled to a large extent and clean tests of the SM are possible. In principle there are three strategies:
- One can focus on inclusive decays modes. These modes are dominated by the partonic contributions because bound-state effects of the final states are eliminated by averaging over a specific sum of hadronic states. Moreover, also long-distance effects of the initial state are accounted for, through the heavy mass expansion in which the inclusive decay rate of a heavy B meson is calculated, using an expansion in inverse powers of the b-quark mass. In fact, one can use quark-hadron duality to derive a heavy mass expansion of the decay rates in powers of Λ_{QCD}/m_b (HME). For example, it turns out that the decay width of the $B \to X_s \gamma$ is well approximated by the partonic decay rate, which can be calculated in renormalization-group-improved perturbation theory:

$$\Gamma(\bar{B} \to X_s \gamma) = \Gamma(b \to X_s^{parton} \gamma) + \Delta^{nonpert.}$$

Non-perturbative corrections occur at the order Λ_{QCD}^2/m_b^2 only. The absence of first-order power corrections is a consequence of the fact that there is no independent gauge-invariant operator of dimension 4 in the operator product expansion because of the equations of motion. The latter fact implies a rather small numerical impact of the non-perturbative corrections to the decay rate of inclusive modes.
Nevertheless, there are additional nonperturbative corrections within inclusive modes due to necessary cuts in the experimental spectra like the photon energy spectrum in $\bar{B} \to X_s \gamma$ (see [9]).
- In exclusive processes, however, one cannot rely on quark-hadron duality and has to face the difficult task of estimating matrix elements between meson states. Therefore, exclusive modes are not well-suited to the new physics search in general. Nevertheless, one can focus on ratios of exclusive decay modes such as asymmetries, where large parts of the hadronic uncertainties partially cancel out. In particular, there are CP asymmetries that are goverend by one weak phase only. In that specific case the hadronic matrix elements cancel out completely.
- There are also specific decays like $K \to \pi \nu \bar{\nu}$ modes where the hadronic uncertainties can be eliminated by experimental data. In these kaon decays the hadronic matrix element can be related to the well-known rare semileptonic K_{l3} decays.

Regarding the hadronic matrix elements of exclusive modes, the method of QCD-improved factorization has been systemized for non-leptonic decays in the heavy-quark limit. This method allows for a perturbative calculation of QCD corrections to naive factorization and is the basis for the up-to-date predictions for exclusive rare B decays in general [10]. However, within this approach, a general, quantitative method to estimate the important $1/m_b$ corrections to the heavy-quark limit is missing.

A more general quantum field theoretical framework for the QCD-improved factorization was proposed – known under the name of SCET – which allows for a deeper understanding of the QCD factorization approach [11, 12]. In contrast to the well-known heavy-quark effective theory (HQET), the recently proposed SCET does not correspond to a local operator expansion. While HQET is only applicable to B decays, when the energy transfer to light hadrons is small, for example to $B \to D$ transitions at small recoil to the D meson, it is not applicable, when some of the outgoing, light particles have

momenta of order m_b; then one faces a multi scale problem:
a) $\Lambda =$ few $\times \Lambda_{QCD}$, the *soft* scale set by the typical energies and momenta of the light degrees of freedom in the hadronic bound states; b) m_b the *hard* scale set by the heavy-b-quark mass (we note, that in the B-meson rest frame, for $q^2 \simeq 0$ also the energy of the final-state hadron is given by $E \simeq m_b/2$); c) the hard-collinear scale $\mu_{hc} = \sqrt{m_b \Lambda}$ appears through interactions between soft and energetic modes in the initial and final states. The dynamics of hard and hard-collinear modes can be described perturbatively in the heavy-quark limit $m_b \to \infty$.

The separation of the two perturbative scales from the non-perturbative hadronic dynamics is formalized, within the framework of SCET, with the small expansion parameter $\lambda = \sqrt{\Lambda/m_b}$. Thus, SCET describes B decays to light hadrons with energies much larger than their masses, assuming that their constituents have momenta collinear to the hadron momentum. On a technical level, the implementation of power counting in λ, at the level of momenta, field and operators, corresponds directly to the well-known method of regions for Feynman diagrams [13].

The large variety of experimental data on those decay modes allows us to test these new tools and perhaps to reach sufficient accuracy for the determination of the CKM parameters and even for the detection of new physics effects.

EXPLORATION OF HIGHER SCALES VIA RARE DECAYS

Rare B and kaon processes often represent flavour changing neutral currents (FCNCs) and occur in the SM only at the loop level. This fact leads to the high sensitivity to potential new degrees of freedom beyond the SM (for a recent review see [23]). Such potential new contributions are not suppressed with respect to the SM contributions (see Fig.1). This indirect search for new physics signatures within flavour physics takes

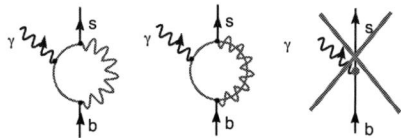

FIGURE 1. Standard and nonstandard contributions to $b \to s\gamma$ at-the-loop level only.

place today in complete darkness, given that we have no direct evidence of new particles beyond the SM. However, the day the existence of new degrees of freedom is established by the LHC, the searches for anomalous phenomena in the flavour sector will become mandatory. The problem then will no longer be to discover new physics, but to measure its (flavour) properties. In this context, the measurement of theoretically clean rare decays, *even when found to be SM-like*, will lead to important and valuable information of the structure of the new physics models and will lead to complementary information to the LHC collider data.

Because new physics effects beyond the SM seem to be rather small it is important to go beyond the pure study of branching ratios and also look at complex kinematic distributions, in particular at CP, forward-backward, isospin and polarization asymmetries. Only the measurements of a large overconstraining set of these observables allow us to detect specific pattern and to distinguish between various new-physics scenarios.

Finally, rare decays are also important tools to analyse the famous flavour problem, namely how FCNCs are suppressed beyond the SM. This problem has to be solved by any viable new physics model. One solution of the flavour problem is given by minimal flavour violation (MFV). In [27], a consistent definition of this scenario was presented, which essentially requires that all flavour and CP-violating interactions be linked to the known structure of Yukawa couplings. The constraint within an effective field approach is introduced with the help of a symmetry concept and can be shown to be renormalization-group-invariant [27].

Perhaps this MFV-based effective field theory approach is too pessimistic from the current point of view. One of the key predictions of the MFV is the direct link between the $b \to s$, $b \to d$, and $s \to d$ transitions. This prediction within the $\Delta F = 1$ sector is definitely not well-tested at the moment and there is still room for new flavour structures to be discovered. Nevertheless, in contrast to the scale of the electroweak symmetry breaking, there is no similarly strong argument that new flavour structures have to appear at the electroweak scale.

$b \to s/d\gamma$ AND $b \to s\ell^+\ell^-$ MODES

The inclusive $b \to s\gamma$ mode is still the most prominent rare decay, because it has already measured by several independent experiments [14, 15, 16, 17, 18, 19] and the present experimental accuracy has reached the 10% level [20]:

$$\text{BR}[\bar{B} \to X_s\gamma] = (3.52 \pm 0.30) \times 10^{-4}.$$

In the near future, more precise data on this mode are expected from the B factories. Thus, it is mandatory to reduce the present theoretical uncertainty accordingly. A systematic improvement certainly consists in performing a complete NNLL calculation which will reduce the well-known large uncertainty due to the definition of the charm mass [21] by a factor 2 as was recently shown [22]. In a recent theoretical update of the NLL prediction of this branching ratio, the uncertainty related to the definition of m_c was taken into account by varying m_c/m_b in the conservative range $0.18 \leq m_c/m_b \leq 0.31$ which covers both, the pole mass (with its numerical error) value and the running mass $\bar{m}_c(\mu_c)$ value with $\mu_c \in [m_c, m_b]$ [24]: $\text{BR}[\bar{B} \to X_s\gamma] = (3.70 \pm 0.35|_{m_c/m_b} \pm 0.02|_{\text{CKM}} \pm 0.25|_{\text{param.}} \pm 0.15|_{\text{scale}}) \times 10^{-4}$. The stringent bounds obtained from the $B \to X_s\gamma$ mode on various non-standard scenarios (see e.g. [25, 26, 27, 28, 29, 30]) are a clear example of the importance of clean FCNC observables in discriminating new-physics models.

Besides the $b \to s\gamma$ mode, also the $b \to s\ell^+\ell^-$ transitions are already accessible at the B factories [31, 32, 33], inclusively and exclusively. Quite recently also the $b \to d\gamma$ transition was measured for the first time [34] (for a recent review see [35]). The inclusive decay $b \to s\ell^+\ell^-$ is particularly attractive because of kinematic observables such as the invariant dilepton mass spectrum and the forward–backward (FB) asymmetry. This inclusive decay is also dominated by perturbative contributions if the $c\bar{c}$ resonances that show up as large peaks in the dilepton invariant mass spectrum are removed by appropriate kinematic cuts. In the 'perturbative windows', namely in the low-s (dilepton mass) region ($0.05 < s = q^2/m_b^2 < 0.25$) and also in the high-$s$ region with $0.65 < s$, a theoretical precision comparable with the one reached in the decay $b \to s\gamma$ is in principle possible. The recently calculated NNLL contributions [36, 37, 38, 39, 40, 41] have significantly improved the sensitivity of the inclusive $B \to X_s \ell^+ \ell^-$ decay in testing extensions of the SM in the sector of flavour dynamics, in particular the value of the dilepton invariant mass (q_0^2), for which the differential forward–backward asymmetry vanishes, is one of the precise predictions in flavour physics

Let us briefly comment on the impact of the exclusive $B \to K^{(*)} \ell^+ \ell^-$ modes. Hadronic uncertainties on these exclusive rates are dominated by the errors on form factors and are much larger than in the corresponding inclusive decays. In fact, following the analysis presented in Ref. [42], we see that inclusive modes already put much stronger constraints on the various Wilson coefficients. Concerning the measurement of a zero in the spectrum of the forward-backward asymmetry, things are different. According to Refs. [43] the value of the dilepton invariant mass (q_0^2), for which the differential forward–backward asymmetry vanishes, can be predicted in quite a clean way. In the QCD factorization approach at leading order in Λ_{QCD}/m_b, the value of q_0^2 is free from hadronic uncertainties at order α_s^0 (a dependence on the soft form factor $\xi_\perp$ and the light cone wave functions of the B and K^* mesons appear at NLL). Within the SM, the authors of Ref. [43] find: $q_0^2 = (4.1 \pm 0.6)\,\text{GeV}^2$. As in the inclusive case, such a measurement will have a huge phenomenological impact.

THERE IS ALSO BEAUTY IN KAON PHYSICS

Although the general focus within flavour physics is at present on B systems, kaon physics offers interesting complementary opportunities in the new physics search, such as the exclusive rare decays $K^+ \to \pi^+ \nu \bar{\nu}$ and $K_L \to \pi^0 \nu \bar{\nu}$. These decay modes are extremely sensitive to possible new degrees of freedom, but they also allow for an accurate determination of the unitarity triangle, which is completely independent from that of the B system (for a recent review, see [44]).

These modes are basically unexplored yet. While there is only an upper limit on the neutral mode, three events were found in the charged mode by the AGS E787 and the E949 Collaborations at Brookhaven [45, 46], leading to

$$BR(K^+ \to \pi^+ \nu \bar{\nu}) = \left(1.47\,^{+1.3}_{-0.9}\right) \times 10^{-10}.$$

Within the experimental and theoretical uncertainties this is fully consistent with the present theory prediction, which is based on a perturbative NNLL QCD analysis [47] and on a recent improvement of the long-distance contributions [48]:

$$BR(K^+ \to \pi^+ \nu \bar{\nu})_{\text{SM}} = (0.80 \pm 0.11) \times 10^{-10}.$$

The error is dominated by parametrical errors on m_c and on CKM matrix elements which can be significantly reduced in the future.

The rare decays $K^+ \to \pi^+ \nu \bar{\nu}$ and $K_L \to \pi^0 \nu \bar{\nu}$ are both exceptionally clean modes, for two reasons essentially. First, the hard (quadratic) GIM mechanism, is active; thus, these decays are dominated by short-distance dynamics. In fact, at the quark level the two processes arise from the $s \to d \nu \bar{\nu}$ process, which originates from a combination of the Z penguin and a double W exchange (see Fig.2) In these graphs the u, c, t quarks

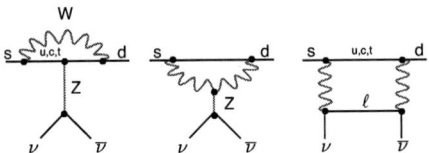

FIGURE 2. Graphs for $s \to d\nu\bar{\nu}$ in the SM

appear as internal lines. The hard GIM mechanism implies on the amplitude level:

$$A_q \sim m_q^2/m_W^2 V_{qs}^* V_{qd}, \quad q = u, c, t.$$

Thus, the top-quark contribution dominates, with a smaller contribution, in the case of the $K^+ \to \pi^+ \nu \bar{\nu}$ decay, from the charm contribution. The up-quark contribution is in both cases negligible, so that $s \to d\nu\bar{\nu}$ is essentially a short-distance process.

Moreover, the short-distance amplitude is governed by a single semileptonic operator whose hadronic matrix element can be determined experimentally by the semileptonic kaon decay $K^+ \to \pi^0 e^+ \nu$ using isospin symmetry; so the main hadronic uncertainties can be eliminated by experimental data. Besides their rich CKM phenomenology, the

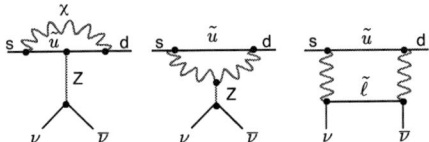

FIGURE 3. Graphs for $s \to d\nu\bar{\nu}$ in supersymmetry

decays $K_L \to \pi^0 \nu \bar{\nu}$ and $K^+ \to \pi^+ \nu \bar{\nu}$, as loop-induced processes, are very sensitive to new physics beyond the SM and are crucial tools to discriminate between various new physics scenarios in the future. Thanks to the cleanliness of the theoretical predictions, the measurement of these decays leads to very accurate constraints on any new physics model and will help us to discriminate between various new physics scenarios in the future, when new physics is discovered within the direct search. Moreover, there are also very interesting and theoretically clean correlations between B and K physics allowing for crucial precision tests of the SM and also of so-called minimal flavour violation scenarios in which the flavour structure is essentially SM-like. These correlations are generally violated in models with new sources of flavor violation. There is also the possibility that these clean rare decay modes themselves lead to direct evidence for new physics, if the measured decay rates are not compatible with the SM. New effects in supersymmetric models, for example, can be induced through new box- and penguin-diagram contributions which involve new particles such as charged Higgs or charginos and stops (Fig. 3), that replace the W boson and the up-type quark of the SM (Fig. 2). Explicit analyses of possible post-SM scenarios, with direct new-physics contributions in the $s \to d\bar{\nu}\nu$ amplitude or in $B\bar{B}$ mixing, can be found in [49] and [44].

Besides the recently finalized Brookhaven experiments on the charged mode [45, 46] and the running experiment on the neutral mode at KEK [51], there are two new proposals for the charged mode[52, 53], one for the neutral [54] currently under discussion.

The novel feature of the proposed measurement of $K^+ \to \pi^+ \nu \bar{\nu}$ at CERN [53] is that, in contrast to previous experiments, one does not use stopped kaons but kaons in flight. One expects 80 $K^+ \to \pi^+ \nu \bar{\nu}$ events in about two years of data taking, starting in 2010, based on the SM predictions. As can be seen in Fig. 4, the main background is the mode

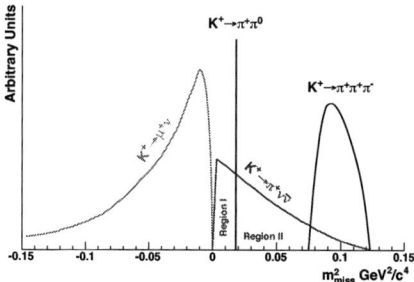

FIGURE 4. Distribution of the missing mass squared for the signal and the most frequent kaon decays. The signal region is divided into region I, which lies between the two prominent two-body decays of the charged kaon, while region II extends to the three-pion kinematical limit [50].

$K^+ \to \pi^+ \pi^0$. Thus, one of the main issues is the vetoing of the photons out of the π^0 decay, which is much easier than in previous experiments because of the high-energy kaon beam.

OUTLOOK

It is expected that the experiments at the LHC will lead to discoveries of new degrees of freedom at the TeV energy scale. The precise nature of this new physics is unknown, but it is strongly expected that it will answer some of the fundamental questions related to the origin of electroweak symmetry breaking. Independently of the nature of the new physics, a flavour physics program parallel to the LHC's will be crucial to disentangle the precise features of the newly uncovered phenomena and to discriminate between different new physics scenarios. In particular, the measurement of theoretically clean loop-induced rare B- and K-meson decays as highly sensitive probes for new degrees of freedom beyond the SM will lead to important information complementary to collider data; there are important fundamental questions that will be addressed exclusively by future flavour experiments, for example how FCNCs are suppressed beyond the SM (flavour problem), if there exist new sources of flavour and CP violation beyond those in the SM, if there is CP violation in the QCD gauge sector, how neutrino masses are generated, and what the relation between the flavour structure in the lepton and quark sectors is. All these questions include exciting options to learn something about physics at a scale much higher than our current experiments. Thus, a diversified and thorough experimental programme in flavour physics will continue to be an essential element for the understanding of nature.

REFERENCES

1. http://www.slac.stanford.edu/BFROOT/
2. http://belle.kek.jp/
3. S. Hashimoto et al., KEK-REPORT-2004-4
4. K. Anikeev et al., arXiv:hep-ph/0201071.
5. P. Ball et al., arXiv:hep-ph/0003238.
6. A. Ali et al., arXiv:hep-ph/0012218.
7. A. G. Akeroyd et al., arXiv:hep-ex/0406071.
8. J. Hewett et al., arXiv:hep-ph/0503261.
9. M. Neubert, Eur. Phys. J. C **40**, 165 (2005)
10. M. Beneke et al., Phys. Rev. Lett. **83**, 1914 (1999)
11. C. W. Bauer et al., Phys. Rev. D **63**, 114020 (2001)
12. M. Beneke et al., Nucl. Phys. B **643**, 431 (2002)
13. M. Beneke and V. A. Smirnov, Nucl. Phys. B **522**, 321 (1998)
14. R. Barate et al.ă [ALEPH Collaboration], Phys. Lett. B **429**, 169 (1998)
15. K. Abe et al.ă [Belle Collaboration], Phys. Lett. B **511**, 151 (2001)
16. S. Chen et al.ă [CLEO Collaboration], Phys. Rev. Lett. **87**, 251807 (2001)
17. B. Aubert et al.ă [BaBar Collaboration], arXiv:hep-ex/0207074.
18. B. Aubert et al.ă [BaBar Collaboration], arXiv:hep-ex/0207076.
19. P. Koppenburg et al. [Belle Collaboration], Phys. Rev. Lett. **93**, 061803 (2004)
20. J. Alexander et al. [Heavy Flavor Averaging Group (HFAG)], arXiv:hep-ex/0412073.
21. P. Gambino and M. Misiak, Nucl. Phys. B **611**, 338 (2001)
22. H. M. Asatrian, C. Greub, A. Hovhannisyan, T. Hurth, V. Poghosyan, Phys. Lett. B **619**, 322 (2005)
23. T. Hurth, Rev. Mod. Phys. **75**, 1159 (2003)
24. T. Hurth, E. Lunghi and W. Porod, Nucl. Phys. B **704** (2005) 56
25. G. Degrassi, P. Gambino and G. F. Giudice, leading JHEP **0012**, 009 (2000)
26. M. Carena, D. Garcia, U. Nierste and C. E. M. Wagner, Phys. Lett. B **499**, 141 (2001)
27. G. D'Ambrosio, G. F. Giudice, G. Isidori and A. Strumia, Nucl. Phys. B **645**, 155 (2002)
28. F. Borzumati, C. Greub, T. Hurth and D. Wyler, corrections Phys. Rev. D **62**, 075005 (2000)
29. T. Besmer, C. Greub and T. Hurth, Nucl. Phys. B **609**, 359 (2001)
30. M. Ciuchini, E. Franco, A. Masiero and L. Silvestrini, Phys. Rev. D **67**, 075016 (2003)
31. M. Iwasaki [Belle Collaboration], arXiv:hep-ex/0503044.
32. B. Aubert et al. [BABAR Collaboration], Phys. Rev. Lett. **93**, 081802 (2004)
33. K. Abe et al. [The Belle Collaboration], arXiv:hep-ex/0508009.
34. K. Abe et al., arXiv:hep-ex/0506079.
35. T. Hurth and E. Lunghi, eConf **C0304052** (2003) WG206
36. H. H. Asatryan, H. M. Asatrian, C. Greub and M. Walker, Phys. Rev. D **65**, 074004 (2002)
37. A. Ghinculov, T. Hurth, G. Isidori and Y. P. Yao, Nucl. Phys. B **648**, 254 (2003)
38. A. Ghinculov, T. Hurth, G. Isidori and Y. P. Yao, Nucl. Phys. B **685**. 351 (2004)
39. H. M. Asatrian, K. Bieri, C. Greub and A. Hovhannisyan, Phys. Rev. D **66** 094013 (2002)
40. C. Bobeth, M. Misiak and J. Urban, Nucl. Phys. B **574**, 291 (2000)
41. C. Bobeth, P. Gambino, M. Gorbahn and U. Haisch, JHEP **0404**, 071 (2004)
42. A. Ali, E. Lunghi, C. Greub and G. Hiller, Phys. Rev. D **66**, 034002 (2002)
43. M. Beneke, T. Feldmann and D. Seidel, Nucl. Phys. B **612**, 25 (2001)
44. A. J. Buras, F. Schwab and S. Uhlig, arXiv:hep-ph/0405132.
45. S. Adler et al. [E787 Collaboration], Phys. Rev. D **70**, 037102 (2004)
46. V. V. Anisimovsky et al. [E949 Collaboration], Phys. Rev. Lett. **93**, 031801 (2004)
47. A. J. Buras et al., arXiv:hep-ph/0508165.
48. G. Isidori, F. Mescia and C. Smith, Nucl. Phys. B **718**, 319 (2005)
49. G. D'Ambrosio and G. Isidori, Phys. Lett. B **530**, 108 (2002)
50. Courtesy of Augusto Ceccucci
51. http://psux1.kek.jp/ e391/
52. http://www-ps.kek.jp/jhf-np/LOIlist/pdf/L04.pdf
53. http://na48.web.cern.ch/NA48/NA48-3/
54. http://www-ps.kek.jp/jhf-np/LOIlist/pdf/L05.pdf

Sum Rules for Leading and Subleading Form Factors in Heavy Quark Effective Theory using the Non-forward Amplitude

F. Jugeau[*], A. Le Yaouanc[†], L. Oliver[†,**] and J.-C. Raynal[†]

[*]Instituto de Física Corpuscular, Valencia, Spain
[†]Laboratoire de Physique Théorique (UMR 8627 CNRS), Université de Paris Sud-XI, Bâtiment 210, 91405 Orsay cedex, France
[**]Speaker at the Workshop

Abstract. Within the OPE, we formulate new sum rules in Heavy Quark Effective Theory in the heavy quark limit and at order $1/m_Q$, using the non-forward amplitude. In the heavy quark limit, these sum rules imply that the elastic Isgur-Wise function $\xi(w)$ is an alternate series in powers of $(w-1)$. Moreover, one gets that the n-th derivative of $\xi(w)$ at $w = 1$ can be bounded by the $(n-1)$-th one, and the absolute lower bound for the n-th derivative $(-1)^n \xi^{(n)}(1) \geq \frac{(2n+1)!!}{2^{2n}}$. Moreover, for the curvature we find $\xi''(1) \geq \frac{1}{5}[4\rho^2 + 3(\rho^2)^2]$ where $\rho^2 = -\xi'(1)$. These results are consistent with the dispersive bounds, and they strongly reduce the allowed region of the latter for $\xi(w)$. The method is extended to the subleading quantities in $1/m_Q$. Concerning the perturbations of the Current, we derive new simple relations between the *functions* $\xi_3(w)$ and $\bar{\Lambda}\xi(w)$ and the sums $\sum_n \Delta E_j^{(n)} \tau_j^{(n)}(1) \tau_j^{(n)}(w)$ ($j = \frac{1}{2}, \frac{3}{2}$), that involve leading quantities, Isgur-Wise functions $\tau_j^{(n)}(w)$ and level spacings $\Delta E_j^{(n)}$. Our results follow because the non-forward amplitude depends on three variables $(w_i, w_f, w_{if}) = (v_i \cdot v', v_f \cdot v', v_i \cdot v_f)$, and we consider the zero recoil frontier $(w, 1, w)$ where only a finite number of j^P states contribute $(\frac{1}{2}^+, \frac{3}{2}^+)$. We also obtain new sum rules involving the elastic subleading form factors $\chi_i(w)$ ($i = 1,2,3$) at order $1/m_Q$ that originate from the $\mathcal{L}_{kin}$ and $\mathcal{L}_{mag}$ perturbations of the Lagrangian. To the sum rules contribute only the same intermediate states $(j^P, J^P) = (\frac{1}{2}^-, 1^-), (\frac{3}{2}^-, 1^-)$ that enter in the $1/m_Q^2$ corrections of the axial form factor $h_{A_1}(w)$ at zero recoil. This allows to obtain a lower bound on $-\delta^{(A_1)}_{1/m^2}$ in terms of the $\chi_i(w)$ and the shape of the elastic IW function $\xi(w)$. An important theoretical implication is that $\chi'_1(1), \chi_2(1)$ and $\chi'_3(1)$ ($\chi_1(1) = \chi_3(1) = 0$ from Luke theorem) must vanish when the slope and the curvature attain their lowest values $\rho^2 \to \frac{3}{4}, \sigma^2 \to \frac{15}{16}$. These constraints should be taken into account in the exclusive determination of $|V_{cb}|$.

Keywords: Heavy Quark Effective Theory;Form factors
PACS: 12.38.-t, 13.20.-v

We will expose the main results that we have obtained in Heavy Quark Effect Theory using the non-forward amplitude and the Operator Product Expansion. We will first examine results in the heavy quark limit on the shape of the Isur-Wise function. The method is then generalized to the study of the $1/m_Q$ perturbations, that are of two types, namely perturbations of the Current, and perturbations of the Lagrangian.

HEAVY QUARK LIMIT

In the leading order of the heavy quark expansion of QCD, Bjorken sum rule (SR) [1] relates the slope of the elastic Isgur-Wise (IW) function $\xi(w)$, to the IW functions of the transitions between the ground state and the $j^P = \frac{1}{2}^+, \frac{3}{2}^+$ excited states, $\tau_{1/2}^{(n)}(w)$, $\tau_{3/2}^{(n)}(w)$, at zero recoil $w = 1$ (n is a radial quantum number). This SR leads to the lower bound $-\xi'(1) = \rho^2 \geq \frac{1}{4}$. Recently, a new SR was formulated by Uraltsev in the heavy quark limit [2] involving also $\tau_{1/2}^{(n)}(1)$, $\tau_{3/2}^{(n)}(1)$, that implies, combined with Bjorken SR, the much stronger lower bound $\rho^2 \geq \frac{3}{4}$, a result that came as a big surprise. In ref. [3], in order to make a systematic study in the heavy quark limit of QCD, we have developed a manifestly covariant formalism within the Operator Product Expansion (OPE). We did recover Uraltsev SR plus a new class of SR.

Using the OPE and the trace formalism in the heavy quark limit, different initial and final four-velocities v_i and v_f, and heavy quark currents $J_1 = \bar{h}_{v'}^{(c)} \Gamma_1 h_{v_i}^{(b)}$, $J_2 = \bar{h}_{v_f}^{(b)} \Gamma_2 h_{v'}^{(c)}$, the following SR can be written [4] :

$$\left\{ \sum_{D=P,V} \sum_n Tr\left[\bar{\mathcal{B}}_f(v_f)\bar{\Gamma}_2 \mathcal{D}^{(n)}(v')\right] Tr\left[\bar{\mathcal{D}}^{(n)}(v')\Gamma_1 \mathcal{B}_i(v_i)\right] \xi^{(n)}(w_i)\xi^{(n)}(w_f) \right.$$

$$\left. + \text{Other excited states} \right\} = -2\xi(w_{if})Tr\left[\bar{\mathcal{B}}_f(v_f)\bar{\Gamma}_2 P'_+ \Gamma_1 \mathcal{B}_i(v_i)\right] . \quad (1)$$

In this formula v' is the intermediate meson four-velocity, $P'_+ = \frac{1}{2}(1 + \slashed{v}')$ comes from the residue of the positive energy part of the c-quark propagator, $\xi(w_{if})$ is the elastic Isgur-Wise function that appears because one assumes $v_i \neq v_f$. $\mathcal{B}_i$ and $\mathcal{B}_f$ are the 4×4 matrices of the ground state B or B^* mesons and $\mathcal{D}^{(n)}$ those of all possible ground state or excited state D mesons coupled to B_i and B_f through the currents. In (1) we have made explicit the $j = \frac{1}{2}^-$ D and D^* mesons and their radial excitations of quantum number n.

The variables w_i, w_f and w_{if} are defined as $w_i = v_i \cdot v'$, $w_f = v_f \cdot v'$, $w_{if} = v_i \cdot v_f$ and are independent within a certain domain. The SR (1) writes $L(w_i, w_f, w_{if}) = R(w_i, w_f, w_{if})$, where $L(w_i, w_f, w_{if})$ is the sum over the intermediate charmed states and $R(w_i, w_f, w_{if})$ is the OPE side. Within the domain one can derive relatively to any of the variables w_i, w_f and w_{if} and obtain different SR taking different limits to the frontiers of the domain.

As in ref. [3] [4], we choose as initial and final states the B meson $\mathcal{B}_i(v_i) = P_{i+}(-\gamma_5)$ $\mathcal{B}_f(v_f) = P_{f+}(-\gamma_5)$ and vector or axial currents projected along the v_i and v_f four-velocities $J_1 = \bar{h}_{v'}^{(c)} \slashed{v}_i h_{v_i}^{(b)}$, $J_2 = \bar{h}_{v_f}^{(b)} \slashed{v}_f h_{v'}^{(c)}$ we obtain SR (1) with the sum of all excited states j^P in a compact form :

$$(w_i + 1)(w_f + 1) \sum_{\ell \geq 0} \frac{\ell+1}{2\ell+1} S_\ell(w_i, w_f, w_{if}) \sum_n \tau_{\ell+1/2}^{(\ell)(n)}(w_i)\tau_{\ell+1/2}^{(\ell)(n)}(w_f)$$

$$+ \sum_{\ell \geq 1} S_\ell(w_i, w_f, w_{if}) \sum_n \tau_{\ell-1/2}^{(\ell)(n)}(w_i)\tau_{\ell-1/2}^{(\ell)(n)}(w_f) = (1 + w_i + w_f + w_{if})\xi(w_{if}) \quad (2)$$

We get, choosing instead the axial currents $J_1 = \bar{h}^{(c)}_{v'} \not{v}_i \gamma_5 h^{(b)}_{v_i}$, $J_2 = \bar{h}^{(b)}_{v_f} \not{v}_f \gamma_5 h^{(c)}_{v'}$

$$\sum_{\ell \geq 0} S_{\ell+1}(w_i, w_f, w_{if}) \sum_n \tau^{(\ell)(n)}_{\ell+1/2}(w_i)\tau^{(\ell)(n)}_{\ell+1/2}(w_f) + (w_i - 1)(w_f - 1)$$

$$\sum_{\ell \geq 1} \frac{\ell}{2\ell-1} S_{\ell-1}(w_i, w_f, w_{if}) \sum_n \tau^{(\ell)(n)}_{\ell-1/2}(w_i)\tau^{(\ell)(n)}_{\ell-1/2}(w_f) = -(1 - w_i - w_f + w_{if})\xi(w_{if}) \quad (3)$$

Following the formulation of heavy-light states for arbitrary j^P given by Falk [5], we have defined in ref. [4] the IW functions $\tau^{(\ell)(n)}_{\ell+1/2}(w)$ and $\tau^{(\ell)(n)}_{\ell-1/2}(w)$, ℓ and $j = \ell \pm \frac{1}{2}$ being the orbital and total angular momentum of the light cloud. S_n is given by [3]

$$S_n = v_{iv_1} \cdots v_{iv_n} v_{f\mu_1} \cdots v_{f\mu_n} \sum_\lambda \varepsilon'^{(\lambda)*v_1 \cdots v_n} \varepsilon'^{(\lambda)\mu_1 \cdots \mu_n}$$

$$= \sum_{0 \leq k \leq \frac{n}{2}} C_{n,k}(w_i^2 - 1)^k (w_f^2 - 1)^k (w_i w_f - w_{if})^{n-2k} \quad (4)$$

with $C_{n,k} = (-1)^k \frac{(n!)^2}{(2n)!} \frac{(2n-2k)!}{k!(n-k)!(n-2k)!}$.

From the sum of (2) and (3) one obtains, differentiating relatively to w_{if} [4] ($\ell \geq 0$):

$$\xi^{(\ell)}(1) = \frac{1}{4}(-1)^\ell \ell! \left\{ \frac{\ell+1}{2\ell+1} 4 \sum_n \left[\tau^{(\ell)(n)}_{\ell+1/2}(1)\right]^2 + \sum_n \left[\tau^{(\ell-1)(n)}_{\ell-1/2}(1)\right]^2 + \sum_n \left[\tau^{(\ell)(n)}_{\ell-1/2}(1)\right]^2 \right\}. \quad (5)$$

Therefore $\xi(w)$ is an alternate series in powers of $(w-1)$. Equation (5) reduces to Bjorken SR [1] for $\ell = 1$. Differentiating (3) relatively to w_{if} one obtains:

$$\xi^{(\ell)}(1) = \ell! (-1)^\ell \sum_n \left[\tau^{(\ell)(n)}_{\ell+1/2}(1)\right]^2 \quad (\ell \geq 0). \quad (6)$$

Combining (5) and (6) one obtains a SR for all ℓ that reduces to Uraltsev SR [2] for $\ell = 1$ and one gets also:

$$(-1)^\ell \xi^{(\ell)}(1) = \frac{1}{4} \frac{2\ell+1}{\ell} \ell! \left\{ \sum_n \left[\tau^{(\ell-1)(n)}_{\ell-1/2}(1)\right]^2 + \sum_n \left[\tau^{(\ell)(n)}_{\ell-1/2}(1)\right]^2 \right\}. \quad (7)$$

$$(-1)^\ell \xi^{(\ell)}(1) \geq \frac{2\ell+1}{4}\left[(-1)^{\ell-1}\xi^{(\ell-1)}(1)\right] \geq \frac{(2\ell+1)!!}{2^{2\ell}} \quad (8)$$

$$-\xi'(1) = \rho^2 \geq \frac{3}{4}, \quad \xi''(1) \geq \frac{15}{16} \quad (9)$$

Considering systematically the derivatives of the SR (2) and (3) relatively to w_i, w_f, w_{if} with the boundary conditions $w_{if} = w_i = w_f = 1$, one obtains a new SR that implies:

$$\sigma^2 \geq \frac{1}{5}\left[4\rho^2 + 3(\rho^2)^2\right]. \quad (10)$$

The result (8), that shows that all derivatives at zero recoil are large, should have important phenomenological implications for the empirical fit needed for the extraction of $|V_{cb}|$ in $B \to D^*\ell\nu$. The fits to extract $|V_{cb}|$ using a linear or linear plus quadratic dependence of $\xi(w)$ are not accurate enough.

$1/M_Q$ PERTURBATIONS OF THE CURRENT

We follow the main lines of our paper [6]. Our starting point is the T-product

$$T_{fi}(q) = i \int d^4x \, e^{-iq \cdot x} < B(p_f)|T[J_f(0)J_i(x)]|B(p_i) > \quad (11)$$

where $J_f(x) = \bar{b}(x)\Gamma_f c(x)$, $J_i(y) = \bar{c}(y)\Gamma_i b(y)$ and p_i is different from p_f.

Inserting in this expression hadronic intermediate states, there are contributions from the direct channel with hadrons with a single heavy quark c, and from hadrons with $b\bar{c}b$ quarks, the Z diagrams. We consider the limit $m_c \gg m_b \gg \Lambda_{QCD}$. The difference between the two energy denominators is large $q^0 - E_f + E_{X_{\bar{c}bb}} - \left(q^0 + E_i - E_{X_c}\right) \sim 2m_c$. In this limit we can neglect the second term, and consider the imaginary part of the direct diagram, the piece proportional to $\delta\left(q^0 + E_i - E_{X_c}\right)$. Our conditions are, in short, as follows : $\Lambda_{QCD} \ll m_b \sim m_c - q^0 \ll q^0 \sim m_c$. To summarize, we are considering the heavy quark limit for the c quark, but we allow for a large finite mass for the b quark. Finally,

$$T_{fi}^{abs} \cong \sum_{D_n} < B_f(v_f)|J_f(0)|D_n(v') > < D_n(v')|J_i(0)|B_i(v_i) > \quad (12)$$

$T_{fi}(q)$ (11) is given, alternatively, in terms of quarks and gluons, by

$$T_{fi}(q) = - \int d^4x \, e^{-iq \cdot x} < B(p_f)|\bar{b}(0)\Gamma_f S_c(0,x)\Gamma_i b(x)|B(p_i) > \quad (13)$$

where $S_c(0,x)$ is the c quark propagator in the background of the soft gluon field [7]. Since we are considering the absorptive part in the c heavy quark limit of the direct graph $S_c(x,0)$ can be replaced by [8]

$$S_c(0,x) \to e^{im_c v' \cdot x} \Phi_{v'}[0,x] D_{v'}(x) \quad (14)$$

where $D_{v'}(x)$ is the *cut* free propagator of a heavy quark ($P'_+ = \frac{1}{2}(1 + \slashed{v}')$)

$$D_{v'}(x) = P'_+ \int \frac{d^4k}{(2\pi)^4} \delta(k \cdot v') e^{ik \cdot x} = P'_+ \int_{-\infty}^{\infty} \frac{dt}{2\pi} \delta^4(x - v't) \quad (15)$$

The eikonal phase $\Phi_{v'}[0,x]$ in (14) corresponds to the propagation of the c quark

$$\Phi_{v'}[0,v't] = P \exp\left(-i \int_0^t ds \, v' \cdot A(v's)\right). \quad (16)$$

and takes care of the dynamics of soft gluons in HQET. One obtains finally the OPE matrix element :

$$T_{fi}^{abs} \cong <B(p_f)|\bar{b}(0)\Gamma_f \frac{1+\slashed{v}'}{2v'^0}\Gamma_i b(0)|B(p_i)> + O(1/m_c). \quad (17)$$

We end up with the SR, valid for *all powers* in $1/m_b$, but only to leading order in $1/m_c$.

$$\sum_{D_n} <B_f(v_f)|J_f(0)|D_n(v')> <D_n(v')|J_i(0)|B_i(v_i)>$$

$$= <B(v_f)|\bar{b}(0)\Gamma_f \frac{1+\slashed{v}'}{2v'^0}\Gamma_i b(0)|B(v_i)> + O(1/m_c) \quad (18)$$

On the other hand, making use of the HQET equations of motion, the field $b(x)$ in (18) can be decomposed into upper and lower components [9] and, keeping the first order in $1/m_b$, the sum rule reads

$$\sum_{D_n} <B_f(v_f)|J_f(0)|D_n(v')> <D_n(v')|J_i(0)|B_i(v_i)>$$

$$= <B(p_f)|\bar{h}_{v_f}(0)\Gamma_f \frac{1+\slashed{v}'}{2v'^0}\Gamma_i h_{v_i}(0)|B(p_i)> \quad (19)$$

$$+ \frac{1}{2m_b} <B(p_f)|\bar{h}_{v_f}(0)\left[(-i\overleftarrow{D})\Gamma_f \frac{1+\slashed{v}'}{2v'^0}\Gamma_i + \Gamma_f \frac{1+\slashed{v}'}{2v'^0}\Gamma_i(i\overrightarrow{D})\right]h_{v_i}(0)|B(p_i)> + \cdots$$

Therefore, in the OPE side we have, besides the leading dimension 3 operator $O^{(3)} = \bar{h}_{v_f}\Gamma_f P'_+ \Gamma_i h_{v_i}$, the dimension 4 operator

$$O^{(4)} = \bar{h}_{v_f}\left[(-i\overleftarrow{D})\Gamma_f P'_+\Gamma_i + \Gamma_f P'_+\Gamma_i(i\overrightarrow{D})\right]h_{v_i}. \quad (20)$$

In the SR we have to compute the l.h.s. including terms of order $1/2m_b$. These terms have been parametrized by Falk and Neubert [10] for the $\frac{1}{2}^-$ doublet and by Leibovich et al. [11] for the transitions between the ground state $\frac{1}{2}^-$ and the $\frac{1}{2}^+, \frac{3}{2}^+$ excited states.

We need also to compute the perturbation of the initial and final wave functions $|B_i(v_i)>$, $|B_f(v_f)>$ due to the kinetic and magnetic perturbations of the Lagrangian. This can be done easily following also the prescriptions of Falk and Neubert to compute these corrections in $1/m_b$ for the leading matrix element $<B_f(v_f)|\bar{h}_{v_f}\Gamma_f P'_+\Gamma_i h_{v_i}|B_i(v_i)>$.

The final result is the following. The subleading quantities, functions of w, $\bar{\Lambda}\xi(w)$ and $\xi_3(w)$ [10] can be expressed in terms of leading quantities, namely the $\frac{1}{2}^- \to \frac{1}{2}^+, \frac{3}{2}^+$ IW functions $\tau_j^{(n)}(w)$ and the corresponding level spacings $\Delta E_j^{(n)}$ ($j = \frac{1}{2}, \frac{3}{2}$) [6]

$$\bar{\Lambda}\xi(w) = 2(w+1)\sum_n \Delta E_{3/2}^{(n)} \tau_{3/2}^{(n)}(1) \tau_{3/2}^{(n)}(w) + 2\sum_n \Delta E_{1/2}^{(n)} \tau_{1/2}^{(n)}(1) \tau_{1/2}^{(n)}(w) \quad (21)$$

$$\xi_3(w) = (w+1)\sum_n \Delta E_{3/2}^{(n)} \tau_{3/2}^{(n)}(1) \tau_{3/2}^{(n)}(w) - 2\sum_n \Delta E_{1/2}^{(n)} \tau_{1/2}^{(n)}(1) \tau_{1/2}^{(n)}(w). \quad (22)$$

These quantities reduce to known SR for $w = 1$, respectively Voloshin SR [12] and a SR for $\xi_3(1)$ [13, 2], and generalizes them for all w.

$1/M_Q$ PERTURBATIONS OF THE LAGRANGIAN

We follow closely our recent work [14]. Instead of using the OPE, we will simply use the definition of the subleading elastic $\frac{1}{2}^- \to \frac{1}{2}^-$ functions $\chi_i(w)$ $(i = 1, 2, 3)$ [10]

$$< D(v')|i \int dx T[J^{cb}(0), \mathcal{L}_{v'}^{(c)}(x)]|B(v) > = \frac{1}{2m_c}\{-2\chi_1(w)Tr[\overline{D}(v')\overline{\Gamma}B(v)]$$
$$-\frac{1}{2}Tr[\overline{A}_{\alpha\beta}(v',v)\overline{D}(v')i\sigma^{\alpha\beta}P'_+\overline{\Gamma}B(v)]\} \qquad (23)$$

with

$$\overline{A}_{\alpha\beta}(v',v) = -2\chi_2(w)(v_\alpha\gamma_\beta - v_\beta\gamma_\alpha) - 4\chi_3(w)i\sigma_{\alpha\beta} \qquad (24)$$

where $\overline{A} = \gamma^0 A^+ \gamma^0$, $J^{cb} = \overline{h}_{v'}^{(c)}\Gamma h_v^{(b)}$ and $\mathcal{L}_v^{(Q)} = \frac{1}{2m_Q}\left[O_{kin,v}^{(Q)} + O_{mag,v}^{(Q)}\right]$ with $O_{kin,v}^{(Q)} = \overline{h}_v^{(Q)}(iD)^2 h_v^{(Q)}$, $O_{mag,v}^{(Q)} = \frac{g_s}{2}\overline{h}_v^{(Q)}\sigma_{\alpha\beta}G^{\alpha\beta}h_v^{(Q)}$.

The $\chi_i(w)$ $(i = 1, 2, 3)$ have dimensions of mass, and correspond to the definition given by Luke [15]. We will now insert intermediate states in the T-products (23). We can separately consider $\mathcal{L}_{kin}^{(b)}$ or $\mathcal{L}_{mag}^{(b)}$. The Z-diagrams involving contributing to the T-products are suppressed by the heavy quark mass. Conveniently choosing initial and final states, we find

$$\chi_1(w) = \frac{1}{2}\sum_{n\neq 0}\frac{1}{\Delta E_{1/2}^{(n)}}\xi^{(n)}(w)\frac{< D^{(n)}(v)|O_{kin,v}^{(c)}(0)|D(v) >}{\sqrt{4m_{D^{(n)}}m_D}} \qquad (25)$$

$$\chi_2(w) = -\frac{3}{4\sqrt{6}}\sum_n\frac{1}{\Delta E_{3/2}^{(n)}}\tau_{3/2}^{(2)(n)}(w)\frac{< D_{3/2}^{*(n)}(v,\varepsilon)|O_{mag,v}^{(c)}(0)|D^*(v,\varepsilon) >}{\sqrt{4m_{D_{3/2}^{*(n)}}m_{D^*}}} \qquad (26)$$

$$\chi_3(w) = -\frac{1}{4}\sum_{n\neq 0}\frac{1}{\Delta E_{1/2}^{(n)}}\xi^{(n)}(w)\frac{< D^{*(n)}(v,\varepsilon)|O_{mag,v}^{(c)}(0)|D^*(v,\varepsilon) >}{\sqrt{4m_{D^{*(n)}}m_{D^*}}}$$
$$-\frac{w-1}{4\sqrt{6}}\sum_n\frac{1}{\Delta E_{3/2}^{(n)}}\tau_{3/2}^{(2)(n)}(w)\frac{< D_{3/2}^{*(n)}(v,\varepsilon)|O_{mag,v}^{(c)}(0)|D^*(v,\varepsilon) >}{\sqrt{4m_{D_{3/2}^{*(n)}}m_{D^*}}} \qquad (27)$$

(i) One should notice that *elastic subleading form factors of the Lagrangian type* are given in terms of *leading IW functions*, namely $\xi^{(n)}(w)$ and $\tau_{3/2}^{(2)(n)}(w)$, and *subleading form factors at zero recoil*.

(ii) $\chi_1(w)$ is given in terms of matrix elements of $\mathcal{L}_{kin}$ and involve transitions $\frac{1}{2}^- \to \frac{1}{2}^-$.

(iii) The *elastic subleading magnetic form factors* $\chi_2(w)$ and $\chi_3(w)$ involve $D^*(1^-) \to D^{*(n)}(1^-)$ transitions $\frac{1}{2}^- \to \frac{1}{2}^-$ and $\frac{1}{2}^- \to \frac{3}{2}^-$.

(iv) $\chi_1(w)$ and $\chi_3(w)$ satisfy $\chi_1(1) = \chi_3(1) = 0$ [15], since $\xi^{(n)}(1) = \delta_{n,0}$.

It is well-known that the determination of $|V_{cb}|$ from the $\overline{B} \to D^*\ell\nu$ differential rate at zero recoil depends on the value of $h_{A_1}(1)$. Precisely the subleading matrix elements of O_{kin} and O_{mag}, that enter in the SR (25)-(27), are related to the quantity $|h_{A_1}(1)|$. The following SR follows from the OPE [16] [11]

$$|h_{A_1}(1)|^2 + \sum_n \frac{|<D^{*(n)}\left(\frac{1}{2}^-,\frac{3}{2}^-\right)(v,\varepsilon)|\vec{A}|B(v)>|^2}{4m_{D^{*(n)}}m_B}$$

$$= \eta_A^2 - \frac{\mu_G^2}{3m_c^2} - \frac{\mu_\pi^2 - \mu_G^2}{4}\left(\frac{1}{m_c^2} + \frac{1}{m_b^2} + \frac{2}{3m_c m_b}\right) \qquad (28)$$

and $\mu_\pi^2 = \frac{1}{2m_B} < B(v)|\overline{h}_v^{(b)}(iD)^2 h_v^{(b)}|B(v)>$, $\mu_G^2 = \frac{1}{2m_B} < B(v)|\overline{h}_v^{(b)}\frac{g_s}{2}\sigma_{\alpha\beta}G^{\alpha\beta}h_v^{(b)}|B(v)>$.

In the l.h.s. of relation (28), $h_{A_1}(1) = \eta_{A_1} + \delta_{1/m^2}^{(A_1)}$ ($\eta_{A_1} = 1+$ radiative corrections) because there are no first order $1/m_Q$ corrections due to Luke theorem. The sum in the l.h.s. of (28) contains two types of contributions, corresponding to $D^{*(n)}\left(\frac{1}{2}^-, 1^-\right)$ ($n \neq 0$), and $D^{*(n)}\left(\frac{3}{2}^-, 1^-\right)$ ($n \geq 0$). One gets,

$$-\delta_{1/m^2}^{(A_1)} = \frac{\mu_G^2}{6m_c^2} + \frac{\mu_\pi^2 - \mu_G^2}{8}\left(\frac{1}{m_c^2} + \frac{1}{m_b^2} + \frac{2}{3m_c m_b}\right) + \frac{1}{2}\sum_n \frac{|<D^{*(n)}\left(\frac{1}{2}^-,\frac{3}{2}^-\right)(v,\varepsilon)|\vec{A}|B(v)>|^2}{4m_{D^{*(n)}}m_B}.$$
(29)

The correction $\delta_{1/m^2}^{(A_1)}$ is therefore negative, both terms being of the same sign.

Leibovich et al. (formulas (4.1) and (4.3)) [11] have expressed
$<D^{*(n)}\left(\frac{1}{2}^-,\frac{3}{2}^-\right)(v,\varepsilon)|\vec{A}|B>$ in terms of $<D^{*(n)}\left(\frac{1}{2}^-\right)(v,\varepsilon)|O_{kin,v}^{(c)}(0)|D^*(v,\varepsilon)>$ and
$<D^{*(n)}\left(\frac{1}{2}^-,\frac{3}{2}^-\right)(v,\varepsilon)|O_{mag,v}^{(c)}(0)|D^*(v,\varepsilon)>$. Hence, $-\delta_{1/m^2}^{(A_1)}$ (29) can be written as

$$-\delta_{1/m^2}^{(A_1)} = \frac{\mu_G^2}{6m_c^2} + \frac{1}{8}\left(\frac{1}{m_c^2} + \frac{1}{m_b^2} + \frac{2}{3m_c m_b}\right)\left(\mu_\pi^2 - \mu_G^2\right)$$

$$+\frac{1}{2}\sum_n\left[\left(\frac{1}{2m_c} + \frac{3}{2m_b}\right)\frac{1}{\Delta E_{1/2}^{(n)}}\frac{<D^{*(n)}\left(\frac{1}{2}^-\right)(v,\varepsilon)|O_{mag,v}^{(c)}(0)|D^*(v,\varepsilon)>}{\sqrt{4m_{D^{*(n)}}m_{D^*}}}\right.$$

$$+\left.\left(\frac{1}{2m_c} - \frac{1}{2m_b}\right)\frac{1}{\Delta E_{1/2}^{(n)}}\frac{<D^{*(n)}\left(\frac{1}{2}^-\right)(v,\varepsilon)|O_{kin,v}^{(c)}(0)|D^*(v,\varepsilon)>}{\sqrt{4m_{D^{*(n)}}m_{D^*}}}\right]^2$$

$$+\frac{1}{2}\sum_n\left[\frac{1}{2m_c}\frac{1}{\Delta E_{3/2}^{(n)}}\frac{<D^{*(n)}\left(\frac{3}{2}^-\right)(v,\varepsilon)|O_{mag,v}^{(c)}(0)|D^*(v,\varepsilon)>}{\sqrt{4m_{D_{3/2}^{*(n)}}m_{D^*}}}\right]^2. \qquad (30)$$

The important point to emphasize is that $<D^{*(n)}\left(\frac{1}{2}^-\right)(v,\varepsilon)|O_{kin,v}^{(c)}(0)|D^*(v,\varepsilon)>$ and $<D^{*(n)}\left(\frac{1}{2}^-,\frac{3}{2}^-\right)(v,\varepsilon)|O_{mag,v}^{(c)}(0)|D^*(v,\varepsilon)>$ are precisely the same matrix elements that enter in the SR (25)-(27). This allows to obtain an interesting lower bound on $-\delta_{1/m^2}^{(A_1)}$.

Taking the relevant linear combinations of the matrix elements suggested by the r.h.s. of (30), and Schwarz inequality

$$\left|\sum_n A_n B_n\right| \leq \sqrt{\left(\sum_n |A_n|^2\right)\left(\sum_n |B_n|^2\right)} \tag{31}$$

one finds the inequality

$$-\delta^{(A_1)}_{1/m^2} \geq \frac{\mu_G^2}{6m_c^2} + \frac{\mu_\pi^2 - \mu_G^2}{8}\left(\frac{1}{m_c^2} + \frac{1}{m_b^2} + \frac{2}{3m_c m_b}\right) + \frac{16}{3}\frac{\left[\frac{1}{2m_c}\chi_2(w)\right]^2}{\sum_n \left[\tau_{3/2}^{(2)(n)}(w)\right]^2}$$

$$+ 2\frac{\left\{\left(\frac{1}{2m_c} - \frac{1}{2m_b}\right)\chi_1(w) - \frac{1}{3}\left(\frac{1}{2m_c} + \frac{3}{2m_b}\right)[-2(w-1)\chi_2(w) + 6\chi_3(w)]\right\}^2}{\sum_{n\neq 0}[\xi^{(n)}(w)]^2} \tag{32}$$

that involves on the r.h.s. *elastic subleading* functions $\chi_i(w)$ ($i = 1, 2, 3$) in the numerator and sums over *inelastic leading IW functions* $\sum_{n\neq 0}[\xi^{(n)}(w)]^2$ and $\sum_n[\tau_{3/2}^{(2)(n)}(w)]^2$ in the denominator. We must emphasize that (32) is valid for all w and constitutes a rigorous constraint between these functions and the correction $-\delta^{(A_1)}_{1/m^2}$. Let us point out that, near $w = 1$, since $\xi^{(n)}(w) \sim (w-1)$ ($n \neq 0$) and, due to Luke theorem $\chi_1(w), \chi_3(w) \sim (w-1)$, the second term on the r.h.s. of (32) is a constant in the limit $w \to 1$. Since $\chi_2(w)$ is not protected by Luke theorem, $\chi_2(1) \neq 0$ and in general $\tau_{3/2}^{(2)}(1) \neq 0$, the last term in the r.h.s. of (32) is also a constant for $w = 1$. The inequality (32) holds in the $w \to 1$ limit :

$$-\delta^{(A_1)}_{1/m^2} \geq \frac{\mu_G^2}{6m_c^2} + \frac{\mu_\pi^2 - \mu_G^2}{8}\left(\frac{1}{m_c^2} + \frac{1}{m_b^2} + \frac{2}{3m_c m_b}\right)$$

$$+ 2\frac{\left\{\left(\frac{1}{2m_c} - \frac{1}{2m_b}\right)\chi_1'(1) - \frac{1}{3}\left(\frac{1}{2m_c} + \frac{3}{2m_b}\right)[-2\chi_2(1) + 6\chi_3'(1)]\right\}^2}{\sum_{n\neq 0}[\xi^{(n)'}(1)]^2} + \frac{16}{3}\frac{\left[\frac{1}{2m_c}\chi_2(1)\right]^2}{\sum_n\left[\tau_{3/2}^{(2)}(1)\right]^2}. \tag{33}$$

On the other hand, we have demonstrated above [4]

$$\sum_n \left[\tau_{3/2}^{(2)}(1)\right]^2 = \frac{4}{5}\sigma^2 - \rho^2 \tag{34}$$

$$\sum_{n\neq 0}[\xi^{(n)'}(1)]^2 = \frac{5}{3}\sigma^2 - \frac{4}{3}\rho^2 - (\rho^2)^2 \tag{35}$$

where ρ^2 and σ^2 are the slope and the curvature of the elastic Isgur-Wise function $\xi(w)$. The positivity of the l.h.s. of (34), (35) yield respectively the lower bounds on the curvature obtained (9) and (10). Relations (33)-(35) give finally the bound

$$-\delta^{(A_1)}_{1/m^2} \geq \frac{\mu_G^2}{6m_c^2} + \frac{\mu_\pi^2 - \mu_G^2}{8}\left(\frac{1}{m_c^2} + \frac{1}{m_b^2} + \frac{2}{3m_c m_b}\right) + \frac{80}{3(4\sigma^2 - 5\rho^2)}\left[\frac{1}{2m_c}\chi_2(1)\right]^2$$

$$+\frac{2}{3[5\sigma^2-4\rho^2-3(\rho^2)^2]}\left\{\left(\frac{1}{2m_c}-\frac{1}{2m_b}\right)3\chi'_1(1)-\left(\frac{1}{2m_c}+\frac{3}{2m_b}\right)[-2\chi_2(1)+6\chi'_3(1)]\right\}^2. \quad (36)$$

(i) The bounds contain an OPE piece, dependent on μ_π^2 and μ_G^2, and a piece that bounds the inelastic contributions, given in terms of the $1/m_Q$ elastic quantities $\chi'_1(1)$, $\chi_2(1)$, $\chi'_3(1)$ and on the slope ρ^2 and curvature σ^2 of the elastic IW function $\xi(w)$.

(ii) Taking roughly constant values for $\chi'_1(1)$, $\chi_2(1)$, $\chi'_3(1)$, independent of ρ^2 and σ^2, as suggested by the QCD Sum Rules calculations (QCDSR) [17] [18] [19], the bounds for the inelastic contributions diverge in the limit $\rho^2 \to \frac{3}{4}$, $\sigma^2 \to \frac{15}{16}$, according to (9).

(iii) Therefore, one should expect that $\chi'_1(1)$, $\chi_2(1)$ and $\chi'_3(1)$ vanish also in this limit. We give a demonstration of this interesting feature below.

(iv) The limit $\rho^2 \to \frac{3}{4}$, $\sigma^2 \to \frac{15}{16}$ seems related to the behaviour of $\chi_i(w)$ near $w=1$.

(v) The feature (iii) does not appear explicitly in the QDCSR approach, where roughly $\rho^2_{ren} \cong 0.7$, and where there is no dependence on ρ^2 of the functions $\chi_i(w)$.

We now demonstrate that indeed $\chi'_1(1)$, $\chi_2(1)$ and $\chi'_3(1)$ vanish in the limit $\rho^2 \to \frac{3}{4}$, $\sigma^2 \to \frac{15}{16}$. Expanding (25)-(27), and using (31), (34) and (35):

$$[\chi'_1(1)]^2 \le \frac{1}{12}[5\sigma^2-4\rho^2-3(\rho^2)^2]\sum_{n\ne 0}\left[\frac{1}{\Delta E^{(n)}_{1/2}}\frac{<D^{(n)}(v)|O^{(c)}_{kin}(0)|D(v)>}{\sqrt{4m_{D^{(n)}}m_D}}\right]^2 \quad (37)$$

$$[\chi_2(1)]^2 \le \frac{1}{480}(4\sigma^2-5\rho^2))\sum_n\left[\frac{1}{\Delta E^{(n)}_{3/2}}\frac{<D^{*(n)}_{3/2}(v,\varepsilon)|O^{(c)}_{mag}(0)|D^*(v,\varepsilon)>}{\sqrt{4m_{D^{*(n)}_{3/2}}m_{D^*}}}\right]^2 \quad (38)$$

$$[-4\chi_2(1)+12\chi'_3(1)]^2 \le \frac{1}{3}[5\sigma^2-4\rho^2-3(\rho^2)^2]\sum_{n\ne 0}\left[\frac{1}{\Delta E^{(n)}_{1/2}}\frac{<D^{(n)}(v)|O^{(c)}_{mag}(0)|D(v)>}{\sqrt{4m_{D^{(n)}}m_D}}\right]^2 \quad (39)$$

Therefore, in the limit $\rho^2 \to \frac{3}{4}$, $\sigma^2 \to \frac{15}{16}$, one gets

$$\chi'_1(1) = \chi_2(1) = \chi'_3(1) = 0 \quad (40)$$

This is a very strong correlation relating the behaviour of the elastic IW function $\xi(w)$ to the elastic subleading IW functions $\chi_i(w)$ ($i=1,2,3$) near zero recoil.

To conclude, we have obtained bounds that relate $1/m_Q^2$ correction of the form factor $h_{A_1}(w)$ to the $1/m_Q$ subleading form factors of the Lagrangian type $\chi_i(w)$ ($i=1,2,3$) and to the shape of the elastic Isgur-Wise $\xi(w)$. This bound should in principle be taken into account in the analysis of the exclusive determination of $|V_{cb}|$ in the channels $\bar{B} \to D(D^*)\ell\nu$. On the other hand, we have demonstrated an important constraint on the behavior of the subleading form factors $\chi_i(w)$: in the limit $\rho^2 \to \frac{3}{4}$, $\sigma^2 \to \frac{15}{16}$, $\chi'_1(1)$, $\chi_2(1)$ and $\chi'_3(1)$ must vanish. It would be very interesting to have a theoretical calculation of the functions $\chi_i(w)$ ($i=1,2,3$) satisfying this constraint. It seems questionable an exclusive determination of $|V_{cb}|$ by fitting the IW function $\xi(w)$ and considering uncorrelated subleading corrections.

In conclusion, using sum rules in HQET, as formulated in ref. [3, 4, 6, 14], we have found lower bounds for the moduli of the derivatives of $\xi(w)$ and non-trivial results on $1/m_Q$ form factors of the Current and Lagrangian types. We have also obtained a lower bound on $-\delta^{(A_1)}_{1/m^2}$. The determination of the CKM matrix element $|V_{cb}|$ in $B \to D^{(*)}\ell\nu$ should satisfy these constraints.

ACKNOWLEDGEMENT

We are indebted to the EC contract HPRN-CT-2002-00311 (EURIDICE) for its support.

REFERENCES

1. J. D. Bjorken, invited talk at Les Rencontres de la Vallée d'Aoste, La Thuile, SLAC-PUB-5278, 1990 ; N. Isgur and M. B. Wise, *Phys. Rev. D* **43**, 819 (1991).
2. N. Uraltsev, *Phys. Lett. B* **501**, 86 (2001).
3. A. Le Yaouanc, L. Oliver and J.-C. Raynal, *Phys. Rev. D* **67**, 114009 (2003).
4. A. Le Yaouanc, L. Oliver and J.-C. Raynal, *Phys. Lett. B* **557**, 207 (2003).
5. A. Falk, *Nucl. Phys. B* **378**, 79 (1992).
6. J. Jugeau, A. Le Yaouanc, L. Oliver and J.-C. Raynal, hep-ph/0407176, *Phys. Rev. D* **71**, 054031 (2005).
7. B. Blok, L. Koryakh, M. Shifman and A. Vainshtein, *Phys. Rev. D* **49**, 3356 (1994).
8. A. Grozin and G. Korchemsky, *Phys. Rev. D* **53**, 1378 (1996).
9. See, for example, M. Neubert, Lectures at the Trieste Summer School in Particle Physics (Part II) (1999), hep-ph/0001334.
10. A. Falk and M. Neubert, *Phys. Rev. D* **47**, 2965 (1993).
11. A. Leibovich et al., *Phys. Rev. D* **57**, 308 (1998).
12. M. Voloshin, *Phys. Rev. D* **46**, 3062 (1992).
13. A. Le Yaouanc et al., *Phys. Lett. B* **480**, 119 (2000).
14. F. Jugeau, A. Le Yaouanc, L. Oliver and J.-C. Raynal, LPT Orsay 05-43, hep-ph/0510178.
15. M. E. Luke, *Phys. Lett. B* **252**, 447 (1990).
16. I. Bigi et al., *Phys. Rev. D* **52**, 196 (1995).
17. M. Neubert, *Phys. Rev. B* **46**, 3914 (1992).
18. M. Neubert, Z. Ligeti and Y. Nir, *Phys. Lett. B* **301**, 101 (1993) ; *Phys. Rev. D* **47**, 5060 (1993).
19. M. Neubert, *Phys. Reports* **245**, 259 (1994).

Non-factorizable contributions to $\overline{B_d^0} \to \overline{D_s^{(*)} D_s^{(*)}}$ from chiral loops and tree level $1/N_c$ terms[1]

J.O. Eeg*, S. Fajfer,† and A. Prapotnik Brdnik**

*Dept. of Physics, Univ. of Oslo, P.O. Box 1048 Blindern, N-0316 Oslo, Norway
†Department of Physics, Univ. of Ljubljana, Jadranska 19, and
J. Stefan Institute, Jamova 39, P.O. Box 3000, 1001 Ljubljana, Slovenia
**Faculty of Civil engineering, Univ. of Maribor, Smetanova ul. 17, 2000 Maribor, Slovenia

Abstract.
We point out that the amplitudes for the decays $\overline{B}^0 \to D_s^+ D_s^-$ and $\overline{B}_s^0 \to D^+ D^-$ have no factorizable contributions. If one or two of the D-mesons in the final state are vectors (i.e D^* 's) there are relatively small factorizable contributions through the annihilation mechanism. The dominant contributions to the decay amplitudes arise from chiral loop contributions and $1/N_c$ suppressed tree level. We predict that the branching ratios for the processes $\overline{B}_d^0 \to D_s^+ D_s^-$, $\overline{B}_d^0 \to D_s^{+*} D_s^-$ and $\overline{B}_d^0 \to D_s^+ D_s^{-*}$ are all of order $(2-3) \times 10^{-4}$, while $\overline{B}_s^0 \to D_d^+ D_d^-$, $\overline{B}_s^0 \to D_d^{+*} D_d^-$ and $\overline{B}_s^0 \to D_d^+ D_d^{-*}$ are of order $(4-7) \times 10^{-3}$. If *both* D-mesons in the final state are D^*'s, we obtain branching ratios of order two times bigger.

Keywords: B-decays, chiral perturbation theory
PACS: 13.20.Hw, 12.39-St, 12.39.Fe,12.39.Hg

INTRODUCTION

Decay modes like $B \to \pi\pi, K\pi$ are intensively studied. From the theoretical side, for instance within QCD-factorization. The decays of a B- meson into two D-mesons are different because the energy release is only of order 1 GeV, and therefore (QCD-) factorization is not expected to hold. $B \to DD$ decays are also studied experimentally [1]. Here we discuss non-factorizable contributions to the decay modes $\overline{B_d^0} \to \overline{D_s^{(*)} D_s^{(*)}}$, where $D_s^{(*)}$ is a pseudoscalar or a vector meson. At quark level such decays occur through the annihilation mechanism $b\bar{q} \to c\bar{c}$, where $q = d, s$ respectively. However, in the factorized limit the annihilation mechanism will give a vanishing amplitude due to current conservation (similar to $D^0 \to K^0 \overline{K^0}$ [2]), unless one or two of the D-mesons in the final state are vectors. The contributions due to the annihilation mechanism are proportional to a numerically non-favored Wilson coefficient. In contrast, the typical factorized decay modes which proceed through the spectator mechanism, $\overline{B^0} \to D^+ D_s^-$ say, are proportional to the numerically favored Wilson coefficient.

In our approach [3, 4], the non-factorizable contributions are coming from chiral loops and from tree level amplitudes generated by soft gluon emission forming a gluon

---

[1] Presented by J.O. Eeg

condensate. The gluon condensate contributions can be calculated within a recently developed Heavy Light Chiral Quark Model (HLχQM) [5]. This model has been applied to processes involving B-mesons in [6, 7]. Both the chiral loop contributions and the gluon condensate contributions are proportional to the numerically favorable Wilson coefficient.

FRAMEWORK

Effective Lagrangian at quark level

The relevant effective Lagrangian at quark level reads:

$$\mathscr{L}_W = -\frac{G_F}{\sqrt{2}} V_{cb} V_{cq}^* \sum_i a_i(\mu) \, Q_i(\mu), \quad (1)$$

where $q = d, s$ and $a_i(\mu)$ are Wilson coefficients that carry all information of the short distance physics above the renormalization scale μ. The matrix elements of $Q_i(\mu)$ contain all non-perturbative, long distance physics below μ. Within Heavy Quark Effective Theory (HQEFT) the effective non-leptonic Lagrangian $\mathscr{L}_W$ can be evolved down to the scale $\mu = \Lambda_\chi \simeq 1$ GeV [8].

The numerically relevant operators in our case are

$$Q_1 = 4(\bar{q}_L \gamma^\mu b_L)(\bar{c}_L \gamma_\mu c_L) \quad , \quad Q_2 = 4(\bar{c}_L \gamma^\mu b_L)(\bar{q}_L \gamma_\mu c_L), \quad (2)$$

where L denotes a left-handed particle. At $\mu = \Lambda_\chi$, which by construction is the matching scale within our approach [3, 5, 6], one finds $a_1 \simeq -0.35 - 0.07i$ and $a_2 \simeq 1.29 + 0.08i$. Note that the Wilson coefficients a_i are complex below $\mu = m_c$ [8]. In the next subsection we will see how the currents in the operators in (2) are bosonized.

In order to obtain all matrix elements of the Lagrangian (1) we need the Fierz transformed version of the operators in (2). To find these, we use the relation:

$$\delta_{ij}\delta_{ln} = \frac{1}{N_c}\delta_{in}\delta_{lj} + 2\, t_{in}^a \, t_{lj}^a, \quad (3)$$

where i, j, l and n are color indices running from 1 to 3 and t^a denotes the color matrices, a being the color octet index. One obtains

$$Q_1^F = \frac{1}{N_c}Q_2 + \widetilde{Q}_2 \, , \quad Q_2^F = \frac{1}{N_c}Q_1 + \widetilde{Q}_1, \quad (4)$$

where the superscript F means "Fierzed", and

$$\widetilde{Q}_1 = 4(\bar{q}_L \gamma^\mu t^a b_L)(\bar{c}_L \gamma_\mu t^a c_L) \quad , \quad \widetilde{Q}_2 = 4(\bar{c}_L \gamma^\mu t^a b_L)(\bar{q}_L \gamma_\mu t^a c_L). \quad (5)$$

These operators generate contributions proportional to the gluon condensate.

Heavy Light Chiral Perturbation Theory

The Heavy Quark Effective Theory (HQEFT) Lagrangian is:

$$\mathscr{L}_{HQEFT} = \pm \overline{Q_v^{(\pm)}} \, iv \cdot D Q_v^{(\pm)} + \mathcal{O}(m_Q^{-1}) \,, \tag{6}$$

where $Q_v^{(+)}(x)$ is a (reduced) heavy quark field (b or c in our case) with velocity v, and $Q_v^{(-)}(x)$ is the field of a heavy anti-quark ($\bar{c}$ in our case). Furthermore, m_Q is the heavy quark mass, and D_μ is the covariant derivative containing the gluon field.

After integrating out the heavy and light quarks, the effective Lagrangian up to $\mathcal{O}(m_Q^{-1})$ can be written as a kinetic term plus a term describing the chiral interaction between heavy and light mesons [5, 9]:

$$\mathscr{L}_\chi = -g_{\mathscr{A}} \, Tr \left[\overline{H_a^{(\pm)}} H_b^{(\pm)} \gamma_\mu \gamma_5 \mathscr{A}_{ba}^\mu \right] \,, \tag{7}$$

where $H_a^{(\pm)}$ is the heavy meson field containing a spin zero and a spin one boson:

$$H_a^{(\pm)} = P_\pm (P_{a\mu}^{(\pm)} \gamma^\mu - i P_{a5}^{(\pm)} \gamma_5) \,. \tag{8}$$

Here a,b are flavor indices and $P_\pm = (1 \pm \gamma \cdot v)/2$ are projecting operators. The axial vector field $\mathscr{A}_\mu$ in (7) is defined as:

$$\mathscr{A}_\mu = -\frac{i}{2}(\xi^\dagger \partial_\mu \xi - \xi \partial_\mu \xi^\dagger) \,, \tag{9}$$

where $\xi \equiv exp[i(\Pi/f)]$. Moreover, f is the bare pion coupling and Π is a 3 by 3 matrix which contains the Goldstone bosons π, K, η in the standard way, and $g_{\mathscr{A}}$ is the axial chiral coupling.

Based on the symmetry of HQEFT, we obtain the bosonized currents. For a decay of the $b\bar{q}$ system (see Fig. 2, left) we have [5]:

$$\overline{q_L} \gamma^\mu Q_{v_b}^{(+)} \longrightarrow \frac{\alpha_H}{2} Tr \left[\xi^\dagger \gamma^\alpha L H_b^{(+)} \right] \,, \tag{10}$$

where (up to QCD and $1/m_Q$ corrections[5]) $\alpha_H = f_H \sqrt{m_H}$ for $H = B, D$. Further, $Q_{v_b}^{(+)}$ is the heavy b-quark field, v_b is its velocity, and $H_b^{(+)}$ is the corresponding heavy meson field. For the W-boson materializing to a $\overline{D}$, the bosonized current $\overline{q_L} \gamma^\mu Q_{v_{\bar{c}}}^{(-)}$ is also given by (10), but with $H_b^{(+)}$ replaced by $H_{\bar{c}}^{(-)}$ representing the $\overline{D}$ meson. $Q_{v_{\bar{c}}}^{(-)}$ is the field of the heavy $\bar{c}$ quark, and $v_{\bar{c}}$ is its velocity (see Fig. 1, right).

The bosonized $b \to c$ transition current in Fig. 1 is given by

$$\overline{Q_{v_b}^{(+)}} \gamma^\mu L Q_{v_c}^{(+)} \longrightarrow -\zeta(\omega) Tr \left[\overline{H_c^{(+)}} \gamma^\alpha L H_b^{(+)} \right] \,, \tag{11}$$

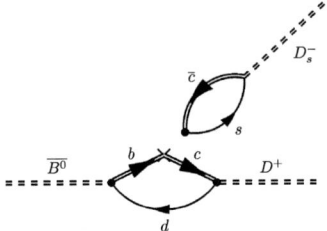

FIGURE 1. Factorized contribution for $\overline{B^0} \to D^+ D_s^-$ through the spectator mechanism, which does not exist for decay mode $\overline{B^0} \to D_s^+ D_s^-$. There are similar diagrams with vector mesons.

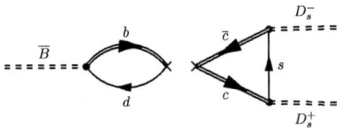

FIGURE 2. Factorized contribution for $\overline{B^0} \to D_s^+ D_s^-$ through the annihilation mechanism, which give zero contributions if both D_s^+ and D_s^- are pseudoscalars.

where $\zeta(\omega)$ is the Isgur-Wise function for the $\bar{B} \to D$ - transition, and v_c is the velocity of the heavy c-quark. Furthermore, $\omega \equiv v_b \cdot v_c = v_b \cdot v_{\bar{c}} = M_B/(2M_D)$. For the weak current for $D\overline{D}$ production (corresponding to the factorizable annihilation mechanism in Fig. 2), the current $Q_{v_c}^{(+)} \gamma^\mu L Q_{v_{\bar{c}}}^{(-)}$ is given by (11) with $H_b^{(+)}$ replaced by $H_{\bar{c}}^{(-)}$, and $\zeta(\omega)$ is replaced by $\zeta(-\lambda)$, where $\lambda = v_{\bar{c}} \cdot v_c = [M_B^2/(2M_D^2) - 1]$. Note that $\zeta(-\lambda)$ is a complex function which is less known than $\zeta(\omega)$.

The factorized contributions for the spectator and annihilation diagrams are shown in the Figs. 1 and 2. The first diagram does not give any (direct) contributions to the class of processes we consider, but is still important because it is the basis of our chiral loops.

The chiral loop amplitudes visualized in Fig. 3 are of order $(g_\mathscr{A} m_K/4\pi f)^2$ compared to typical factorizable amplitudes in processes where these exist. For instance, the ratio between the chiral loop amplitude for $\overline{B^0} \to D_s^+ D_s^-$ and the factorized amplitude for $\overline{B^0} \to D^+ D_s^-$ is $\simeq -0.20 + 0.26i$ (before the difference in KM structure is taken into account). For vectors (i.e. D^*'s) in the final state there are similar diagrams with various combinations of pseudoscalars and vectors in the loop, but the diagrams in Fig. 3 constitute the two classes of diagrams [4].

In a complete analysis, counterterms to the chiral loops has to be included. These counterterms are not considered here (or in [3, 4]) and has to be be considered together with the constant (non-logarithmic) chiral loop terms which we also have dropped in this analysis. The inclusion of counterterms and constant chiral loops terms will be discussed elsewhere.

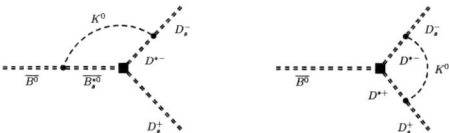

FIGURE 3. Non-factorizable chiral loops for $\overline{B^0} \to D_s^+ D_s^-$. There are similar diagrams for vector mesons in the final state. The two diagrams illustrate the two classes of diagrams.

One might also write down possible terms consistent with HQEFT and chiral symmetry, for instance the following three terms:

$$Tr\left[\xi^\dagger \sigma^{\mu\alpha} LH_b^{(+)}\right] \cdot Tr\left[\overline{H_c^{(+)}} \gamma_\alpha LH_{\tilde{c}}^{(-)} \gamma_\mu\right] \quad , \quad Tr\left[\xi^\dagger LH_b^{(+)}\right] \cdot Tr\left[\overline{H_c^{(+)}} \gamma^\alpha LH_{\tilde{c}}^{(-)} \gamma_\alpha\right],$$

$$\varepsilon^{\mu\nu\alpha\lambda}(v_c + \bar{v})_\nu Tr\left[\xi^\dagger \gamma^\mu LH_b^{(+)}\right] \cdot Tr\left[\overline{H_c^{(+)}} \gamma^\alpha LH_{\tilde{c}}^{(-)} \gamma_\lambda\right] . \tag{12}$$

Such terms do not appear in the factorized limit, and will correspond to (at least) $1/N_c$ suppressed terms. Within pure chiral perturbation theory their coefficients are unknown, but they might be calculated within the HLχQM, as described in the next subsection.

The Heavy Light Chiral Quark Model

The HLχQM Lagrangian is

$$\mathscr{L}_{\text{HL}\chi\text{QM}} = \mathscr{L}_{HQEFT} + \mathscr{L}_{\chi QM} + \mathscr{L}_{Int} . \tag{13}$$

The first term is given in (6) and the second term is described by the Chiral Quark Model of the light sector [10] involving interactions between quarks and (Goldstone) mesons:

$$\mathscr{L}_{\chi QM} = \bar{\chi}\left[\gamma^\mu(i\mathscr{D}_\mu + \gamma_5 \mathscr{A}_\mu) - m\right]\chi . \tag{14}$$

Here $m = (230 \pm 20)$MeV is the $SU(3)$ invariant constituent light quark mass, and χ is the flavor rotated quark field given by $\chi_L = \xi^\dagger q_L$ and $\chi_R = \xi q_R$, where $q^T = (u,d,s)$ is the light quark field. The covariant derivative $\mathscr{D}_\mu$ contains the soft gluon field forming the gluon condensates (besides some chiral interactions) [5, 6, 10].

The interaction between heavy meson fields and quarks is described by [5]:

$$\mathscr{L}_{Int} = -G_H \left[\bar{\chi}_a \overline{H_a^{(\pm)}} Q_v^{(\pm)} + \overline{Q_v^{(\pm)}} H_a^{(\pm)} \chi_a\right], \tag{15}$$

where the coupling constant $G_H = \sqrt{2m\rho}/f$, and ρ is a hadronic parameter depending on m (numerically ρ is of order one). For further details, see ref. [5].

The gluon condensate amplitudes can be written, within the framework presented in the previous section, in a quasi-factorized way as a product of matrix elements of colored

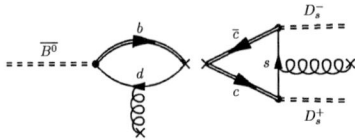

FIGURE 4. Non-factorizable contribution for $\overline{B^0} \to D_s^+ D_s^-$ through the annihilation mechanism with additional soft gluon emission. The wavy lines represent soft gluons ending in vacuum to make gluon condensates.

currents in (5), as visualized in Fig. 4. The left part of Fig. 4 corresponds to the bosonized colored current:

$$\left(\overline{q_L} t^a \gamma^\alpha Q_{v_b}^{(+)}\right)_{1G} \longrightarrow -\frac{G_H g_s}{64\pi} G_{\mu\nu}^a$$
$$\times Tr\left[\xi^\dagger \gamma^\alpha L H_b^{(+)} (\sigma^{\mu\nu} - F\{\sigma^{\mu\nu}, \gamma \cdot v_b\})\right], \quad (16)$$

where $G_{\mu\nu}^a$ is the octet gluon tensor, and $F \equiv 2\pi f^2/(m^2 N_c)$ is a dimensionless quantity of the order 1/3. The symbol $\{\,,\,\}$ denotes the anti-commutator. For the creation of a $D\overline{D}$ pair in the right part of Fig. 4, the colored current $\left(\overline{Q_{v_c}^{(+)}} t^a \gamma^\alpha L Q_{v_{\bar{c}}}^{(-)}\right)_{1G}$ is bosonized similarly to (16), but involves $H_c^{(+)}$ and $H_{\bar{c}}^{(-)}$.

Multiplying the two colored currents, and introducing gluon condensate contributions by the replacement:

$$g_s^2 G_{\mu\nu}^a G_{\alpha\beta}^a \to 4\pi^2 \langle \frac{\alpha_s}{\pi} G^2 \rangle \frac{1}{12}(g_{\mu\alpha} g_{\nu\beta} - g_{\mu\beta} g_{\nu\alpha}), \quad (17)$$

we obtain a bosonized effective Lagrangian which is $1/N_c$ suppressed compared to the factorized contributions. This effective Lagrangian corresponds to a certain linear combination of a priori possible $1/N_c$ suppressed terms at tree level (in the chiral perturbation theory sense). Among these are the three terms in (12).

$1/m_b$ suppressed terms seem to be negligible. In order to include $1/m_c$ terms, one must for instance consider the $c\bar{c}$ production current to this order:

$$\Delta J_\mu(\bar{c}c) = \frac{1}{m_c} \overline{Q_{v_c}^{(+)}} \left(i\gamma \cdot \overleftarrow{D}_\perp(v_c) \gamma^\mu L + \gamma^\mu L i\gamma \cdot D_\perp(v_{\bar{c}})\right) Q_{v_{\bar{c}}}^{(-)}, \quad (18)$$

where $D_\perp^\mu(v) = (g^{\mu\nu} - v^\mu v^\nu) D_\nu$. Within the HL$\chi$QM one may estimate both factorizable and non-factorizable $1/m_c$ corrections due to this current There are also other operators. Compared to other contributions studied here, the relative size of $1/m_c$ suppressed contributions will be of order $\tilde{m}/m_c$, where $\tilde{m}$ is some hadronic parameter within the HLχQM with dimension mass, such as linear commbinations of m and $\langle \bar{q}q \rangle / f^2$. The total contributions from such terms are significant and will be studied elsewhere.

RESULTS AND DISCUSSION

In our calculation we used the following input parameters: $\alpha_B \simeq \alpha_D \simeq 0.33 \text{ GeV}^{-3/2}$, $G_H = 7.5 \text{ GeV}^{-1/2}$ and $\langle \frac{\alpha_s}{\pi} G^2 \rangle = [(315 \pm 20) \text{ MeV}]^4$ [5, 6], $g_{\mathcal{A}} = 0.6$, $f_\pi = 93 \text{ MeV}$. We find [4] the following branching ratios

$$Br(\bar{B}^0 \to D_s^+ D_s^-) = 2.5 \times 10^{-4}, \quad Br(\bar{B}_s^0 \to D^+ D^-) = 4.5 \times 10^{-3}, \quad (19)$$
$$Br(\bar{B}^0 \to D_s^{+*} D_s^-) = 3.3 \times 10^{-4}, \quad Br(\bar{B}_s^0 \to D^{+*} D^-) = 6.8 \times 10^{-3}, \quad (20)$$
$$Br(\bar{B}^0 \to D_s^+ D_s^{-*}) = 2.0 \times 10^{-4}, \quad Br(\bar{B}_s^0 \to D^+ D^{-*}) = 4.3 \times 10^{-3}, \quad (21)$$
$$Br(\bar{B}^0 \to D_s^{*+} D_s^{-*}) = 5.4 \times 10^{-4}, \quad Br(\bar{B}_s^0 \to D^{*+} D^{-*}) = 9.1 \times 10^{-3}. \quad (22)$$

The contribution of the constant term and the corresponding counterterm can change the branchig ratio for B-meson decaying into two pseudoscalars by about 10%, while in the case of decay into one pseudoscalar and one vector D-meson, this contribution is in the range of 20 – 40%. In the case of B-meson decaying into two vector mesons, the constant term is estimated to be 2-8 times larger than the logaritmic contribution, depending on the choice of the scheme in which the products of two Levi-Civita terms are considered. The uncertainty in input parameters can result in an additional error for the branching ratios. We estimate that it can be of the order of 20%. Within our approach the $1/m_Q$ corrections, with $Q = c, b$ have so far been omitted. At least the $1/m_c$ corrections will be numerically significant.

ACKNOWLEDGMENTS

J.O. Eeg is supported in part by the Norwegian research council and by the European Union RTN network, Contract No. HPRN-CT-2002-00311 (EURIDICE). S. Fajfer is supported in part by the Ministry of Education, Science and Sport of the Republic of Slovenia.

REFERENCES

1. For a recent study, see K. Abe et al., e-Print Archive: hep-ex/0508040.
2. J. O. Eeg, S. Fajfer, J. Zupan, *Phys. Rev.* **D 64**, 034010 (2001).
3. J.O. Eeg, S. Fajfer, A. Hiorth, *Phys. Lett.* **B 570** (2003) 46-52.
4. J.O. Eeg, S. Fajfer, A. Prapotnik, *Eur.Phys.J.* **C42** (2005) 29-36.
5. A. Hiorth and J. O. Eeg, *Phys. Rev.* **D 66** (2002) 074001, and references therein.
6. A. Hiorth and J. O. Eeg, *Eur.Phys.J.direct* **C30** (2003) 006.
7. J.O. Eeg, A. Hiorth,and A.D. Polosa *Phys.Rev.* **D65** (2002) 054030.
   J.O. Eeg, K. Kumericki, and I. Picek, e-Print Archive: hep-ph/0506152.
8. B. Grinstein, W. Kilian, T. Mannel, and M.B. Wise, *Nucl. Phys.* **B363** (1991) 19.
   R. Fleischer, *Nucl. Phys.* **B 412** (1994) 201.
9. R. Casalbuoni, A. Deandrea, N. Di Bartelomeo, R. Gatto, F. Feruglio and G. Nardulli,
   Phys. Rep. **281**, 145 (1997);
10. See for example: A. Pich and E. de Rafael, *Nucl. Phys.* **B 58**, 311 (1991), S. Bertolini, J.O. Eeg and M. Fabbrichesi, *Nucl. Phys.* **B449** (1995) 197, V. Antonelli, S. Bertolini, J.O. Eeg, M. Fabbrichesi and E.I. Lashin, *Nucl. Phys.* **B469** (1996) 143, S. Bertolini, J.O. Eeg, M. Fabbrichesi, and E.I. Lashin, *Nucl.Phys.* **B514** (1998) 63-92, and references therein.

Relations for Direct CP asymmetries in $B \to PP$ and $B \to PV$ decays

T. N. Pham

Centre de Physique Théorique,
Centre National de la Recherche Scientifique, UMR 7644,
Ecole Polytechnique, 91128 Palaiseau Cedex, France

Abstract. The presence of additional strong phase from power corrections and other chirally enhanced terms makes it more difficult to predict direct CP asymmetries in two-body charmless B decays. In this talk, I would like to report on a recent work on QCD Factorisation and Power Corrections in Charmless B Decays. Using the measured branching ratios for $B \to PV$, it is shown that power corrections in charmless B decays are probably large, at least for penguin dominated PV channels. Since the tree-penguin interference responsible for direct CP asymmetries in two-body charmless B decays are related by CKM factors and $SU(3)$ symmetry, we find that, if power corrections other than the chirally enhanced power corrections and annihilation topology were negligible, QCD Factorisation would predict the direct CP asymmetry of $B \to \pi^+\pi^-$ to be about 3 times larger than that of $B \to \pi^\pm K^\mp$, with opposite sign, in agreement with the latest measurement from Belle. Similar relations are also given for direct CP asymmetries in $B \to PV$.

Keywords: CP violation; B physics
PACS: 13.25.Hw 12.38.Bx

INTRODUCTION

The large direct CP asymmetries observed at BaBar and Belle in $B \to K\pi$ and $B \to \pi\pi$ decays [1, 2] indicate a large strong phase in these decay amplitudes. Since in general QCD Factorisation(QCDF) predicts a much smaller strong phase and a small CP asymmetry, one would then need important power correction terms or other power-suppressed non-factorisable term (e.g FSI effects etc.) to generate a large strong phase and a large CP asymmetry in these decays. In fact the charmed meson inelastic rescattering FSI effects [3, 4] or charming penguin [5] are able to produce a large absorptive part and therefore a large strong phase for $B \to PP$ and $B \to PV$ amplitudes. Though the presence of these power-suppressed terms makes it difficult to predict the amount of CP asymmetry, because of $SU(3)$ symmetry and the CKM factor, one can derive however relations between CP asymmetries in $B \to PP$ and $B \to PV$. In the following I shall first present an analysis showing possible evidence for power-suppressed terms in charmless B decays.

In QCD Factorization (QCDF) [6], the $O(1/m_b)$ power corrections in penguin matrix elements and other chirally enhanced corrections could make important contributions to the penguin-dominated charmless B decays as in $B \to \pi K$ decays. Other power corrections terms such as annihilation contributions may also be present in PP and PV decays as first noticed in the perturbative QCD method for charmless B decays [7] and indicated by recent analysis of charmless two-body non-leptonic B decays [8, 9, 10]. In a recent work [11], we have shown that in QCDF, it is possible to consider certain ratios of the $B \to PV$ amplitudes which depend only on the Wilson coefficients and the known hadronic parameters. The discrepancy between prediction and experiment for the ratio would be a clear evidence for annihilation or other non-factorisable contributions. We find that annihilation topology likely plays an indispensable role at least for penguin-dominated PV channels. Including the annihilation terms in QCDF, we find that the direct CP asymmetry of $B \to \pi^+\pi^-$ to be about 3 times larger than that of $B \to K^{\mp}\pi^{\pm}$, with opposite sign, in agreement with experiment [2].

QCD FACTORIZATION FOR CHARMLESS B DECAYS

The effective Lagrangian for non-leptonic B decays can be obtained from operator product expansion and renormalization group equation, in which short-distance effects involving large virtual momenta of the loop corrections from the scale M_W down to $\mu = \mathcal{O}(m_b)$ are integrated into the Wilson coefficients. The amplitude for the decay $B \to M_1 M_2$ can be expressed as

$$\mathscr{A}(B \to M_1 M_2) = \frac{G_F}{\sqrt{2}} \sum_{i=1}^{6} \sum_{q=u,c} \lambda_q C_i(\mu) \langle M_1 M_2 | O_i(\mu) | B \rangle \qquad (1)$$

λ_q is a CKM factor, $C_i(\mu)$ are the Wilson coefficients perturbatively calculable from first principles and O_i are the tree and penguin operators given by(neglecting other operators):

$$O_1 = (\bar{s}u)_L(\bar{u}b)_L \quad , \quad O_4 = \sum_q (\bar{s}q)_L(\bar{q}b)_L$$
$$O_6 = -2\sum_q (\bar{s}_L q_R)(\bar{q}_R b_L) \qquad (2)$$

The hadronic matrix elements $\langle M_1 M_2 | O_i(\mu) | B \rangle$ contain the physics effects from the scale $\mu = \mathcal{O}(m_b)$ down to Λ_{QCD}. In the heavy quark limit, QCD Factorisation [6] allows the decay amplitude $\langle M_1 M_2 | O_i(\mu) | B \rangle$ to be factorized into hard radiative corrections and non-perturbative matrix elements which can be parametrized by the semi-leptonic decay form factors and meson light-cone distribution amplitudes (LCDAs).

Power corrections in $1/m_b$ come from penguin matrix elements, chirally enhanced corrections and annihilation contributions. For example, in the $B \to \pi K$ amplitude, the matrix element of O_6 is of the order $O(1/m_b)$ compared to the $(V-A) \times (V-A)$ O_1 and O_4 matrix elements, since $< K|\bar{s}_L d_R|0 >$ is proportional to $m_K^2/m_s \approx 2.5\,\text{GeV}$ while $< K|\bar{s}_L d_L|0 >$ is proportional to K momentum which is $O(m_b)$, thus numerically, the matrix element of O_6 has a factor

$$r_\chi^K = \frac{2m_K^2}{m_b(m_s + m_d)} \approx O(1) \qquad (3)$$

and is comparable to that of O_4. For penguin-dominated decays, the O_4 and O_6 matrix element are of the same sign in PP channnel, while in PV channel they are of opposite sign. Thus in QCDF one expects a small $B \to K\rho$ branching ratio relative to $B \to \pi K$.

Because of a cancellation between the O_4 and O_6 contributions, the $B \to K\rho$ decay is more sensitive to other power corrections and non-factorisable contributions. Including the chirally-enhanced corrections in terms of two quantities $X_{A,H}$ and a strong phase, the $B \to M_1 M_2$ decay amplitudes in QCDF can be thus be written as [12, 13]:

$$\mathcal{A}(B \to M_1 M_2) = \frac{G_F}{\sqrt{2}} \sum_{p=u,c} V_{pb} V_{ps}^* \left(-\sum_{i=1}^{6} a_i^p \langle M_1 M_2 | O_i | B \rangle_f + \sum_j f_B f_{M_1} f_{M_2} b_j \right). \quad (4)$$

POWER CORRECTIONS IN $B \to$ PV DECAYS

Consider the ratio of $A(B^+ \to \pi^+ K^{*0})$ to $A(B^0 \to \rho^+ \pi^-)$ amplitudes. If the power corrections were negligible, this ratio would be theoretically very clean where the form factors cancel out, furthermore it is almost independent of the CKM angle γ and the strange-quark mass:

$$\left| \frac{\mathcal{A}(B^+ \to \pi^+ K^{*0})}{\mathcal{A}(B^0 \to \rho^+ \pi^-)} \right| \simeq \left| \frac{V_{cb} V_{cs}^*}{V_{ub} V_{ud}^*} \right| \frac{f_{K^*}}{f_\rho} \left| \frac{a_4^c(\pi K^*) + r_\chi^{K^*} a_6^c(\pi K^*)}{a_1^u} \right| \quad (5)$$

$|(a_4^c(\pi K^*) + r_\chi^{K^*} a_6^c(\pi K^*))/a_1^u|$ should be about or less than 0.04 in QCDF. ($f_{K^*}/f_\rho \approx 1$). The ratio $|V_{ub}/V_{cb}|$ is not very well determined experimentally, but a stringent lower limit can be obtained from the unitarity of the CKM matrix. Since [14, 15]:

$$\left| \frac{V_{ub}}{V_{cb}} \right| = \lambda \sin\beta \sqrt{1 + \frac{\cos^2\alpha}{\sin^2\alpha}} \geq \lambda \sin\beta. \quad (6)$$

and from the current Babar and Belle measured values : $\sin 2\beta = 0.725 \pm 0.037$ [16], we have

$$\left| \frac{V_{ub}}{V_{cb}} \right| \geq \lambda \sin\beta = 0.090 \pm 0.007 > 0.078 \quad (7)$$

Eq.(5) implies the following inequality :

$$0.53 > \left| \frac{\mathcal{A}(B^+ \to \pi^+ K^{*0})}{\mathcal{A}(B^0 \to \rho^+ \pi^-)} \right| = 0.77 \pm 0.09, \quad (8)$$

where the number on the rhs is from the measured branching ratios [17, 18]. The lhs would be reduced further to 0.46 ± 0.04, if in Eq.(6) one neglects a small $\cos^2\alpha$ term according to a recent determined value $\alpha = (101^{+16}_{-9})°$ [16].

Since the chirally enhanced corrections for penguin-dominated decays are not expected to be large, this large discrepancy is strong indication that annihilation topology and/or other sources of power corrections might play an important role at least in $B \to$ PV decays. There is similar disagreement between theory and experiment in another ratio, the branching fraction of $B^0 \to K^+ \rho^-$ to that of $B^0 \to \rho^- \pi^+$, though with large theoretical uncertainties. For $\gamma = 70°$, $V_{ub}/V_{cb} = 0.09$, $a_4^c(\rho K) - r_\chi^K a_6^c(\rho K) = 0.037 + 0.003i$, $m_s = 90$ MeV, we find

$$\frac{\mathcal{B}(B^0 \to K^+ \rho^-)}{\mathcal{B}(B^0 \to \rho^- \pi^+)} = 0.38 \quad (9)$$

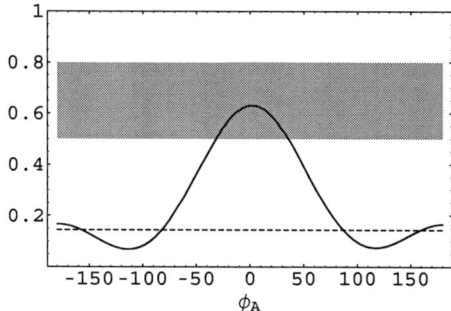

FIGURE 1. The ratio $\mathcal{B}(B^+ \to \pi^+ K^{*0})/\mathcal{B}(B^0 \to \rho^+\pi^-)$ versus the weak annihilation phase ϕ_A. The default parameters are used but letting the annihilation parameter $\rho_A = 1$. The dashed lines show the ratios without weak annihilation contributions. The gray areas denote the experimental measurements with 1σ error.

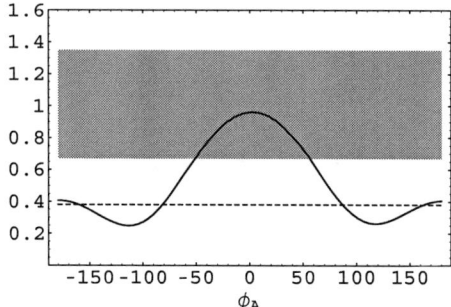

FIGURE 2. The ratio $\mathcal{B}(B^0 \to K^+\rho^-)/\mathcal{B}(B^0 \to \rho^-\pi^+)$ versus the weak annihilation phase ϕ_A. The default parameters are as in Fig.1.

far below the measured value of 1.01 ± 0.34, though, this ratio could be increased to 0.69, if m_s is lowered to 70 MeV. The discrepancy could be greater according to recent Belle measurements [17] which give a very large $B^0 \to K^+\rho^-$ branching ratio, $(15.1^{+3.4+2.4}_{-1.5-2.6}) \times 10^{-6}$ which could be obtained by a large annihilation contributions [13]. Recent charmed meson inelastic rescattering FSI calculation [4] also produces a large branching ratio. Taken together, these results indicate that the penguin-dominated $B \to$ PV decay amplitudes are consistently underestimated without annihilation contributions. Including the annihilation terms, from Eq. (4), we have

$$A(B^+ \to \pi^+ K^{*0}) = f_{K^*} F^{B\pi} m_B^2 a_4 + b_3(V,P)$$
$$A(B^0 \to K^+\rho^-) = f_K A_0^{B\rho} m_B^2 (a_4 - r_\chi^K a_6) + b_3(P,V) \quad (10)$$

$$b_3(M_1,M_2) = \frac{C_F}{N_c^2}\{C_3 A_1^i(M_1,M_2) + C_5 A_3^i(M_1,M_2) + (C_5 + N_c C_6) A_3^f(M_1,M_2)\} \quad (11)$$

With the penguin terms $a_4 \simeq -0.03$ and $a_4 - r_\chi^K a_6 \simeq 0.037$ having opposite sign, the key observation is that $b_3(V,P)$ and $b_3(P,V)$, which get most of the contribution from $(C_5+N_cC_6)A_3^f$ term, are also roughly of the opposite sign since $A_3^f(P,V)=-A_3^f(V,P)$. Thus QCDF can easily enhance both ratios without fine tuning (no large strong phase) as can be seen in Fig.2 .

DIRECT CP VIOLATIONS

We now turn to the CP asymmetries in QCDF with annihilation terms included. Because of the CKM factor and $SU(3)$ symmetry for the tree and penguin matrix elements in $B^0 \to \pi^+\pi^-$ and $B^0 \to K^+\pi^-$ decays, one can derive a relation between direct CP asymmetries in these two channels. With the CP asymmetry given as:

$$A_{\pi\pi} = \frac{4|V_{ub}V_{ud}V_{cb}V_{cd}T_{\pi\pi}P_{\pi\pi}|\sin\gamma\sin\delta}{2\mathcal{B}(B\to\pi^+\pi^-)},$$

$$A_{\pi K} = -\frac{4|V_{ub}V_{us}V_{cb}V_{cs}T_{\pi K}P_{\pi K}|\sin\gamma\sin\tilde{\delta}}{2\mathcal{B}(B\to\pi^+K^-)}, \quad (12)$$

($\delta = \delta_P - \delta_T$ = strong phases difference between the penguin and tree amplitudes), we find

$$\frac{A_{\pi\pi}}{A_{\pi K}} = -\frac{f_\pi^2}{f_K^2}\frac{\mathcal{B}(B\to\pi^+K^-)}{\mathcal{B}(B\to\pi^+\pi^-)}\left|\frac{T_{\pi\pi}P_{\pi\pi}}{T_{\pi K}P_{\pi K}}\right|\frac{\sin\delta}{\sin\tilde{\delta}}$$

$$\simeq (-2.7\pm 0.3)\frac{\sin\delta}{\sin\tilde{\delta}} \quad (13)$$

a consequence of the fact that $T_{\pi\pi}P_{\pi\pi}/T_{\pi K}P_{\pi K}$ is close to 1, a reasonable approximation in QCDF, at about 10 percent level uncertainty. A previous derivation of this relation is given in [19, 20]. Belle has claimed large direct CP asymmetry observed in $B^0 \to \pi^+\pi^-$ decay while BaBar has not confirmed it yet, but both of them are close in measurements on $A_{CP}(\pi^-K^+)$ [2, 1, 21]

$$A_{\pi\pi} = \begin{cases} 0.56\pm 0.12\pm 0.06 & \text{(Belle)}, \\ 0.09\pm 0.15\pm 0.04 & \text{(BaBar)}, \\ 0.37\pm 0.10 & \text{(Average)}. \end{cases} \quad (14)$$

$$A_{\pi K} = \begin{cases} -0.101\pm 0.025\pm 0.005 & \text{(Belle)}, \\ -0.133\pm 0.030\pm 0.009 & \text{(BaBar)}, \\ -0.114\pm 0.020 & \text{(Average)}. \end{cases} \quad (15)$$

We thus expect very naturally a larger direct CP violation for $\pi^+\pi^-$ decay compared with π^-K^+ decay, since the $\pi^+\pi^-$ decay rate is smaller than the π^-K^+ decay rate by factor $3-4$.

Experimentally,

$$\frac{A_{\pi\pi}}{A_{\pi K}} = \frac{0.37\pm 0.10}{-0.11\pm 0.03} = -3.4\pm 1.5 , \quad (16)$$

still consistent with the theoretical estimation of -2.7 ± 0.3.

Similar relation between CP asymmetries for the $B \to PV$ decays for which the CP-violating interference terms are essentially of the same magnitude, but with opposite sign:

$$\frac{A_{\mathrm{CP}}(B^0 \to \rho^+ \pi^-)}{A_{\mathrm{CP}}(B^0 \to K^{*+} \pi^-)} \simeq \frac{\mathscr{B}(B^0 \to K^{*+} \pi^-)}{\mathscr{B}(B^0 \to \rho^+ \pi^-)} \frac{f_\rho^2 \sin \delta_{\pi\rho}}{f_K^2 \sin \delta_{\pi K^*}}$$
$$\frac{A_{\mathrm{CP}}(B^0 \to \rho^- \pi^+)}{A_{\mathrm{CP}}(B^0 \to \rho^- K^+)} \simeq \frac{\mathscr{B}(B^0 \to \rho^- K^+)}{\mathscr{B}(B^0 \to \rho^- \pi^+)} \frac{f_\pi^2 \sin \delta_{\rho\pi}}{f_K^2 \sin \delta_{\rho K}}. \qquad (17)$$

In the presence of charming penguin or charmed meson inelastic rescattering FSI effects, the above CP asymmetry relation applies since the tree-penguin interference terms are related by $SU(3)$ symmetry and CKM factor [3]. Thus any significant deviation from theoretical estimation would suggest either different strong phases, e.g between $\pi\pi$ and πK decays, or possible new physics contributions.

CONCLUSION

Power corrections in charmless B decays are probably large, at least for the penguin-dominated PV channel. The key observation is that QCDF predicts the annihilation terms for $B^+ \to \pi^+ K^{*0}$ and $B^0 \to K^+ \rho^-$ to be almost equal in magnitude but opposite in sign and thus enhance the decay rates for these two modes to accommodate the experimental data. The relation for the direct CP asymmetry would naturally implies a large CP asymmetry for $B \to \pi^+ \pi^-$, about 3 times larger than that of $B \to \pi^\pm K^\mp$ with opposite sign.

I would like to thank G. Nardulli, P. Colangelo, F. De Fazio and the organisers of QCD@Work for the warm hospitality extended to me at Conversano.

REFERENCES

1. BaBar Collaboration, B.Aubert *et al*, hep-ex/0501071.
2. Belle Collaboration, K. Abe *et al*., hep-ex/0502035.
3. C. Isola, M. Ladisa, G. Nardulli, T. N. Pham and P. Santorelli, Phys. Rev. D **64**, 014029 (2001); ibid, **65**, 094005 (2002).
4. C. Isola, M. Ladisa, G. Nardulli, and P. Santorelli, Phys. Rev. D **68**, 114001 (2003).
5. M. Ciuchini, E. Franco, G. Martinelli, and L. Silvestrini, Phys. Lett. B **515**, 33 (2001); and previous works cited therein.
6. M. Beneke, G. Buchalla, M. Neubert and C. T. Sachrajda, Phys. Rev. Lett. **83**, 1914 (1999); Nucl. Phys. B **591**, 313 (2000).
7. Y. Y. Keum, H. N. Li and A. I. Sanda, Phys. Rev. D **63**, 054008 (2001); Phys. Lett. B **504**, 6 (2001).
8. D. S. Du, J. F. Sun, D.S. Yang and G. H. Zhu, Phys. Rev. D **67**, 014023 (2003).
9. R. Aleksan, P. F. Giraud, V. Morenas, O. Pene and A. S. Safir, Phys. Rev. D **67**, 094019 (2003).
10. N. de Groot, W. N. Cottingham and I. B. Whittingham, Phys. Rev. D **68**, 113005 (2003).
11. T. N. Pham and Guohuai Zhu, Phys. Rev. D **69**, 114016 (2004).
12. M. Beneke, G. Buchalla, M. Neubert and C. T. Sachrajda, Nucl. Phys. B **606**, 245 (2001).
13. M. Beneke and M. Neubert, Nucl. Phys. B **675**, 333 (2003).
14. T. N. Pham, invited talk at the 2nd Workshop on the CKM Unitarity Triangle, Durham, April 2003, hep-ph/0306271.

15. A. J. Buras, F. Parodi and A. Stocchi, JHEP **0301**, 029 (2003).
16. J. Smith, Talk given at the 3rd Workshop on the CKM Unitarity Triangle, San Diego, USA, March 2005.
17. Heavy Flavor Averaging Group(HFAG), hep/hep-ph/0505100.
18. Heavy Flavor Averaging Group(HFAG),
    `www.slac.stanford.edu/xorg/hfag/triangle/ichep2004/index.shtml`.
19. R. Fleischer, Phys. Lett. B **459**, 306 (1999); A. J. Buras, R. Fleischer, S. Recksiegel and F. Schwab, hep-ph/0402112.
20. M. Dariescu, N. G. Deshpande, X. G. He, and G. Valencia, Phys. Lett. B **557**, 60 (2003).
21. Heavy Flavor Averaging Group(HFAG),
    `www.slac.stanford.edu/xorg/hfag/triangle/summer2005/index.shtml`.

Final state interactions in the B meson decay into two pions

Aldo Deandrea*, Massimo Ladisa†, Vincenzo Laporta†, Giuseppe Nardulli†,** and Pietro Santorelli‡

*Université de Lyon 1, Institut de Physique Nucléaire, Villeurbanne Cedex, France
†Dipartimento di Fisica dell'Università di Bari, Italy and INFN, Sezione di Bari, Italy
**PH Department, TH Unit, CERN, 1211 Geneva 23, Switzerland
‡Dipartimento di Scienze Fisiche, Università di Napoli "Federico II", Italy and INFN, Sezione di Napoli, Italy

Abstract. We estimate final state interactions in the B-meson decays into two pions by the Regge model. We take into account Pomeron exchange and the leading Regge trajectories that can be related to the final state. In the $B \to \pi^0 \pi^0$ and $B \to \pi^+ \pi^-$ channels the effect produces a better agreement between theory and experiment.

Keywords: Decays of bottom mesons
PACS: 13.25.Hw

INTRODUCTION

In the Standard Model the $B \to \pi\pi$ processes are expected to be dominated by the *tree* amplitudes and so one can write the relation

$$\mathcal{B}(\pi^0 \pi^0) \ll \mathcal{B}(\pi^+ \pi^-) \approx 2 \mathcal{B}(\pi^+ \pi^0) , \qquad (1)$$

which is violated by the present experimental data, as we will see in the next sections. There can be several factors leading to violations of the expectation (1). First of all the role of *penguin* amplitudes should not be neglected. Moreover, (1) does not take into account final state interactions (FSI). To this issue was devoted the paper [1] we discuss here. An alternative approach is to go beyond *naive* factorization and use more sophisticated schemes taking into account QCD in factorization, for example the BBNS approach [2, 3] or the Soft-Collinear-Effective-Theory (SCET) [4, 5, 6, 7].

It is well known that FSI play a significant role in B decays into two light mesons where the charming penguin diagrams may produce the discrepancy between experimental data and the *naive* factorization findings. The effects of charming penguins on charmless B meson decays were estimated in ref. [8, 9, 10, 11, 12, 13]. In particular, for the final states with a strange light meson, e.g. $B \to \pi K$, these long-distance contributions are not numerically suppressed. In fact, the Cabibbo–Kobayashi–Maskawa (CKM) matrix elements produce an enhancement $\sim |V_{cb}V_{cs}^*|/|V_{ub}V_{us}^*|$ which can compensate the parametric suppression predicted by QCD factorization. In $B \to \pi\pi$ processes the lack of the above-mentioned enhancement should reduce the charming contributions. On the other hand in [14] their role is found to be significant. This matter should be settled, but in any case FSI must be taken into account, be they dominated by charming penguins

TABLE 1. Bare amplitudes for $B \to \pi\pi$ and $B \to \rho\rho$ processes. $\lambda = \pm 1, 0$ refers to the helicities of the vector particles. Results in 10^{-8} GeV.

Process	A_b	Process	$A_b\ (\lambda = +1)$	$A_b\ (\lambda = -1)$	$A_b\ (\lambda = 0)$
$B^+ \to \pi^+\pi^0$	$+2.02 - 1.24i$	$B^+ \to \rho^+\rho^0$	$-0.02 + 0.01i$	$-1.1 + 0.65i$	$+4.49 - 2.76i$
$B^0 \to \pi^0\pi^0$	$-0.41 + 0.053i$	$B^0 \to \rho^0\rho^0$	$+0.004 - 0.001i$	$+0.20 - 0.07i$	$-0.83 + 0.31i$
$B^0 \to \pi^+\pi^-$	$+2.43 - 1.74i$	$B^0 \to \rho^+\rho^-$	$-0.02 + 0.02i$	$-1.31 + 0.85i$	$+5.53 - 3.59i$

or by other rescattering processes, involving non-charmed particles in the intermediate state.

In ref. [1], we take into account FSI in hadronic B decays by means of the Regge model of high energy scattering processes, which can be applied to hadronic B decays due to the rather large value of $s = m_B^2$. The model evaluates FSI by unitarity diagrams and the Watson's theorem [15]. Some studies on the application of the Regge model to B decays are in refs. [16, 17, 18]. Elastic contributions to high energy scattering are dominated by the Pomeron exchange, while the inelastic channels get contributions from both Pomeron and Regge trajectories. Also charming penguins find a place in this scheme, provided one introduces also charmed Regge trajectories, as, for example, in the study performed in [19] for the charmless B decay into two light vector mesons.

RELEVANT BARE AMPLITUDES

Many intermediate states can rescatter in the $\pi\pi$ states. Clearly one should select the most prominent channels. Among the inelastic ones we single out the decays $B \to \rho\rho$ and $B \to a_1\pi$ for their large branching ratios: $\mathcal{B}(B^+ \to \rho^0\rho^+) = (26.4 \pm 6.1) \times 10^{-6}$; $\mathcal{B}(B^0 \to \rho^-\rho^+) = (30.0 \pm 6.0) \times 10^{-6}$; $\mathcal{B}(B^0 \to a_1^-\pi^+) = (42.6 \pm 5.9) \times 10^{-6}$ [20]. Obviously, we have to add the elastic $B \to \pi\pi$ channels, though they have smaller branching ratios. Finally, to take into account the charming penguins contributions we will consider the $D^{(*)}\bar{D}^{(*)}$ intermediate states. In conclusion the final state interactions that we consider are the elastic scattering $\pi\pi \to \pi\pi$, and the $\rho\rho \to \pi\pi$ and $a_1\pi \to \pi\pi$ and the $D^{(*)}\bar{D}^{(*)} \to \pi\pi$ inelastic channels.

We evaluate the bare amplitudes for the relevant decay processes in the *naive* factorization approximation. The effective non-leptonic hamiltonian is well known and we do not write it here, see e.g. [21]. For the Wilson coefficients we use: $a_1 = 1.029$, $a_2 = 0.140$, and $(a_3, a_4, a_5, a_6, a_7, a_8, a_9, a_{10})$ $= (33.33, -246.66, -10, -300, 1.95, 4.81, -93.30, -12.63) \times 10^{-4}$ [22]. The relevant CKM matrix elements are taken from [23]. As for the form factors and constant decay we use $f_\pi = 0.132$ GeV, $f_\rho = 0.210$ GeV, $f_{a_1} \approx 0.21$ GeV (see the discussion in [24]) $F_1^{B \to \pi}(0) = 0.26$, $A_1^{B \to \rho}(0) = 0.26$, $A_2^{B \to \rho}(0) = 0.23$, $V_0^{B \to a_1}(0) = A_0^{B \to \rho}(0) = 0.39$, where we use the notations of [25] for the $B \to \rho$ transition and the parameterization of ref. [26] for the $B \to a_1\pi$ matrix element. All the other parameters are taken from [23]. We get in this way the results of tables 1-3.

TABLE 2. Bare amplitudes for $B \to a_1\pi$; a_1 with longitudinal polarization. Units are 10^{-8} GeV.

Process	A_b
$B^+ \to a_1^+ \pi^0$	$+3.4 - 2.0\,i$
$B^+ \to a_1^0 \pi^+$	$+2.2 - 1.6\,i$
$B^0 \to a_1^0 \pi^0$	$-0.60 + 0.20\,i$
$B^0 \to a_1^+ \pi^-$	$+4.2 - 2.7\,i$
$B^0 \to a_1^- \pi^+$	$+3.4 - 2.4\,i$

TABLE 3. Bare amplitudes for $B \to D^{(*)}\bar{D}^{(*)}$; vector particles have longitudinal polarization. Results in 10^{-7} GeV.

Process	A_b	Process	A_b
$B^+ \to D^+ \bar{D}^0$	$-2.8\,i$	$B^+ \to D^{*+} \bar{D}^0$	$+2.5\,i$
$B^0 \to D^+ D^-$	$-2.8\,i$	$B^0 \to D^{*+} D^-$	$+2.5\,i$
$B^+ \to D^{*+} \bar{D}^{*0}$	$-2.9\,i$	$B^+ \to D^+ \bar{D}^{*0}$	$+2.5\,i$
$B^0 \to D^{*+} D^{*-}$	$-2.9\,i$	$B^0 \to D^+ D^{*-}$	$+2.5\,i$

FSI AND REGGE THEORY

The bare amplitudes, A_b, are modified by final state interactions in agreement with the Watson's theorem [15]:

$$A = \sqrt{S} A_b, \qquad (2)$$

where S is the S-matrix, and A are the full amplitudes. An application of the Watson's theorem was first discussed in [17] and subsequently applied to other decay channels in [18] and [19, 1]. Here we review only the relevant points.

In our approach we will include, besides the Pomeron, the ρ and a_2 (almost) exchange-degenerate trajectories and π Regge trajectories. Furthermore, as discussed before, we add charmed Regge trajectories in parameterizing charming penguins.

For the Pomeron and leading Regge trajectories we write:

$$A(B \to \pi\pi)^{(I)} \approx \sqrt{1 + 2iT^{\mathscr{P}}} A_b^{(I)} + \frac{1}{2\sqrt{1 + 2iT^{\mathscr{P}}}} \sum_k \sum_{\mathscr{R}} (\mathscr{R})^{(k,I)} A_b^{(k)}. \qquad (3)$$

Here the sum over k refers to the various intermediate states contributing to the final state $\pi\pi$; I is the isospin index. For $B \to \pi\pi$ we only have the $I = 0$ and $I = 2$ transition amplitudes; the decay amplitude $B \to \pi^+\pi^0$ is only $I = 2$. The expressions for $T^{\mathscr{P}}$ and $\mathscr{R}^{(k,I)}$ in eq. (3) can be found in ref. [1] together with a detailed discussion on the values of the parameters used in the calculations.

The calculation of the residues performed in [1] gives a small value for the $\beta^{\rho}_{a_1\pi}$ allowing us to neglect the contribution of this channel. Similarly, the numerical calculation of the contribution from the $D^{(*)}\bar{D}^{(*)}$ intermediate states to $B \to \pi\pi$ produces a negligible result as a consequence of the negative intercept of the corresponding Regge trajectories.

TABLE 4. Theoretical branching ratios for $B \to \pi\pi$ decay channels with and without final state interactions and their comparison with experimental data. The column FSI is computed with $\beta^{a_2}_{\rho^+\pi^0}$ in the range given in eq. (4). Units 10^{-6}.

Process	$\mathscr{B}$ (without FSI)	$\mathscr{B}$ (with FSI)	$\mathscr{B}$ (exp.)	
$B^0 \to \pi^0\pi^0$	0.08	$0.10 - 0.65$	$1.17 \pm 0.32 \pm 0.10$	[29, 30]
			$2.3^{+0.4+0.2}_{-0.5-0.3}$	[31, 32]
			1.45 ± 0.29	[20]
$B^0 \to \pi^+\pi^-$	8.1	$3.8 - 4.4$	$4.7 \pm 0.6 \pm 0.2$	[29, 30]
			$4.4 \pm 0.6 \pm 0.3$	[31, 32]
			4.5 ± 0.4	[20]
$B^+ \to \pi^+\pi^0$	5.0	$3.6 - 5.0$	$5.8 \pm 0.6 \pm 0.4$	[29, 30]
			$5.0 \pm 1.2 \pm 0.5$	[31, 32]
			5.5 ± 0.6	[20]

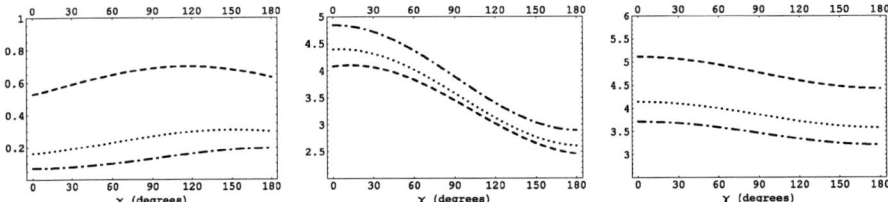

FIGURE 1. Branching ratios (units 10^{-6}) for $B \to \pi\pi$ as functions of the angle γ (degrees). From left to right the decays $B^0 \to \pi^0\pi^0$, $B^0 \to \pi^+\pi^-$ and $B^+ \to \pi^+\pi^0$. Dashed, dotted, and dot-dashed lines refer to the upper, central and lower values of the Regge residue in eq. (4).

Moreover, the discrepancy between our calculation of $\beta^{a_2}_{\rho^+\pi^0}$ from the $a_2 \to \rho\pi$ process and the value obtained in [27, 28] suggest us to assume (see [1])

$$\beta^{a_2}_{\rho^+\pi^0} = 13.1 \times (1 \pm 0.50). \qquad (4)$$

RESULTS AND DISCUSSION

Our numerical results on $B \to \pi\pi$ branching ratios are plotted in figure 1 by taking the γ angle as a parameter and allowing $\beta^{a_2}_{\rho^+\pi^0}$ to assume the upper, central and lower values in eq. (4). A survey of the experimental results is in table 4. Here we have also reported our results for $\beta^{a_2}_{\rho^+\pi^0}$ in the range of values given in eq. (4). We see that the role of FSI is especially important for the $B \to \pi^0\pi^0$ channel. Nevertheless also for the $B^0 \to \pi^+\pi^-$ channel we can see that FSI contribution produce a better agreement with the data.

Although our analysis cannot be considered complete, our results give a strong indication of the relevant role played by the rescattering effects when the bare amplitudes are for some reason small.

We have also evaluated the CP integrated asymmetries defined by

$$\mathscr{A}_{ij} = \frac{\Gamma(\bar{B}^0 \to \pi^i \pi^j) - \Gamma(B^0 \to \pi^i \pi^j)}{\Gamma(\bar{B}^0 \to \pi^i \pi^j) + \Gamma(B^0 \to \pi^i \pi^j)} \qquad (i,j) = \{(0,0),(+,-),(-,0)\} \quad (5)$$

and we report the results in Fig. 2. For $\mathscr{A}_{00}$ the HFAG group reports the average of the BaBar and Belle Collaborations as follows [20]: $\mathscr{A}_{00} = 0.28 \pm 0.39$. For $\gamma \simeq 60°$ our result is compatible, within error, with the experiment.

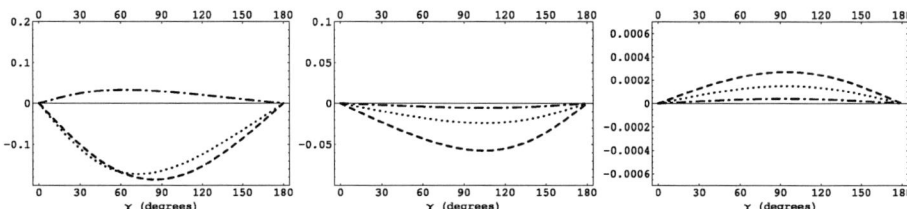

FIGURE 2. Time integrated asymmetries as defined in eq. (5) as functions of the angle γ (degrees). From the left to right $\mathscr{A}_{00}$, $\mathscr{A}_{+-}$ and $\mathscr{A}_{-0}$. Dashed, dotted, and dot-dashed lines refer to the upper, central and lower values of the Regge residue in eq. (4).

REFERENCES

1. A. Deandrea, M. Ladisa, V. Laporta, G. Nardulli and P. Santorelli, hep-ph/0508083.
2. M. Beneke, G. Buchalla, M. Neubert, and C. T. Sachrajda, Phys. Rev. Lett. **83**, 1914 (1999), hep-ph/9905312.
3. M. Beneke, G. Buchalla, M. Neubert, and C. T. Sachrajda, Nucl. Phys. **B591**, 313 (2000), hep-ph/0006124.
4. C. W. Bauer, S. Fleming, and M. E. Luke, Phys. Rev. **D63**, 014006 (2001), hep-ph/0005275.
5. C. W. Bauer, S. Fleming, D. Pirjol, and I. W. Stewart, Phys. Rev. **D63**, 114020 (2001), hep-ph/0011336.
6. C. W. Bauer and I. W. Stewart, Phys. Lett. **B516**, 134 (2001), hep-ph/0107001.
7. C. W. Bauer, D. Pirjol, and I. W. Stewart, Phys. Rev. **D65**, 054022 (2002), hep-ph/0109045.
8. P. Colangelo, G. Nardulli, N. Paver, and Riazuddin, Z. Phys. **C45**, 575 (1990).
9. M. Ciuchini, E. Franco, G. Martinelli, and L. Silvestrini, Nucl. Phys. **B501**, 271 (1997), hep-ph/9703353.
10. C. Isola, M. Ladisa, G. Nardulli, T. N. Pham, and P. Santorelli, Phys. Rev. **D64**, 014029 (2001), hep-ph/0101118.
11. C. Isola, M. Ladisa, G. Nardulli, T. N. Pham, and P. Santorelli, Phys. Rev. **D65**, 094005 (2002), hep-ph/0110411.
12. C. Isola, M. Ladisa, G. Nardulli, and P. Santorelli, Phys. Rev. **D68**, 114001 (2003), hep-ph/0307367.
13. P. Colangelo, F. De Fazio, and T. N. Pham, Phys. Lett. **B597**, 291 (2004), hep-ph/0406162.
14. C. W. Bauer, D. Pirjol, I. Z. Rothstein, and I. W. Stewart, Phys. Rev. **D70**, 054015 (2004), hep-ph/0401188.
15. K. M. Watson, Phys. Rev. **88**, 1163 (1952).
16. H.-q. Zheng, Phys. Lett. **B356**, 107 (1995), hep-ph/9504360.
17. J. F. Donoghue, E. Golowich, A. A. Petrov, and J. M. Soares, Phys. Rev. Lett. **77**, 2178 (1996), hep-ph/9604283.
18. G. Nardulli and T. N. Pham, Phys. Lett. **B391**, 165 (1997), hep-ph/9610525.
19. M. Ladisa, V. Laporta, G. Nardulli, and P. Santorelli, Phys. Rev. **D70**, 114025 (2004), hep-ph/0409286.

20. J. Alexander et al. (Heavy Flavor Averaging Group (HFAG)) (2005), hep-ex/0412073.
21. A. Ali, G. Kramer, and C.-D. Lu, Phys. Rev. **D58**, 094009 (1998), hep-ph/9804363.
22. A. J. Buras, hep-ph/9806471.
23. S. Eidelman et al. (Particle Data Group), Phys. Lett. **B592**, 1 (2004).
24. G. Nardulli and T. N. Pham (2005), hep-ph/0505048.
25. P. Ball, ECONF **C0304052**, WG101 (2003), hep-ph/0306251.
26. A. Deandrea, R. Gatto, G. Nardulli and A. D. Polosa, Phys. Rev. D **59**, 074012 (1999) hep-ph/9811259.
27. A. C. Irving and R. P. Worden, Phys. Rept. **34**, 117 (1977).
28. M. Gell-Mann, D. Sharp, and W.G. Wagner, Phys. Rev. Lett. **8**, 261 (1962).
29. B. Aubert et al. (BABAR), Phys. Rev. Lett. **89**, 281802 (2002), hep-ex/0207055.
30. B. Aubert et al. (BABAR), Phys. Rev. Lett. **94**, 181802 (2005), hep-ex/0412037.
31. Y. Chao et al. (Belle), Phys. Rev. **D69**, 111102 (2004), hep-ex/0311061.
32. K. Abe et al. (Belle), Phys. Rev. Lett. **94**, 181803 (2005), hep-ex/0408101.

Charm meson resonances in D semileptonic decays

Svjetlana Fajfer[*,†] and Jernej Kamenik[*]

[*]*J. Stefan Institute, Jamova 39, P. O. Box 3000, 1001 Ljubljana, Slovenia*
[†]*Department of Physics, University of Ljubljana, Jadranska 19, 1000 Ljubljana, Slovenia*

Abstract. Motivated by recent experimental results we reconsider semileptonic $D \to P\ell\nu_\ell$ and $D \to V\ell\nu_\ell$ decays within a model which combines heavy quark symmetry and properties of the chiral Lagrangian. Using limits of soft collinear effective theory and heavy quark effective theory we parametrize the semileptonic form factors. We include excited charm meson states in our Lagrangians and determine their impact on the charm meson semileptonic form factors. Then we calculate branching ratios for all $D \to P\ell\nu_\ell$ and $D \to Vl\nu_l$ decays.

PACS: 13.20.Fc,12.39.Hg,12.39.Fe,14.40.Lb

The knowledge of the form factors which describe the weak *heavy* → *light* semileptonic transitions is very important for the accurate determination of the CKM parameters from the experimentally measured exclusive decay rates. Usually, the attention has been devoted to B decays and the determination of the phase of the V_{ub} CKM matrix element. At the same time in the charm sector, the most accurate determination of the size of V_{cs} and V_{cd} matrix elements is not from a direct measurement, mainly due to theoretical uncertainties in the calculations of the relevant form factors' shapes.

Recently, there have been new interesting results on D-meson semileptonic decays. The CLEO and FOCUS collaborations have studied semileptonic decays $D^0 \to \pi^- \ell^+ \nu$ and $D^0 \to K^- \ell^+ \nu$ [1, 2]. Their data provide new information on the $D^0 \to \pi^- \ell^+ \nu$ and $D^0 \to K^- \ell^+ \nu$ form factors. Usually in D semileptonic decays a simple pole parametrization was used in the past. The results of Refs. [1, 2] for the single pole parameters required by the fit of their data, however, suggest pole masses, which are inconsistent with the physical masses of the lowest lying charm meson resonances. In their anlyses they also utilized a modified pole fit as suggested in [3] and their results indeed suggest the existence of contributions beyond the lowest lying charm meson resonances [1].

In addition to these results new experimental studies of charm meson resonances have provided a lot of new information on the charm sector [4, 5, 6, 7] which we can now apply to D and D_s semileptonic decays.

The purpose of our studies [8, 9] is to accommodate contributions of the newly discovered and theoretically predicted charm mesons in form factors which are parametrized using constraints coming from heavy quark effective theory (HQET) limit for the region of q^2_{max} and in the $q^2 \simeq 0$ region using results of soft collinear effective theory (SCET). We restrain our discussion to the leading chiral and $1/m_H$ terms in the expansion.

The standard decomposition of the current matrix elements relevant to semileptonic decays between a heavy pseudoscalar meson state $|H(p_H)\rangle$ with momentum p_H^ν and a light pseudoscalar meson state $|P(p_P)\rangle$ with momentum p_P^μ is in terms of two scalar functions of the exchanged momentum squared $q^2 = (p_H - p_P)^2$ – the form factors $F_+(q^2)$ and $F_0(q^2)$. Here F_+ denotes the vector form factor and it is dominated by vector meson resonances, while F_0 denotes the scalar form factor and is expected to be dominated by scalar meson resonance exchange [10, 11]. In order that the matrix elements are finite at $q^2 = 0$, the form factors must also satisfy the relation $F_+(0) = F_0(0)$.

The transition of $|H(p_H)\rangle$ to light vector meson $|V(p_V, \varepsilon_V)\rangle$ with momentum p_V^ν and polarization vector ε_V^ν is similarly parameterized in terms of four form factors V, A_0, A_1 and A_2, again functions of the exchanged momentum squared $q^2 = (p_H - p_V)^2$. Here V denotes the vector form factor and is expected to be dominated by vector meson resonance exchange, the axial A_1 and A_2 form factors are expected to be dominated by axial resonances, while A_0 denotes the pseudoscalar form factor and is expected to be dominated by pseudoscalar meson resonance exchange [11]. As in previous case in order that the matrix elements are finite at $q^2 = 0$, the form factors must also satisfy the well known relation $A_0(0) + A_1(0)(m_H + m_V)/2m_V - A_2(0)(m_H - m_V)/2m_V = 0$.

Next we follow the analysis of Ref. [3], where the F_+ form factor in $H \to P$ transitions is given as a sum of two pole contributions, while the F_0 form factor is written as a single pole. This parametrization includes all known properties of form factors at large m_H. Using a relation which connects the form factors within large energy release approach [12] the authors in Ref. [3] propose the following form factor parametrization

$$F_+(q^2) = \frac{c_H(1-a)}{(1-x)(1-ax)}, \quad F_0(q^2) = \frac{c_H(1-a)}{1-bx}, \qquad (1)$$

where $x = q^2/m_{H^*}^2$.

Utilizing the same approach we propose a general parametrization of the heavy to light vector form factors, which also takes into account all the known scaling and resonance properties of the form factors. As already mentioned, there exist the well known HQET scaling laws in the limit of zero recoil [13] while in the SCET limit $q^2 \to 0$ one obtains that all four $H \to V$ form factors can be related to only two universal SCET scaling functions [12].

The starting point is the vector form factor V, which is dominated by the pole at $t = m_{H^*}^2$ when considering the part of the phase space that is close to the zero recoil. For the *heavy $\to$ light* transitions this situation is expected to be realized near the zero recoil where also the HQET scaling applies. On the other hand, in the region of large recoils, SCET dictates the scaling described in [12]. In the full analogy with the discussion made in Refs. [3, 14], the vector form factor consequently receives contributions from two poles and can be written as

$$V(q^2) = \frac{c'_H(1-a)}{(1-x)(1-ax)}, \qquad (2)$$

where $x = q^2/m_{H^*}^2$ ensures, that the form factor is dominated by the physical H^* pole, while a measures the contribution of higher states which are parametrized by another effective pole at $m_{\text{eff}}^2 = m_{H^*}^2/a$.

An interesting and useful feature one gets from the SCET is the relation between V and A_1 [12, 15, 16, 17] at $q^2 \approx 0$. When combined with our result (2), it imposes a single pole structure on A_1. We can thus continue in the same line of argument and write

$$A_1(q^2) = \xi \frac{c'_H(1-a)}{1-b'x}. \tag{3}$$

Here $\xi = m_H^2/(m_H + m_V)^2$ is the proportionality factor between A_1 and V from the SCET relation, while b' measures the contribution of resonant states with spin-parity assignment 1^+ which are parametrized by the effective pole at $m_{H'^*_{\text{eff}}}^2 = m_{H^*}^2/b'$. It can be readily checked that also A_1, when parametrized in this way, satisfies all the scaling constraints.

Next we parametrize the A_0 form factor, which is completely independent of all the others so far as it is dominated by the pseudoscalar pole and is proportional to a different universal function in SCET. To satisfy both HQET and SCET scaling laws we parametrize it as

$$A_0(q^2) = \frac{c''_H(1-a')}{(1-y)(1-a'y)}, \tag{4}$$

where $y = q^2/m_H^2$ ensures the physical 0^- pole dominance at small recoils and a' again parametrizes the contribution of higher pseudoscalar states by an effective pole at $m_{H'_{\text{eff}}}^2 = m_H^2/a'$. The resemblance to V is obvious and due to the same kind of analysis [3] although the parameters appearing in the two form factors are completely unrelated.

Finally for the A_2 form factor, due to the pole behavior of the A_1 form factor on one hand and different HQET scaling at q_{max}^2 on the other hand, we have to go beyond a simple pole formulation. Thus we impose

$$A_2(q^2) = \frac{(m_H + m_V)\xi c'_H(1-a) + 2m_V c''_H(1-a')}{(m_H - m_V)(1-b'x)(1-b''x)}, \tag{5}$$

which again satisfies all constraints. Due to the relations between the form factors we only gain one parameter in this formulation, b''. This however causes the contribution of the 1^+ resonances to be shared between the two effective poles in this form factor.

At the end we have parametrized the four $H \to V$ vector form factors in terms of the six parameters c'_H, a, a', b', c''_H and b''.

In our heavy meson chiral theory (HMχT) calculations we use the leading order heavy meson chiral Lagrangian in which we include additional charm meson resonances. The details of this framework are given in [8] and [9]. We first calculate values of the form factors in the small recoil region. The presence of charm meson resonances in our Lagrangian affects the values of the form factors at q_{max}^2 and induces saturation of the second poles in the parameterizations of the $F_+(q^2)$, $V(q^2)$ and $A_0(q^2)$ form factors by the next radial excitations of $D^*_{(s)}$ and $D_{(s)}$ mesons respectively. Although the D mesons mat not be considered heavy enough, we employ these parameterizations with model matching conditions at q_{max}^2. Using HQET parameterization of the current matrix elements [8, 9], which is especially suitable for HMχT calculations of the form factors near zero recoil, we are able to extract consistently the contributions of individual resonances from our Lagrangian to the various $D \to P$ and $D \to V$ form factors. We use physical pole masses of excited state charmed mesons in the extrapolation, giving for the pole parameters $a = m_{H^*}^2/m_{H'^*}^2$, $a' = m_H^2/m_{H'}^2$, $b' = m_{H^*}^2/m_{H_A}^2$. Although in the general parameterization of the form factors the extra poles in F_+, V and $A_{0,1,2}$ parametrized all the neglected higher resonances beyond the ground state heavy meson spin doublets $(0^-, 1^-)$, we are here saturating those by a single nearest resonance. The single pole q^2 behavior of the $A_1(q^2)$ form factor is explained by the presence of a single 1^+ state relevant to each decay, while in $A_2(q^2)$ in addition to these states one might also account for their next radial excitations. However, due to the lack of data on their presence we assume their masses being much higher than the first 1^+ states and we neglect their effects, setting effectively $b'' = 0$.

The values of the new model parameters appearing in $D \to P l \nu_l$ decay amplitudes [8] are determined by fitting the model predictions to known experimental values of branching ratios $\mathscr{B}(D^0 \to K^- \ell^+ \nu)$, $\mathscr{B}(D^+ \to \bar{K}^0 \ell^+ \nu)$, $\mathscr{B}(D^0 \to \pi^- \ell^+ \nu)$, $\mathscr{B}(D^+ \to \pi^0 \ell^+ \nu)$, $\mathscr{B}(D_s^+ \to \eta \ell^+ \nu)$ and $\mathscr{B}(D_s^+ \to \eta' \ell^+ \nu)$ [18]. In our calculations of decay widths we neglect the lepton mass, so the form factor F_0, which is proportional to q^μ, does not contribute. For the decay width we then use the integral formula proposed in [19] with the flavor mixing parametrization of the weak current defined in [8].

Similarly in the case of $D \to V l \nu_l$ transitions we have to fix additional model parameters [9] and we again use known experimental values of branching ratios $\mathscr{B}(D_0 \to K^{*-} \ell^+ \nu)$, $\mathscr{B}(D_s^+ \to \Phi \ell^+ \nu)$, $\mathscr{B}(D^+ \to \rho^0 \ell^+ \nu)$, $\mathscr{B}(D^+ \to K^{*0} \ell^+ \nu)$, as well as partial decay width ratios $\Gamma_L/\Gamma_T(D^+ \to K^{*0} \ell^+ \nu)$ and $\Gamma_+/\Gamma_-(D^+ \to K^{*0} \ell^+ \nu)$ [18]. We calculate the decay rates for polarized final light vector mesons using helicity amplitudes $H_{+,-,0}$ as in for example [20]. By neglecting the lepton masses we again arrive at the integral expressions from [19] with the flavor mixing parametrization of the weak current defined in [9].

We first draw the q^2 dependence of the F_+ and F_0 form factors for the $D^0 \to K^-$, $D^0 \to \pi^-$ and $D_s \to K^0$ transitions. The results are depicted in Fig. 1. Our model results, when extrapolated with the double pole parameterization, agree well with previous theoretical [21, 22] and experimental [1, 2] studies whereas the single pole extrapolation does not give satisfactory results. Note that without the scalar resonance, one only gets a soft pion contribution to the F_0 form factor. This gives for the q^2 dependence of F_0 a constant value for all transitions, which largely disagrees with lattice QCD results [22] as well as heavily violates known form factor relations.

We also calculate the branching ratios for all the relevant $D \to P$ semileptonic decays and compare the predictions of our model with experimental data from PDG. The results are summarized in Table 1. For comparison we also include the results for the rates obtained with our approach for $F_+(q_{max}^2)$ but using a single pole fit. It is very interesting that our model extrapolated with a double pole gives branching ratios for $D \to P \ell \nu_\ell$ in rather good agreement with experimental results for the already measured decay rates. It is also obvious that the single pole fit gives the rates up to a factor of two larger than the experimental results. Only for decays to η and η' as given in Table 1, an agreement with experiment of the double pole version of the model is not better but worse than for the single pole case.

We next draw the q^2 dependence of all the form factors for the $D^0 \to K^{*-}$, $D^0 \to \rho^-$ and $D_s \to \phi$ transitions. The results are depicted in Fig. 2. Our extrapolated results for the shapes of the $D \to V$ semileptonic form factors agree well with existing theoretical studies [20, 21, 23, 24], while currently no experimental determination of the form

TABLE 1. The branching ratios for the $D \to P$ semileptonic decays. Comparison of our model fit with experiment as explained in the text.

Decay	$\mathscr{B}$ (model, double pole) [%]	$\mathscr{B}$ (model, single pole) [%]	$\mathscr{B}$ (Exp. [18]) [%]
$D^0 \to K^-$	3.4	4.9	3.43 ± 0.14
$D^0 \to \pi^-$	0.27	0.56	0.36 ± 0.06
$D_s^+ \to \eta$	1.7	2.5	2.5 ± 0.7
$D_s^+ \to \eta'$	0.61	0.74	0.89 ± 0.33
$D^+ \to \bar{K}^0$	9.4	12.4	6.8 ± 0.8
$D^+ \to \pi^0$	0.33	0.70	0.31 ± 0.15
$D^+ \to \eta$	0.10	0.15	< 0.5
$D^+ \to \eta'$	0.016	0.019	< 1.1
$D_s^+ \to K^0$	0.20	0.32	

factors' shapes in these decays exists.

We complete our study by calculating branching ratios and partial decay width ratios also for all relevant $D \to V l \nu_l$ decays. They are listed in Table 2 together with known experimentally measured values.

Finally, we summarize our results: We have investigated semileptonic form factors for $D \to P$ and $D \to V$ decays within an approach which combines heavy meson and chiral symmetry. The form factors are parametrized to satisfy all constraints coming from HQET and SCET. The contributions of excited charm meson states are included

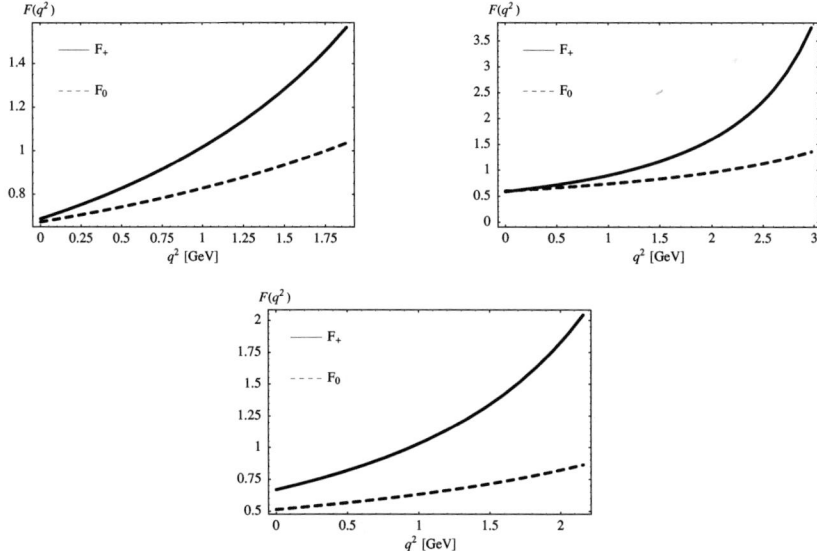

FIGURE 1. q^2 dependence of the $D^0 \to K^-$ (upper left), $D^0 \to \pi^-$ (upper right) and $D_s \to K^0$ (lower) transition form factors.

TABLE 2. The branching ratios and partial decay width ratios for the $D \to V$ semileptonic decays. Comparison of our model fit with experiment as explained in the text.

Decay	$\mathscr{B}$ (Mod.) [%]	$\mathscr{B}$ (Exp.) [%]	Γ_L/Γ_T (Mod.)	Γ_+/Γ_- (Mod.)
$D_0 \to K^*$	2.2	2.15 ± 0.35 [18]	1.14	0.22
$D_0 \to \rho$	0.20	$0.194 \pm 0.039 \pm 0.013$ [25]	1.11	0.14
$D^+ \to K_0^*$	5.6	5.73 ± 0.35 [18]	1.13 *	0.22 †
$D^+ \to \rho_0$	0.25	0.25 ± 0.08 [18]	1.11	0.14
$D^+ \to \omega$	0.25	$0.17 \pm 0.06 \pm 0.01$ [25]	1.10	0.14
$D_s \to \Phi$	2.4	2.0 ± 0.5 [18]	1.08	0.21
$D_s \to K_0^*$	0.22		1.03	0.13

* Exp. 1.13 ± 0.08 [18]
† Exp. 0.22 ± 0.06 [18]

into analysis. The values of form factors at q^2_{max} are calculated using heavy meson chiral Lagrangian. The second poles of the $F_+(q^2)$, $V(q^2)$ and $A_0(q^2)$ form factors are saturated by the presence of the next radial excitations of $D^*_{(s)}$ and $D_{(s)}$. The single pole q^2 behavior of the $A_1(q^2)$ form factor is explained by the presence of a single 1^+ state relevant to each decay, while in $A_2(q^2)$ in addition to 1^+ states one might include their next radial excitations.

The obtained q^2 dependence of the form factors is in good agreement with recent experimental results and existing theoretical studies. The calculated branching ratios are

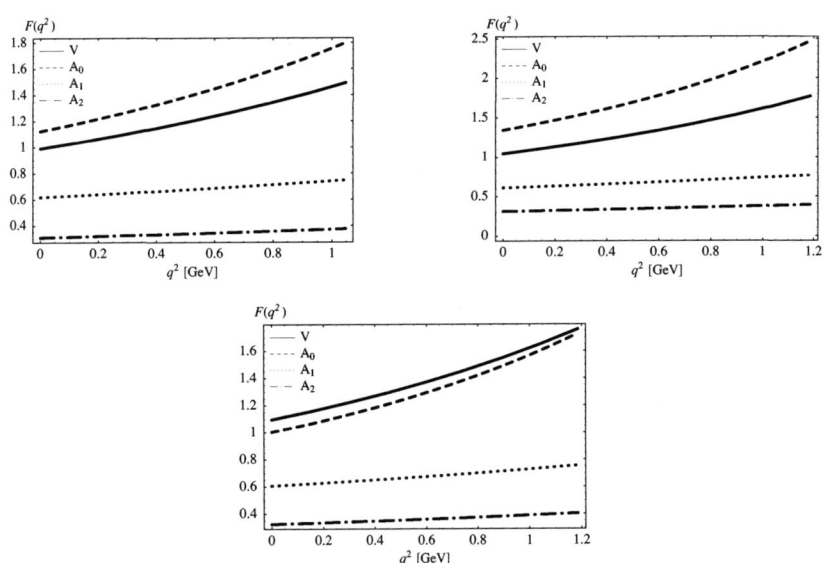

FIGURE 2. q^2 dependence of the $D^0 \to K^{*-}$ (upper left), $D^0 \to \rho^-$ (upper right) and $D_s \to \phi$ (lower) transition form factors.

close to the experimental ones. We hope that the ongoing experimental studies will help to shed more light on the shapes of the $D \to P, V$ form factors.

ACKNOWLEDGMENTS

We thank D. Bećirević for many very fruitful discussions on this subject. S. F. thanks Alexander von Humboldt foundation for financial support and A. J. Buras for his warm hospitality during her stay at the Physik Department, TU München, where part of this work has been done. This work is supported in part by the Ministry of Education, Science and Sport of the Republic of Slovenia.

REFERENCES

1. G. S. Huang, et al., *Phys. Rev. Lett.* **94**, 011802 (2005), hep-ex/0407035.
2. J. M. Link, et al., *Phys. Lett.* **B607**, 233–242 (2005), hep-ex/0410037.
3. D. Becirevic, and A. B. Kaidalov, *Phys. Lett.* **B478**, 417–423 (2000), hep-ph/9904490.
4. B. Aubert, et al., *Phys. Rev. Lett.* **90**, 242001 (2003), hep-ex/0304021.
5. E. W. Vaandering (2004), hep-ex/0406044.
6. D. Besson, et al., *AIP Conf. Proc.* **698**, 497–502 (2004), hep-ex/0305017.
7. A. V. Evdokimov, et al., *Phys. Rev. Lett.* **93**, 242001 (2004), hep-ex/0406045.
8. S. Fajfer, and J. Kamenik, *Phys. Rev.* **D71**, 014020 (2005), hep-ph/0412140.
9. S. Fajfer, and J. Kamenik, *Phys. Rev.* **D72**, 034029 (2005), hep-ph/0506051.
10. R. E. Marshak, Riazuddin, and C. P. Ryan, *Theory of Weak Interactions in Particle Physics*, vol. XXIV of *Interscience Monographs and Texts in Physics and Astronomy*, Wiley-Interscience, New York, 1969.
11. M. Wirbel, B. Stech, and M. Bauer, *Z. Phys.* **C29**, 637 (1985).
12. J. Charles, A. Le Yaouanc, L. Oliver, O. Pene, and J. C. Raynal, *Phys. Rev.* **D60**, 014001 (1999), hep-ph/9812358.
13. N. Isgur, and M. B. Wise, *Phys. Rev.* **D42**, 2388–2391 (1990).
14. R. J. Hill (2005), hep-ph/0505129.
15. D. Ebert, R. N. Faustov, and V. O. Galkin, *Phys. Rev.* **D64**, 094022 (2001), hep-ph/0107065.
16. G. Burdman, and G. Hiller, *Phys. Rev.* **D63**, 113008 (2001), hep-ph/0011266.
17. R. J. Hill (2004), hep-ph/0411073.
18. S. Eidelman, et al., *Phys. Lett.* **B592**, 1 (2004).
19. B. Bajc, S. Fajfer, and R. J. Oakes, *Phys. Rev.* **D53**, 4957–4963 (1996), hep-ph/9511455.
20. P. Ball, V. M. Braun, and H. G. Dosch, *Phys. Rev.* **D44**, 3567–3581 (1991).
21. D. Melikhov, and B. Stech, *Phys. Rev.* **D62**, 014006 (2000), hep-ph/0001113.
22. C. Aubin, et al. (2004), hep-ph/0408306.
23. A. Abada, et al., *Nucl. Phys. Proc. Suppl.* **119**, 625–628 (2003), hep-lat/0209116.
24. P. Ball, *Phys. Rev.* **D48**, 3190–3203 (1993), hep-ph/9305267.
25. S. Blusk (2005), hep-ex/0505035.

How Charm can still be charming: some recent results from FOCUS

Sandra Malvezzi

INFN, Sezione di Milano, Via Celoria 16, 20133 Milano, Italy

Abstract. Charm physics is a paradigm of the way in which precise measurements have led to a revival of the sector, allowing for New Physics searches through mixing, CP violation, and measurements of rare and forbidden decays. New vigorous spectroscopy studies of high-mass states (the so-called "Renaissance of spectroscopy") complement the scenario. These promising investigations, which are typical of a mature field under study for several decades, require knowledge and control of QCD effects. Recent studies of charm weak decays in hadronic and semileptonic processes through Dalitz-plot analyses and form-factor measurements respectively, have revealed limits in the generally adopted approaches for treating strong dynamics effects. FOCUS has performed pioneering analyses, suggesting new directions for strong decay dynamics investigation; a few examples will be discussed in this paper.

Keywords: Semileptonic and Hadronic Charmed Decay, Multichannel Scattering
PACS: 13.20Fc, 13:25Ft, 11.80Gw

CHARM DECAY DYNAMICS INVESTIGATION IN FOCUS

The semileptonic sector

Semileptonic decays have always been considered the best candidates to study charm phenomenology. Decay rates can be calculated from first principles, e.g., Feynman diagrams; owing to the presence of only one hadron in the final state we do not have to worry much about final-state interaction complications; the hadronic part of the decay can be embodied in proper form factors, which, in turn can be predicted by theory via approaches such as HQET, LGT and quark models. However, recent FOCUS studies have shown that decays in the semileptonic sector also reveal the presence of quantum-mechanical effects and hadronic complications have to be dealt with in the analysis. Of particular relevance is the study of $D^+ \to K^-\pi^+\mu^+\nu$ [1]. The FOCUS mass spectrum appears, as expected, dominated by $\bar{K}^{*0}(892)$; yet an unexpected forward–backward asymmetry was exposed in the $\cos\theta_V$ variable, defined in Fig. 1. The asymmetry is striking below the pole and essentially absent above the pole; it thus suggests quantum-mechanical interference between the Breit–Wigner amplitude describing the $\bar{K}^{*0}(892)$ and a broad or nearly constant S-wave amplitude. FOCUS has tried to describe this behavior of the decay distribution for $D^+ \to K^-\pi^+\mu^+\nu$ through a simple model. This has been written in terms of three helicity-basis form factors: H_+, H_0 and H_-, with the addition of an S-wave amplitude, modeled as a constant of modulus A and phase δ.

The shape of the $\cos\theta_V$ term versus $m_{K\pi}$ turns out to be a strong function of the interfering S-wave amplitude phase δ. Figure 2 shows how the $\cos\theta_V$ distribu-

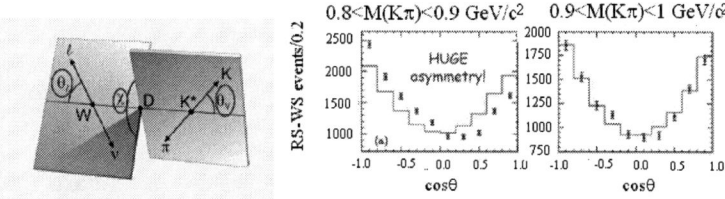

FIGURE 1. The definition of the kinematic variables for the $D^+ \to \bar{K}^{*0}\mu^+\nu$ and $\cos\theta_V$ distribution.

tion (properly weighted) is consistent with a constant S-wave amplitude of the form $0.36\exp(i\pi/4)\,\text{GeV}^{-1}$. At this stage of the analysis, the solution is was not unique; al-

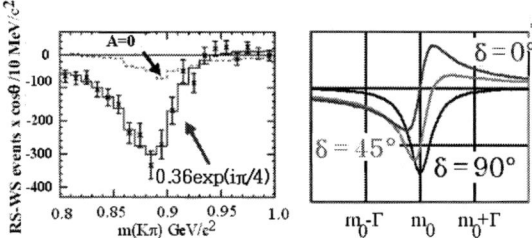

FIGURE 2. The asymmetry distribution in $K\pi$ invariant mass. The dashed line is the MC with no S-wave amplitude, the solid line is the MC with an S-wave amplitude of $0.36\exp(i\pi/4)\,(\text{GeV})^{-1}$. Different modelings of the S-wave are also shown.

ternative modelings of the S-wave amplitude, as shown in the same Fig. 2, could also fit the data. However, it is already interesting to note that the measured phase of $\pi/4$ is consistent with that found by LASS from a $K\pi$ phase-shift analysis and that the hypothesis of the broad scalar resonance $\kappa(900)$ is disfavored since it would imply a phase-shift of approximately $90°$ with respect to $\bar{K}^{*0}(892)$, unlikely to be manifest in the semileptonic sector, where FSI are not expected to be wild [2]. More recently FOCUS has performed some additional checks on the decay, essentially confirming that the scalar $\kappa(900)$ is not necessary to fit their data [3]. It has to be noted that the inclusion of the S-wave amplitude dramatically improves the quality of the form factor fits [4].

The hadronic sector: Dalitz-plot analysis of heavy flavor decays

Over the last decade, the Dalitz plot has emerged as a powerful analysis tool to study charm-meson hadronic decay dynamics. It provides a "complete observation" of the three-body decay, allowing, in principle, for a rich variety of measurements: from the dynamics features of the heavy flavor (HF) decay mechanism and the relative importance of non-spectator processes, up to CP-violating phases and mixing, such as the determination of the angle α in the decay $B^0 \to \rho\pi$ and of the angle $\gamma(\phi_3)$ in $B^\pm \to D(^*)K^\pm$. Just recently we have witnessed what I like to call the Dalitz plot revenge, i.e., a vast application of this analysis technique in all the major experiments for sophisticated investigations of both charm and beauty studies. It is instructive to refer the reader to some of the latest related papers [5–9].

We have already learnt a lot about charm and experienced the limits of the formalism that has been traditionally adopted for treating resonances. FOCUS has performed seminal work, suggesting a method to perform the analysis in the D^+ and $D_s \to \pi^+\pi^-\pi^+$ channels [10]. In particular, the need to model intermediate scalar particles contributing to charm-meson decays into three-body hadronic channels has caused experimentalists in the field to question the validity of the Breit–Wigner (BW) approximation for the description of the relevant scalar resonances [11, 12]. A formalism for studying overlapping and many-channel resonances was proposed long ago and is based on the *K-matrix* [13, 14] parametrization. This formalism, with origins in the context of two-body scattering, can be generalized to cover the case of production of resonances in more complex reactions [15], with the assumption that the two-body system in the final state is an isolated one and that the two particles do not simultaneously interact with the rest of the final state in the production process [14]. The *K-matrix* approach allows for including the positions of the poles in the complex plane directly into the analysis, incorporating the results from light spectroscopy experiments [16, 17] into the charm analysis. In addition, the *K-matrix* formalism provides a direct way of imposing the two-body unitarity constraint, which is not explicitly guaranteed in the simple Breit-Wigner sum (here called the "traditional isobar model").

For a well-defined wave of specific isospin and spin, *IJ*, characterized by narrow and isolated resonances, the propagator is, as anticipated, of the simple BW form and the isobar model can represent the data reasonably. In contrast, when the specific wave *IJ* is characterized by large and heavily overlapping resonances, just as the scalars, the propagation is no longer dominated by a single resonance, but is the result of complicated interplay among the various resonances. In this case, it can be demonstrated on very general grounds that the propagator may be written in the context of the *K-matrix* approach as

$$(I - iK \cdot \rho)^{-1} \qquad (1)$$

where K is the scattering matrix and ρ is the phase-space matrix. In this picture, the production process is viewed as consisting of an initial preparation of several states, which then propagate via the term $(I - iK\rho)^{-1}$ into the final state. In particular, the three-pion final state can be fed by an initial formation of $(\pi\pi)\pi$, $(K\bar{K})\pi$, $(\eta\eta)\pi$, $(\eta\eta')\pi$ and multi-meson states (mainly four-pion states at $\sqrt{s} < 1.6\,\text{GeV}$). Indeed the $\pi\pi$ scalar resonances are large and overlap each other in such a way that it is impossible to single out the effect of any one of them on the real axis. In order to write down the propagator, we need the scattering matrix. At the time of this analysis a self-consistent description of *S*-wave isoscalar scattering was given in the *K-matrix* representation by Anisovich and Sarantsev in [17] through a global fit of the available scattering data from the $\pi\pi$ threshold up to 1900 MeV. FOCUS has performed the first fit to charm data with the *K-matrix* formalism in the D^+ and $D_s^+ \to \pi^+\pi^-\pi^+$ channels [10]. The *K-matrix* used is that of [17]. As anticipated, the *K-matrix* formalism, having origins within the context of two-body scattering, can be generalized to deal with formation of resonances in more complex reactions through the *P-vector* approach [15]. The decay amplitude for the *D* meson into three-pion final state, where $\pi^+\pi^-$ are in a $(IJ^{PC} = 00^{++})$-wave, can thus be written as

$$A(D \to (\pi^+\pi^-)_{00^{++}}\pi^+) = F_1 = (I - iK\rho)^{-1}_{1j}$$

$$\times \left\{ \sum_\alpha \frac{\beta_\alpha g_j^{(\alpha)}}{m_\alpha^2 - m^2} + f_{1j}^{prod} \frac{1\,\text{GeV}^2 - s_0^{prod}}{s - s_0^{prod}} \right\} \times \frac{(s - s_A m_\pi^2/2)(1 - s_{A0})}{(s - s_{A0})}, \quad (2)$$

where β_α is the coupling to the pole α in the "initial" production process, f_{1j}^{prod} and s_0^{prod} are the P-vector slowly varying (SVP) parameters. The factor $\frac{(s-s_A m_\pi^2/2)(1-s_{A0})}{(s-s_{A0})}$ is the possible Adler zero term. Finally, the complete decay amplitude of the D meson into a three-pion final state is [10]:

$$A(D) = a_0 e^{i\delta_0} + \sum_i a_i e^{i\delta_i} A_i + F_1 \quad (3)$$

where the first term represents the direct, non-resonant, three-body decay; the index i only runs over the vector and tensor resonances, which can be safely treated as simple Breit–Wigner's. In the fit to the data, the K-matrix parameters are fixed to the values of [17], which consistently reproduce measured S-wave isoscalar scattering. The only free parameters are those peculiar to the P-vector, i.e., β_α, f_{1j}^{prod} and s_0^{prod}, and those in the remaining isobar part of the amplitude, a_i and δ_i. The three-pion selected samples (Fig. 3) consist of 1527 ± 51 and 1475 ± 50 events for the D^+ and D_s respectively.

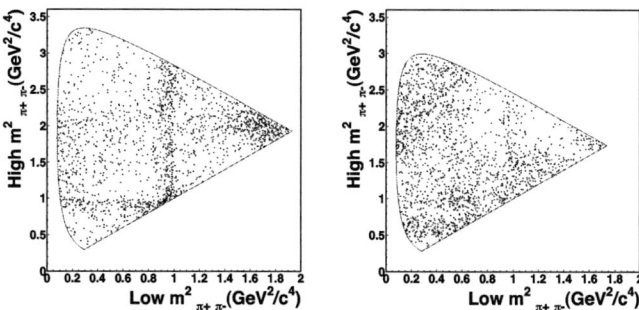

FIGURE 3. a) D_s^+ and b) D^+ Dalitz plots.

Results for the D_s^+ and $D^+ \to \pi^+\pi^-\pi^+$ decays

The resulting fit fractions [1], phases and amplitude coefficients for D_s^+ are quoted in Table 1. The D_s^+ Dalitz projections of FOCUS data are shown in Fig. 4 with final fit projections superimposed. The fit C.L. is 3%. The $D^+ \to \pi^+\pi^-\pi^+$ Dalitz plot shows

---

[1] The quoted fit fractions are defined as the ratios between the intensity for single amplitudes integrated over the Dalitz plot and that of the total amplitude with all the modes and interference present.

TABLE 1. Fit results from the K-matrix model for D_s^+.

decay channel	fit fraction (%)	phase (deg)	amplitude coefficient
$(S\text{-wave})\pi^+$	$87.04 \pm 5.60 \pm 4.17$	0 (fixed)	1 (fixed)
$f_2(1270)\pi^+$	$9.74 \pm 4.49 \pm 2.63$	$168.0 \pm 18.7 \pm 2.5$	$0.165 \pm 0.033 \pm 0.032$
$\rho^0(1450)\pi^+$	$6.56 \pm 3.43 \pm 3.31$	$234.9 \pm 19.5 \pm 13.3$	$0.136 \pm 0.030 \pm 0.035$
Fit C.L	3.0%		

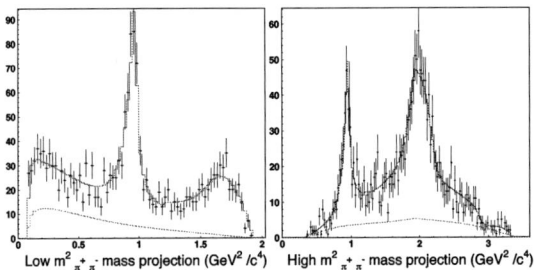

FIGURE 4. D_s^+ Dalitz-plot projections with the K-matrix fit superimposed. The background shape under the signal is also shown.

an excess of events at low $\pi^+\pi^-$ mass (Fig. 3), which cannot be explained in the context of the simple isobar model with the usual mixture of well established resonances along with a constant, non-resonant amplitude. A new scalar resonance, the $\sigma(600)$, has been previously proposed [18] to describe this excess. However, we know that complex structure can be generated by the interplay among the S-wave resonances and the underlying non-resonant S-wave component that cannot be properly described in the context of a simple isobar model. It is therefore interesting to study this channel with the present formalism, which embeds all the experimental knowledge about the S-wave $\pi^+\pi^-$ scattering dynamics. Beside the S-wave component, the decay appears to be dominated by the $\rho^0(770)$ plus a $f_2(1270)$ component. In analogy with the D_s^+, the direct three-body non-resonant component was not necessary since the SVP of the S-wave could reproduce the entire non-resonant portion of the Dalitz plot. The complete fit results are reported in Table 2. The D^+ Dalitz projections are shown in Fig. 5. The fit C.L. is 7.7%. This analysis suggests that any σ-like object in the D decay should be consistent with the same σ-like object measured in the $\pi^+\pi^-$ scattering. Higher statistics is needed to understand the σ nature.

TABLE 2. Fit results from the K-matrix model fit for D^+.

decay channel	fit fraction (%)	phase (deg)	amplitude coefficient
$(S\text{-wave})\pi^+$	$56.00 \pm 3.24 \pm 2.08$	0 (fixed)	1 (fixed)
$f_2(1270)\pi^+$	$11.74 \pm 1.90 \pm 0.23$	$-47.5 \pm 18.7 \pm 11.7$	$1.147 \pm 0.291 \pm 0.047$
$\rho^0(770)\pi^+$	$30.82 \pm 3.14 \pm 2.29$	$-139.4 \pm 16.5 \pm 9.9$	$1.858 \pm 0.505 \pm 0.033$
Fit C.L.	7.7%		

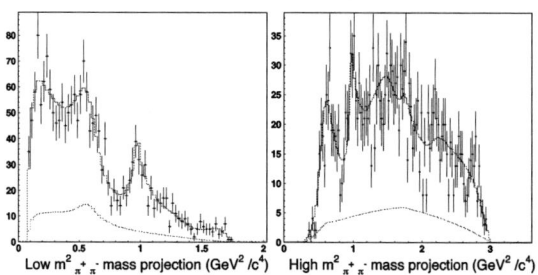

FIGURE 5. D^+ Dalitz-plot projections with the final fit superimposed. The background shape under the signal is also shown.

Interpretation of the D_s^+ and $D^+ \to \pi^+\pi^-\pi^+$ results

The *K-matrix* formalism, which has been applied for the first time to the charm sector in the FOCUS Dalitz-plot analyses of the D_s^+ and $D^+ \to \pi^+\pi^-\pi^+$ final states appears to give a coherent picture of both two-body scattering measurements in light-quark experiments *as well as* charm-meson decay. This result was not obvious beforehand. In addition, the non-resonant component of each decay seems to be described by known two-body S-wave dynamics without the need to include constant amplitude contributions. The *K-matrix* treatment of the S-wave component of the decay amplitude allows for a direct interpretation of the decay mechanism in terms of the five virtual channels considered: $\pi\pi$, $K\bar{K}$, $\eta\eta$, $\eta\eta'$ and 4π. By inserting KK^{-1} in the decay amplitude, F,

$$F = (I - iK\rho)^{-1}P = (I - iK\rho)^{-1}KK^{-1}P = TK^{-1}P = TQ \qquad (4)$$

we can view the decay as consisting of an initial production of the five virtual states, which then scatter via the physical T into the final state. The Q-vector contains the production amplitude of each virtual channel in the decay. The resulting picture, for both D_s^+ and D^+ decay, is that the S-wave decay is dominated by an initial production of $\eta\eta$, $\eta\eta'$ and $K\bar{K}$ states. Dipion production is always much smaller. This suggests that in both cases the S-wave decay amplitude primarily arises from a $s\bar{s}$ contribution such as that produced by the Cabibbo-favored weak diagram for the D_s^+ and one of the two possible singly Cabibbo-suppressed diagrams for the D^+. For the D^+, the $s\bar{s}$ contribution competes with a $d\bar{d}$ contribution. That the $f_0(980)$ appears as a peak in the $\pi\pi$ mass distribution in D^+ decay, as it does in D_s decay, shows that for the S-wave component the $s\bar{s}$ contribution dominates [12]. Comparing the relative S-wave fit fractions observed for D_s^+ and D^+ reinforces this picture. The S-wave decay fraction is larger for the D_s^+ (87%) than for the D^+ (56%). Rather than coupling to an S-wave dipion, the $d\bar{d}$ piece prefers to couple to a vector state such as $\rho^0(770)$, which alone accounts for $\sim 30\%$ of D^+ decay. This interpretation also bears on the role of the annihilation diagram in the $D_s^+ \to \pi^+\pi^-\pi^+$ decay. This study suggests that the S-wave annihilation contribution is negligible over much of the dipion mass spectrum. It might be interesting to search for annihilation contributions in higher spin channels, such as $\rho^0(1450)\pi$ and $f_2(1270)\pi$.

CONCLUSIONS

Systematic investigation of charm decay dynamics is producing interesting results in both semileptonic and hadronic sectors. In particular, Dalitz-plot analysis is and will be a crucial tool to extract physics from heavy flavor decays. Nevertheless, to fully exploit this unlimited potential, a systematic revision of the amplitude formalization is required: FOCUS has applied the *K-matrix* approach for the first time to the heavy flavor sector. The *K-matrix formalism* also allows for a rigorous coupled-channel analysis; this will be the next step in the Dalitz analysis of HF decays, as for instance, D^+ and $D_s^+ \to f_0 \pi$ amplitudes, which can feed both 3π and $KK\pi$ final channels. Strong dynamics effects in D decays now seem under control and fully consistent with those measured by light-quark experiments, without requiring *ad hoc* resonances. The new scenario is very promising for the future measurements of the CP violating phases in the B sector, where a proper description of the different amplitudes is essential [7]. FOCUS is now studying the high statistics channel $D^+ \to K^- \pi^+ \pi^+$, which, with more than 50K events, will provide a test of the model assumption, such as the quasi two-body nature of the decay.

Let me close by indicating some other recent FOCUS measurements regarding D^* spectroscopy [19], lifetime [20], mixing [21] and charmed pentaquark searches [22].

REFERENCES

1. J. Link *et al.*, *Phys. Lett.B*, **535**, 43, (2002).
2. D. Kim for the FOCUS collaboration Proc. of the XXXVIIIth Rencontres de Moriond, QCD amd High Energy Hadronic Interactions Les Arcs , March, 2003, also hep-ex/0305038.
3. J. Link *et al.*, *Phys. Lett. B* **72**, 80 (2005).
4. J. Link *et al.*, *Phys. Lett. B* **544**, 89 (2002).
5. D. Cronin-Hennessy, *et al.*, *Phys. Rev. D* **72**, 031102 (2005).
6. D. M. Asner *et al.*, *Phys. Rev. D* **2**, 012001 (2005).
7. B. Aubert *et al.*, accepted by *Phys. Rev. Lett Report-no: SLAC-PUB-11127*.
8. K. Abe *et al.*,contributed to *Moriond Electroweak 2005 and CKM 2005 Workshop Report-no: BELLE-CONF-0502*.
9. B. Aubert *et al.*, contributed to *the 32nd. International Conference on High-Energy Physics, ICHEP 04, 16-22 August 2004, Beijing, China Report-no: BABAR-CONF-04/38, SLAC-PUB-10658*.
10. J. Link *et al.*, *Phys. Lett. B* **585**, 200 (2004).
11. S. Spanier and N. A. Törnqvist, Scalar Mesons (rev.), Particle Data Group, *Phys. Rev. D* **66**, 010001-450 (2002).
12. M. R. Pennington, "In Search of Hadrons Beyond the Quark Model", *Proc. of Oxford Conf. in honour of R. H. Dalitz*, Oxford, July, 1990, Ed. by I. J. R. Aitchison, *et al.*, (World Scientific) pp. 66–107; *Proc. of Workshop on Hadron Spectroscopy* (WHS 99), Rome, March 1999, Ed. by T. Bressani *et al.*, (INFN, Frascati).
13. E. P. Wigner, *Phys. Rev.* **70**,15 (1946).
14. S. U. Chung *et al.*, *Ann. Physik* **4**, 404 (1995).
15. I. J. R. Aitchison, *Nucl. Phys. A* **189**, 417 (1972).
16. K. L. Au, D. Morgan, and M. R. Pennington, *Phys. Rev. D***35**, 1633 (1987).
17. V. V. Anisovich and A. V. Sarantsev, *Eur. Phys. J. A* **16**, 229 (2003).
18. E. M. Aitala *et al.*, *Phys. Rev. Lett.* **86**, 770 (2001).
19. J. M. Link *et al.*, *Phys. Lett. B*, **586**, 11 (2004).
20. J. M. Link *et al.*, submitted to *Phys. Rev. Lett.*, e-Print Archive: hep-ex/0504056 .
21. J. M. Link *et al.*, *Phys. Lett. B*, **618**, 23 (2005).
22. J. M. Link *et al.*, *Phys. Lett. B*, **622**, 229 (2005).

Radiative decays of excited charm mesons: a light-cone QCD sum rule analysis

P. Colangelo, F. De Fazio and A. Ozpineci[1]

INFN, Sezione di Bari, Bari, Italy

Abstract. The radiative decay modes of the newly discovered $D_{sJ}^*(2317)$ and $D_{sJ}(2460)$ states are important to shed light on the quark structure of these mesons. If they are studied within the framework of light cone QCD sum rules, assuming the D_{sJ} mesons as made of a quark-anti-quark pair, one obtains that the radiative decay widths follow the present experimentally observed pattern.

Keywords: Charm Mesons, Radiative Decays
PACS: 12.38.Lg, 13.20.Fe

INTRODUCTION

The last few years represented an interesting period for the hadron spectroscopy due to the observation, for example, of excitations of heavy quarkonium states [1] and of doubly charmed baryons such as Ξ_{cc} [2]. Claims for the observations of five-quark configurations were also reported; a discussion of the experimental status of such pentaquark states can be found in these proceedings [3].

Another interesting discovery for the hadron spectroscopy was the observation of a narrow resonance, which is called $D_{sJ}^*(2317)$, in the $D_s^+\pi^0$ invariant mass distribution by the BaBar collaboration at a mass of 2317 MeV [4]. The quantum numbers are consistent with $J^P = 0^+$. The existence of the resonance was later on confirmed by Belle and CLEO (for a review see [5] and references therein). Moreover, CLEO also reported another narrow resonance, $D_{sJ}(2460)$, with a mass 2460 MeV and with the quantum numbers consistent with $J^P = 1^+$. Both of these states have widths consistent with the experimental resolution, thus $\Gamma < 10\ MeV$.

One of the surprises regarding these resonances is the low value of their masses. Predictions about the masses for the scalar and pseudo vector charmed-strange mesons were higher then the $D^{(*)}K$ threshold. Hence, these resonances would decay into $D^{(*)}K$ and thus have a large decay width. But experimentally their masses are measured to be lower then this threshold, leading to a small width. They have to decay through other modes. An important decay channel is the isospin violating decay into $D_s^{(*)}\pi$. This decay could be described by a two stage process: first the D_{sJ} meson decays into $D_s^{(*)}\eta$. Then using the mixing between η and π^0 due to isospin violation, η is converted into π^0.

Another surprise about the masses of these mesons was that, in all previous studies, the non-strange D-mesons were always found to be lighter then the strange D-mesons.

---

[1] Speaker at the Workshop. Present address: Middle East Technical University, Ankara, Turkey

TABLE 1. Measurements and 90% CL upper limits of ratios of $D_{sJ}^*(2317)$ and $D_{sJ}(2460)$ decay widths.

Decay Channel	Belle [6]	BaBar [7]	CLEO [8]
$\frac{\Gamma(D_{sJ}^*(2317)\to D_s^*\gamma)}{\Gamma(D_{sJ}^*(2317)\to D_s\pi^0)}$	< 0.18	–	< 0.059
$\frac{\Gamma(D_{sJ}(2460)\to D_s\gamma)}{\Gamma(D_{sJ}(2460)\to D_s^*\pi^0)}$	$0.55 \pm 0.13 \pm 0.08$ $0.38 \pm 0.11 \pm 0.04$	$0.375 \pm 0.054 \pm 0.057$ $0.274 \pm 0.045 \pm 0.020$	< 0.49
$\frac{\Gamma(D_{sJ}(2460)\to D_s^*\gamma)}{\Gamma(D_{sJ}(2460)\to D_s^*\pi^0)}$	< 0.31	–	< 0.16
$\frac{\Gamma(D_{sJ}(2460)\to D_{sJ}^*(2317)\gamma)}{\Gamma(D_{sJ}(2460)\to D_s^*\pi^0)}$	–	< 0.23	< 0.58

But experimental results show that they have more or less equal masses. These surprises led to possible interpretations of these mesons as meson molecules or 4-quark states.

In order to shed light on the structure of these mesons, one strategy is to assume a specific structure for the mesons and study various decays to see if the assumption reproduces the experimental results. Various theoretical estimates for the partial decay widths into $D_s\pi$ assuming a $c\bar{s}$ picture give widths in the range $6-20$ KeV, which is well below the experimental limit.

Another interesting decay mode which can shed light on the structure of these mesons is the radiative decay mode. The only observed radiative decay is $D_{sJ}(2460) \to D_s\gamma$, while for the other modes there are only upper bounds. The present experimental situation is summarized in Table 1.

A classification of the D mesons can be obtained in the heavy quark limit. In this limit, the spin of the heavy quark s_Q decouples from the total angular momentum of the light degrees of freedom s_ℓ. Hence they both become conserved quantities. As a consequence, to a given value of s_l corresponds a degenerate doublet the members of which differ only for the orientation of the heavy quark spin. Finite quark mass effects lead to mass splitting of the doublet.

When the orbital angular momentum is $L=0$ we have $s_l = \frac{1}{2}$ and hence the $(0^-, 1^-)$ doublet, which in case of charm is composed by the states $(D_{(s)}, D_{(s)}^*)$. For $L=1$, s_l can take two values: $s_l = \frac{1}{2}$ or $s_l = \frac{3}{2}$. For $s_l = \frac{1}{2}$, we obtain the doublet $(0^+, 1^+)$ denoted by $(D_{(s)0}^*, D_{(s)1}')$; and for $s_l = \frac{3}{2}$, there corresponds the doublet $(1^+, 2^+)$, denoted by $(D_{(s)1}, D_{(s)2}^*)$. If the mass of the quark is finite, then in general it is expected that there should be a mixing between the 1^+ member of the $s_l = \frac{1}{2}$ doublet and the 1^+ member of the $s_l = \frac{3}{2}$ doublet. But experimentally, this mixing is found to be small for the non-strange case [9]. We neglect this mixing also in the strange D-meson case.

In order to study these decays, one needs a non-perturbative method. One is the sum rule approach proposed by Shifman, Vainshtein and Zakharov [10]. This method has been applied successfully to many problems involving hadrons. In QCD sum rules, one expresses hadronic parameters in terms of the properties of the vacuum parameterized by the vacuum condensates. There had been many modification of the sum rules to improve in various ways the original sum rules. Light cone QCD sum rules is one of these modifications [11, 12].

Here we present the results obtained in [13], where light cone QCD sum rules are used to study the radiative decay modes assuming that $(D_{sJ}^*(2317), D_{sJ}(2460))$ is identified with the doublet $(0^+, 1^+)$. In the following section we collect results for all the decay modes and compare them with the predictions of other methods. Details can be found in [13].

RESULTS AND DISCUSSIONS

$$D_{sJ}^*(2317) \to D_s^* \gamma$$

The relevant transition amplitude can be parameterized as:

$$\langle D_s^*(p,\eta)\gamma(q,\varepsilon)|D_{sJ}^*(p+q)\rangle = ed\left((\varepsilon^*\eta^*)(pq) - (\eta^*p)(\varepsilon^*q)\right)$$

where η is the polarization vector of the photon, ε is the polarization of the vector meson, and gauge invariance has been used to reduce the number of parameters to one, i.e. d.

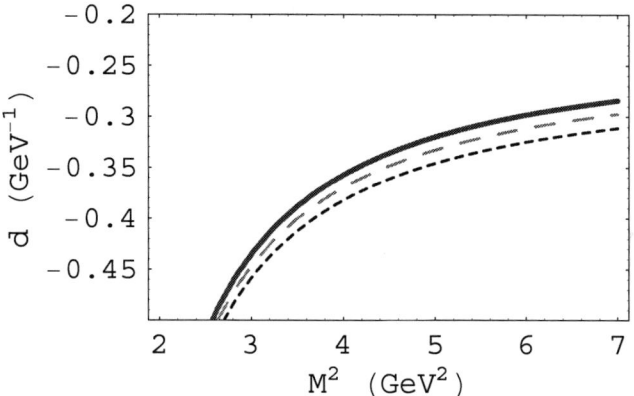

FIGURE 1. The parameter d in the $D_{s0}^* \to D_s^* \gamma$ decay amplitude versus the Borel parameter M^2. The curves correspond to the thresholds $s_0 = 2.45^2$ GeV^2 (continuous line), $s_0 = 2.5^2$ GeV^2 (long-dashed line) and $s_0 = 2.55^2$ GeV^2 (dashed line).

In Fig. 1, the dependence of d on the Borel parameter M^2 is depicted for various values of the continuum threshold s_0 separating the QCD continuum from the considered resonances. The Borel parameter is introduced as an external, auxiliary quantity; since the physical quantiy d should be independent of the external parameter M^2, a suitable region for M^2 has to be chosen such that within this region, the physical predictions are independent of the precise value of M^2. From Fig. 1 one can see that such a region exists for $5\ GeV^2 < M^2 < 7\ GeV^2$. This region corresponds to $-0.35\ GeV^{-1} < d < -0.28\ GeV^{-1}$. Once d is calculated, the width is obtained from

$$\Gamma = \frac{\alpha}{8m_{D_{sJ}^*}^3}(m_{D_{sJ}^*}^2 - m_{D_s^*}^2)^3|d|^2 \qquad (1)$$

which turns out to be $\Gamma = 5 \pm 1$ *KeV*, 4 – 5 times bigger then other predictions which are obtained using heavy quark theory.

The uncertainty quoted in the result is due to the residual dependence on the Borel parameter M^2 and the threshold s_0. Another source of uncertainty is due to the variation of the parameters entering the sum rules. One such parameter is the quark magnetic susceptibility χ which we fix to $\chi = -3.15\ GeV^{-2}$ [14].

In order to study the possible reason of the discrepancy with the other predictions which use the heavy quark limit, we also derived d in the same limit, obtaining $d = -0.14 \pm 0.2\ GeV^{-1}$, consistent with the predictions of the VMD model in the heavy quark limit: $d \simeq -0.15\ GeV^{-1}$ [15, 5].

$$D_{sJ}(2460) \to D_s \gamma$$

The transition amplitude for the decay $D_{sJ}(2460) \to D_s \gamma$ can be written as,

$$\langle \gamma(q,\varepsilon) D_s(p) | D'_{s1}(p+q,\eta) \rangle = eg_1 [(\varepsilon^* \cdot \eta)(p \cdot q) - (\varepsilon^* \cdot p)(\eta \cdot q)] \ . \tag{2}$$

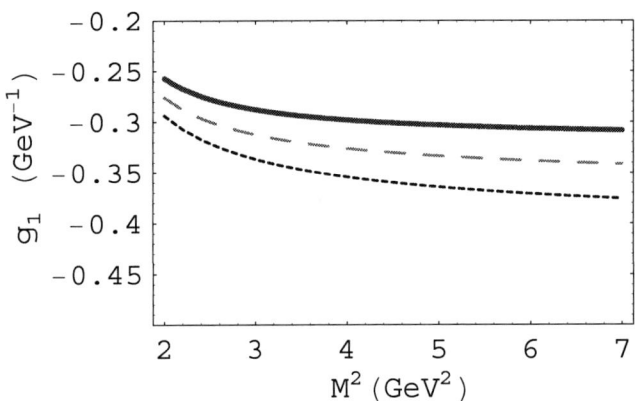

FIGURE 2. g_1 as a function of the Borel parameter M^2 for the threshold $s_0 = 2.5^2$ GeV^2 (continuous), $s_0 = 2.55^2$ GeV^2 (long-dashed) and $s_0 = 2.6^2$ GeV^2 (dashed line).

In Fig. 2 we present the dependence of g_1 on the Borel parameter M^2 for various values of the threshold that separates the QCD continuum from the the considered resonances. We quote the result: $g_1 = -0.33 \pm 0.04\ GeV^{-1}$ corresponding to $\Gamma(D'_{s1} \to D_s^* \gamma) = 24 \pm 5\ KeV$.

$D_{sJ}(2460) \to D_s^* \gamma$

The transition amplitude for the $D_{sJ}(2460) \to D_s^* \gamma$ transition can be parameterized by:

$$\langle \gamma(q,\varepsilon) D_s^*(p,\tilde{\eta}) | D'_{s1}(p+q,\eta) \rangle = ie g_2 \varepsilon_{\alpha\beta\sigma\tau} \eta^\alpha \tilde{\eta}^{*\beta} \varepsilon^{*\sigma} q^\tau$$

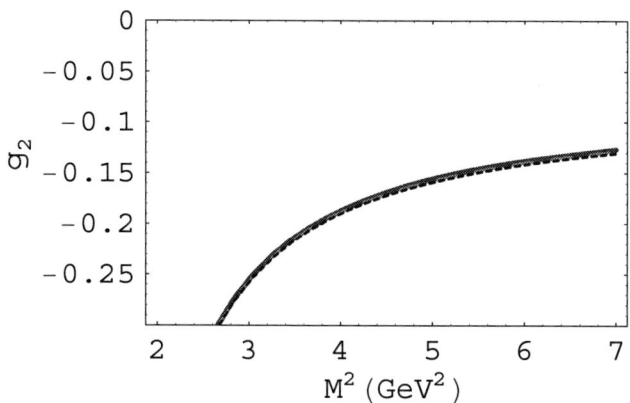

FIGURE 3. g_2 versus the Borel parameter M^2. The continuous, long-dashed and dashed lines refer to $s_0 = 2.5^2$ GeV^2, $s_0 = 2.55^2$ GeV^2 and $s_0 = 2.6^2$ GeV^2, respectively.

In Fig. 3, the dependence of the parameter g_2 on the Borel parameter is depicted for various values of the continuum threshold. The result $g_2 = -0.15 \pm 0.03$ corresponds to $\Gamma(D'_{s1} \to D_s^* \gamma) = 0.8 \pm 0.3$ KeV.

$D_{sJ}(2460) \to D_{sJ}^*(2317) \gamma$

The parameterization of the transition matrix element for the decay $D_{sJ}(2460) \to D_{sJ}^*(2317) \gamma$ is given by:

$$\langle \gamma(q,\varepsilon) D_{s0}(p) | D'_{s1}(p+q,\eta) \rangle = ie g_3 \varepsilon_{\alpha\beta\sigma\tau} \varepsilon^{*\alpha} \eta^\beta p^\sigma q^\tau .$$

In Fig. 4, we present the dependence of g_3 on the Borel parameter for various values of the continuum threshold. In the stability region we get $g_3 = -0.31 \pm 0.04$ GeV^{-1} corresponding to $\Gamma(D'_{s1} \to D_{s0} \gamma) = 0.7 \pm 0.2$ KeV.

In Table 1, we summarize our results for the decay widths of the radiative modes.

From Table 2 it is seen that, except for the decay $D_{sJ}(2460) \to D_s^* \gamma$, the predictions of the light cone sum rules with finite quark mass for the branching ratios are larger then the predictions of the other methods. Another interesting feature is that the prediction for the decay width of $D_{sJ}(2460) \to D_s \gamma$ is larger then for the other channels: this might be the explanation why it is only this mode which is observed experimentally. There are no measurements of the individual radiative decay widths. Experimental data are available

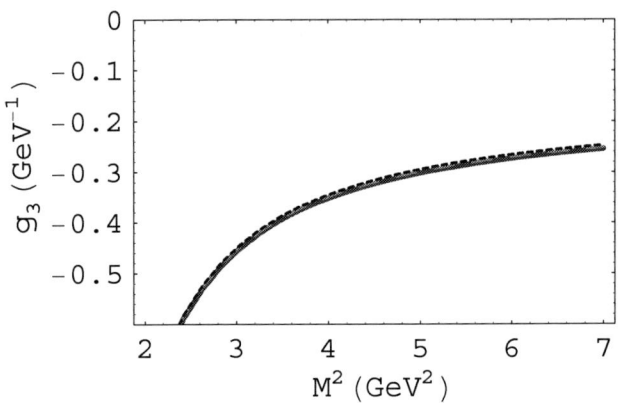

FIGURE 4. The parameter g_3 in the $D'_{s1} \to D^*_{s0}\gamma$ amplitude, versus the Borel parameter M^2.

TABLE 2. Radiative decay widths (in KeV) of $D^*_{sJ}(2317)$ and $D_{sJ}(2460)$ obtained by light-cone sum rules (LCQSR). Vector Meson Dominance (VMD) and constituent quark model (QM) results are also reported.

Initial state	Final state	LCQSR	VMD [5, 15]	QM [16]	QM [17]
$D^*_{sJ}(2317)$	$D^*_s \gamma$	4-6	0.85	1.9	1.74
$D_{sJ}(2460)$	$D_s \gamma$	19-29	3.3	6.2	5.08
	$D^*_s \gamma$	0.6-1.1	1.5	5.5	4.66
	$D^*_{sJ}(2317)\gamma$	0.5-0.8	–	0.012	2.74

only for the relative decay widths of the radiative decay to the pionic mode. VMD predictions in the heavy quark limit for these ratios are consistent with the experiment. In our study, we have found that the radiative mode is enhanced when finite quark mass effects are taken into account. It is reasonable that such an enhancement would also occur for the pionic modes, however further studies are required on this point. In any case, the results of our analysis based on the quark-antiquark picture of the D_{sJ} mesons are consistent with the experimentally observed pattern.

REFERENCES

1. See T. Skwarnicki, Int. J. Mod. Phys. **A19**, 1030 (2004) and references therein.
2. M. Mattson, et al. [SELEX Collaboration], Phys. Rev. Lett. **89**, 112001 (2002).
3. M. Battaglieri, these proceedings.
4. B. Aubert et al. [BaBar Collaboration], Phys. Rev. Lett. **90**, 242001 (2003).
5. P. Colangelo, F. De Fazio, R. Ferrandes, Mod. Phys. Lett. **A19**, 2083 (2004).
6. P. Krokovny et al. [Belle Collaboration], Phys. Rev. Lett. **91**, 262002 (2003). K. Abe et al., Phys. Rev. Lett. **92**, 012002 (2004).
7. B. Aubert et al. [BABAR Collaboration], Phys. Rev. Lett. **93**, 181801 (2004).
8. D. Besson et al. [CLEO Collaboration], Phys. Rev. D **68**, 032002 (2003).
9. K. Abe et al. [Belle Collaboration], Phys. Rev. D **69**, 112002 (2004).

10. M. A. Shifman, V. I. Vainshtein, V. I. Zakharov, Nucl. Phys. **B147**, 385 (1979).
11. N. S. Craigie and J. Stern, Nucl. Phys. B **216**, 209 (1983); V. M. Braun and I. E. Filyanov, Z. Phys. C **44**, 157 (1989); V. L. Chernyak and I. R. Zhitnitsky, Nucl. Phys. B **345**, 137 (1990); V. M. Belyaev, V. M. Braun, A. Khodjamirian and R. Ruckl, Phys. Rev. D **51**, 6177 (1995).
12. For a reviews see: P. Colangelo and A. Khodjamirian, in 'At the Frontier of Particle Physics/Handbook of QCD', ed. by M. Shifman (World Scientific, Singapore, 2001), page 1495 (arXiv:hep-ph/0010175).
13. P. Colangelo, F.De Fazio, and A. Ozpineci, Phys. Rev. **D72**, 074004 (2005).
14. P. Ball, V. M. Braun, N. Kivel, Nucl. Phys. **B649**, 263 (2003).
15. P. Colangelo, F. De Fazio, Phys. Lett. **B570**, 180 (2003).
16. S. Godfrey, Phys. Lett. **B 568**, 254 (2003).
17. W. A. Bardeen, E. J. Eichten and C. T. Hill, Phys. Rev. D **68**, 054024 (2003).

Heavy quarkonium decays and transitions in the language of effective field theories

Antonio Vairo

Dipartimento di Fisica dell'Università di Milano and INFN, via Celoria 16, 20133 Milano, Italy

Abstract. Heavy quarkonium decays and transitions are discussed in the framework of non-relativistic effective field theories. Emphasis is put on the matching procedure in the non-perturbative regime. Some exact results valid for the magnetic dipole couplings are discussed.

Keywords: NRQCD, pNRQCD, heavy quarkonium, decay, radiative transitions
PACS: 12.38.-t, 12.39.Hg, 13.25.Gv

INTRODUCTION

In the last years the B factories, CLEO and BES have produced a large amount of new data for heavy quarkonium observables [1]. These data are not only interesting because they may signal new states or new decay or production mechanisms, but also because heavy quarkonium is a system that to a large extent can rigorously be studied in QCD. Therefore, any new understanding of it may potentially provide new insight on the non-perturbative dynamics of QCD.

Heavy quarkonium, as a non-relativistic bound state, is characterized by a hierarchy of energy scales: m, mv and mv^2, where m is the heavy-quark mass and $v \ll 1$ the heavy-quark relative velocity. Whenever a system is described by a hierarchy of scales, observables may be calculated by expanding one scale with respect to the other. An effective field theory (EFT) is a field theory that makes this expansion explicit at the Lagrangian level. To be more precise, let us call H a system described by a fundamental Lagrangian $\mathscr{L}$ and suppose it characterized by 2 scales: $\Lambda \gg \lambda$. The EFT Lagrangian, $\mathscr{L}_{\text{EFT}}$, suitable to describe H at scales lower than Λ, is characterized by (1) a cut off $\Lambda \gg \mu \gg \lambda$; (2) some degrees of freedom that exist at scales lower than μ. The Lagrangian $\mathscr{L}_{\text{EFT}}$ is then made of all operators O_n that may be built from the effective degrees of freedom and are consistent with the symmetries of the original Lagrangian $\mathscr{L}$:

$$\mathscr{L}_{\text{EFT}} = \sum_n c_n(\Lambda/\mu) \frac{O_n(\mu,\lambda)}{\Lambda^n}. \qquad (1)$$

The advantage is that, once the scale μ has been run down to λ, the power counting is homogeneous $\langle O_n \rangle \sim \lambda^n$, so that the EFT is, indeed, organized as an expansion in λ/Λ. Despite the EFT not being renormalizable in the traditional sense, it is renormalizable order by order in λ/Λ. The matching coefficients $c_n(\Lambda/\mu)$ encode the non-analytic behaviour in Λ. They are calculated by imposing that $\mathscr{L}_{\text{EFT}}$ and $\mathscr{L}$ describe the same physics at any finite order in the expansion. The procedure is called matching. Finally, we note that if $\Lambda \gg \Lambda_{QCD}$, $c_n(\Lambda/\mu)$ may be calculated in perturbation theory, if $\Lambda \sim \Lambda_{QCD}$, the matching must rely on non-perturbative methods.

Several effective field theories for heavy quarkonium that take full advantage of the non-relativistic hierarchy of scales have been developed and used over the last decade. For a recent review we refer to [2]. NRQCD is the EFT that exploits the hierarchy $\Lambda = m \gg \lambda = mv$ [3, 4]. Since $m \gg \Lambda_{QCD}$, the matching coefficients of NRQCD may be calculated in perturbation theory. pNRQCD is the EFT that exploits the hierarchy $\Lambda = mv \gg \lambda = mv^2$ [5, 6]. If $\Lambda_{QCD} \sim mv^2$, then the matching to pNRQCD may be still done in perturbation theory. We call weak coupling this regime, which may be suited to describe ground-state quarkonium. If $\Lambda_{QCD} \sim mv$, then the matching to pNRQCD is non perturbative. We call strong coupling this regime, which may be suited to describe excited quarkonium states.

The fact that EFTs may be built to describe heavy quarkonium in the strong-coupling regime follows from the observation that the non-relativistic hierarchy of scales survives also below Λ_{QCD} [7, 8]. The complication of the strong-coupling regime comes from the non-perturbative matching and from new scales that may arise in loops sensitive to Λ_{QCD}. An example is the three-momentum scale $\sqrt{m\Lambda_{QCD}}$ discussed in [9]. Nevertheless, many advantages remain in treating even strongly-coupled heavy quarkonium in an EFT framework.

In the following we shall sketch a unified framework for the description of inclusive and electromagnetic decays, and radiative transitions in the framework of strongly coupled pNRQCD. For a treatment of inclusive and electromagnetic decay widths in the weak-coupling regime we refer to [2] and references therein. For a treatment of magnetic dipole transitions in the weak-coupling regime we refer to [10].

NRQCD

NRQCD is the EFT that follows from QCD when modes of energy or momentum m are integrated out. The structure of the EFT Lagrangian is like Eq. (1) with $\Lambda = m$ and $\lambda = mv \sim \Lambda_{QCD}$. The scale mv is sometimes called soft. The degrees of freedom of the EFT Lagrangian are quarks, antiquarks and gluons with energy and momentum lower than m (we neglect light quarks).

The NRQCD Lagrangian may be written as

$$\mathscr{L}_{NRQCD} = \mathscr{L}_{2-f} + \mathscr{L}_{4-f} + \mathscr{L}_{light}, \qquad (2)$$

where

$$\mathscr{L}_{2-f} = \psi^\dagger \left(iD_0 + \frac{\mathbf{D}^2}{2m}\right)\psi + \frac{c_F}{2m}\psi^\dagger \sigma \cdot g\mathbf{B}\psi - \frac{2c_F - 1}{8m^2}\psi^\dagger \sigma \cdot [-i\mathbf{D}\times, g\mathbf{E}]\psi + \dots$$
$$+ [\psi \to i\sigma^2 \chi^*], \qquad (3)$$

$$\mathscr{L}_{light} = -\frac{1}{4}F^{\mu\nu a}F^a_{\mu\nu}. \qquad (4)$$

ψ is the Pauli spinor field that annihilates a heavy quark of mass m, χ is the corresponding one that creates a heavy antiquark, $iD_0 = i\partial_0 - gT^a A_0^a$ and $i\mathbf{D} = i\nabla + gT^a \mathbf{A}^a$. The term $\mathscr{L}_{4-f}$ stands for the 4-fermion part of the NRQCD Lagrangian (for an explicit expression see, for instance, [4]).

The coefficient c_F is a matching coefficient of the EFT. In Eq. (3), we have made use of reparameterization invariance to reduce the other matching coefficients to this one. It is known at two loops and may be found in [11]. The 4-fermion matching coefficients encode the contribution of the annihilation graphs. As a consequence they develop an imaginary part. We refer to [12] for an updated list of them and for references.

Let us give some definitions concerning the Fock space of NRQCD. If H_{NRQCD} is the Hamiltonian of NRQCD, we call $|\underline{n};\mathbf{r},\mathbf{R}\rangle$ the eigenstates of H_{NRQCD}, and E_n the corresponding eigenvalues. $\mathbf{r}$ stands for the relative distance of the two heavy quarks and $\mathbf{R}$ for their centre-of-mass coordinate. Both are good quantum numbers in the static limit. With n we indicate a generic set of conserved quantum numbers. $|\underline{n};\mathbf{r},\mathbf{R}\rangle$ and $E_n(\mathbf{r},\mathbf{R};\nabla_r,\nabla_R)$ satisfy the system of equations:

$$H_{\text{NRQCD}}|\underline{n};\mathbf{r},\mathbf{R}\rangle = \int d^3 r' d^3 R' |\underline{n};\mathbf{r}',\mathbf{R}'\rangle E_n(\mathbf{r}',\mathbf{R}';\nabla_{r'},\nabla_{R'})\delta^3(\mathbf{r}'-\mathbf{r})\delta^3(\mathbf{R}'-\mathbf{R}), \quad (5)$$

$$\langle \underline{m};\mathbf{r},\mathbf{R}|\underline{n};\mathbf{r}',\mathbf{R}'\rangle = \delta_{nm}\delta^3(\mathbf{r}-\mathbf{r}')\delta^3(\mathbf{R}-\mathbf{R}'). \quad (6)$$

INCLUSIVE DECAYS IN PNRQCD

pNRQCD is the EFT that follows from NRQCD when gluons of energy or momentum and quarks of energy larger than mv^2 and quarks of momentum larger than mv are integrated out. The structure of the EFT Lagrangian is like Eq. (1) with $\Lambda = mv \sim \Lambda_{\text{QCD}}$ and $\lambda = mv^2$. The scale mv^2 is sometimes called ultrasoft. In the strong-coupling regime, if the gluonic excitations between the two heavy quarks develop a mass gap of order Λ_{QCD}, then they are all integrated out from the theory. Therefore, the degrees of freedom of the EFT Lagrangian are only singlet quarkonium fields. The Lagrangian of pNRQCD is very simple:

$$\mathscr{L}_{\text{pNRQCD}} = \int d^3 r \, \text{Tr}\left\{ S^\dagger \left(i\partial_0 + \frac{\nabla_R^2}{4m} + \frac{\nabla_r^2}{m} - V_S \right) S \right\}, \quad (7)$$

S is a non-local field, function of $\mathbf{r}$, $\mathbf{R}$ and t, $2 \otimes 2$ in spin space and a $3 \otimes 3$ singlet in colour space. The trace is taken over colour and spin indices.

All complications go into the potential V_S, which is a non-perturbative function of $\mathbf{r}$ to be determined by a non-perturbative matching procedure. In general V_S contains also an imaginary part inherited from the matching coefficients of the 4-fermion operators of NRQCD. The matching condition reads

$$\langle \underline{0};\mathbf{r}',\mathbf{R}'|H_{\text{NRQCD}}|\underline{0};\mathbf{r},\mathbf{R}\rangle = \left(-\frac{\nabla_R^2}{4m} - \frac{\nabla_r^2}{m} + V_S\right)\delta^3(\mathbf{r}'-\mathbf{r})\delta^3(\mathbf{R}'-\mathbf{R}). \quad (8)$$

The matching condition determines V_S as a function of quantities defined in NRQCD (the left-hand side of Eq. (8)). Once V_S has been determined, one may calculate the solutions $\Phi_H(\mathbf{r})$ and E_H of the Schrödinger equation

$$\left(-\frac{\nabla_r^2}{m} + V_S\right)\Phi_H(\mathbf{r}) = E_H\Phi_H(\mathbf{r}). \tag{9}$$

Using the optical theorem, the inclusive decay width to light particles (l.p.) is given by

$$\Gamma_{H\to\text{l.p.}} = -2\,\text{Im}\,\langle H(0)| - \mathscr{L}_{\text{pNRQCD}}|H(0)\rangle. \tag{10}$$

$|H(0)\rangle$ stands for a quarkonium state in the rest-frame ($\mathbf{P} = 0$):

$$|H(\mathbf{0})\rangle = \int d^3r \int d^3R\,\text{Tr}\left\{\Phi_H(\mathbf{r})\text{S}^\dagger(\mathbf{r},\mathbf{R})\right\}|0\rangle, \tag{11}$$

where $|0\rangle$ is the Fock subspace containing no heavy quarks but an arbitrary number of ultrasoft particles. Note that, since Φ_H and $\mathscr{L}_{\text{pNRQCD}}$ have been calculated through the matching procedure, Eq. (10) provides, indeed, a practical tool to calculate the inclusive decay width. Explicit applications of Eq. (10) have been worked out in [13, 14, 15].

RADIATIVE TRANSITIONS IN PNRQCD

Radiative transitions may be described in the same EFT framework that we have discussed so far by enlarging the gauge group to $SU_c(3) \times U_{\text{em}}(1)$. This means that more degrees of freedom have to be taken into account (photons) and more operators added to the EFT Lagrangians. We will concentrate in the following on magnetic dipole transitions [10].

At the level of the NRQCD Lagrangian, magnetic transitions are accounted for by replacing $iD_0 \to iD_0 - ee_Q A_0^{\text{em}}$ and $i\mathbf{D} \to i\mathbf{D} + ee_Q \mathbf{A}^{\text{em}}$ in Eq. (3),

$$\begin{aligned}\mathscr{L}_{2-f} \to \mathscr{L}_{2-f} &+ \frac{c_F^{\text{em}}}{2m}\psi^\dagger\sigma\cdot ee_Q\mathbf{B}^{\text{em}}\psi - \frac{2c_F^{\text{em}}-1}{8m^2}\psi^\dagger\sigma\cdot[-i\mathbf{D}\times,ee_Q\mathbf{E}^{\text{em}}]\psi \\ &+ \frac{c_{W1}^{\text{em}}}{8m^3}\psi^\dagger\{\mathbf{D}^2,\sigma\cdot ee_Q\mathbf{B}^{\text{em}}\}\psi - \frac{c_{W1}^{\text{em}}-1}{4m^3}\psi^\dagger D^i\sigma\cdot ee_Q\mathbf{B}^{\text{em}}D^i\psi \\ &+ \frac{c_F^{\text{em}}-1}{8m^3}\psi^\dagger\left(\sigma\cdot\mathbf{D}\,ee_Q\mathbf{B}^{\text{em}}\cdot\mathbf{D} + \mathbf{D}\cdot ee_Q\mathbf{B}^{\text{em}}\,\sigma\cdot\mathbf{D}\right)\psi + \cdots + [\psi \to i\sigma^2\chi^*],\end{aligned} \tag{12}$$

and $\mathscr{L}_{\text{light}} \to \mathscr{L}_{\text{light}} - \frac{1}{4}F^{\mu\nu\,\text{em}}F_{\mu\nu\,\text{em}}$, where the gauge fields with upperscript "em" are electromagnetic fields and ee_Q stands for the charge of the quark of flavour Q.

The coefficients c_F^{em} and c_{W1}^{em} are new matching coefficients of the EFT associated with the electromagnetic couplings. Again, we have made use of reparameterization invariance to reduce their number. All coefficients are known at least at one-loop level

[16]. In particular, we have[1]

$$c_F^{em} \equiv 1 + \kappa_Q^{em} = 1 + \frac{4}{3}\frac{\alpha_s}{2\pi} + \mathcal{O}(\alpha_s^2), \tag{13}$$

$$c_{W1}^{em} = 1 + \frac{4}{3}\frac{\alpha_s}{\pi}\left(\frac{1}{12} + \frac{4}{3}\ln\frac{m}{\mu}\right) + \mathcal{O}(\alpha_s^2), \tag{14}$$

κ_Q^{em} is usually identified with the anomalous magnetic moment of the heavy quark.

At the level of pNRQCD, magnetic transitions involving ultrasoft photons are described by adding to the Lagrangian (7) the electromagnetic Lagrangian $-F^{\mu\nu\,em} \times F_{\mu\nu\,em}/4$ and a term $\mathscr{L}_{\gamma\mathrm{pNRQCD}}$ responsible for the coupling of the quarkonium to the electromagnetic field:

$$\begin{aligned}
\mathscr{L}_{\gamma\mathrm{pNRQCD}} = \int d^3r \, \mathrm{Tr}\Bigg\{ & V_A^{em} \, \mathrm{S}^\dagger \mathbf{r} \cdot ee_Q \mathbf{E}^{em} \mathrm{S} \\
& + \frac{V_S^{\sigma\cdot B}/m}{2m}\{\mathrm{S}^\dagger, \sigma \cdot ee_Q \mathbf{B}^{em}\}\mathrm{S} + \frac{V_S^{(r\cdot\nabla)^2 \sigma\cdot B/m}}{16m}\{\mathrm{S}^\dagger, r^i r^j (\nabla_R^i \nabla_R^j \sigma \cdot ee_Q \mathbf{B}^{em})\}\mathrm{S} \\
& + \frac{V_S^{\sigma\cdot(r\times r\times B)/m^2}}{4m^2 r}\{\mathrm{S}^\dagger, \sigma \cdot [\hat{\mathbf{r}} \times (\hat{\mathbf{r}} \times ee_Q \mathbf{B}^{em})]\}\mathrm{S} + \frac{V_S^{\sigma\cdot B/m^2}}{4m^2 r}\{\mathrm{S}^\dagger, \sigma \cdot ee_Q \mathbf{B}^{em}\}\mathrm{S} \\
& - \frac{V_S^{\sigma\cdot \nabla\times E/m^2}}{16m^2}[\mathrm{S}^\dagger, \sigma \cdot [-i\nabla_R \times, ee_Q \mathbf{E}^{em}]]\mathrm{S} - \frac{V_S^{\sigma\cdot \nabla_r\times r\cdot\nabla E/m^2}}{16m^2}[\mathrm{S}^\dagger, \sigma \cdot [-i\nabla_r \times, r^i (\nabla_R^i ee_Q \mathbf{E}^{em})]]\mathrm{S} \\
& + \frac{V_S^{\nabla_r^2 \sigma\cdot B/m^3}}{4m^3}\{\mathrm{S}^\dagger, \sigma \cdot ee_Q \mathbf{B}^{em}\}\nabla_r^2\mathrm{S} + \frac{V_S^{(\nabla_r\cdot\sigma)(\nabla_r\cdot B)/m^3}}{4m^3}\{\mathrm{S}^\dagger, \sigma^i ee_Q \mathbf{B}^{em\,j}\}\nabla_r^i \nabla_r^j \mathrm{S} \\
& + \frac{V_S^{\sigma\cdot(r\times r\times B)/m^3}}{4m^3 r^2}\{\mathrm{S}^\dagger, \sigma \cdot [\hat{\mathbf{r}} \times (\hat{\mathbf{r}} \times ee_Q \mathbf{B}^{em})]\}\mathrm{S} + \frac{V_S^{\sigma\cdot B/m^3}}{4m^3 r^2}\{\mathrm{S}^\dagger, \sigma \cdot ee_Q \mathbf{B}^{em}\}\mathrm{S} + \cdots \Bigg\}. \tag{15}
\end{aligned}$$

All gauge fields are calculated in the centre-of-mass coordinate $\mathbf{R}$. The field S is understood as a singlet also under $U_{em}(1)$ gauge transformations.

In the centre-of-mass of the initial quarkonium state, the power counting goes as follows: $\nabla_r \sim mv$, $r \sim 1/mv$, the electromagnetic fields associated to the external photons go like $\mathbf{E}^{em}, \mathbf{B}^{em} \sim k_\gamma^2$. The centre-of-mass derivative ∇ acting on the recoiling final quarkonium state or emitted photon is of order k_γ, where k_γ is the energy and momentum of the emitted photon.

The coefficients V in Eq. (15) are the matching coefficients of pNRQCD. They encode high-energy contributions to the electromagnetic couplings and are of the same nature as V_S in Eq. (7). In the strong-coupling regime they are determined by non-perturbative matching of 5-points Green functions involving two external quarks, two external antiquarks and an external photon. Let us consider the matching condition for

---

[1] The coefficients get also QED corrections, but they are numerically negligible.

the $1/m$ operators, it reads

$$\langle \underline{0}; \mathbf{r}', \mathbf{R}'| \otimes \langle \gamma | \left(\frac{c_F^{em}}{2m} \int d^3x \, \psi^\dagger \sigma \cdot ee_Q \mathbf{B}^{em} \psi + [\psi \to i\sigma^2 \chi^*] \right) |0\rangle \otimes |\underline{0}; \mathbf{r}, \mathbf{R}\rangle =$$

$$\left(\frac{V_S^{\frac{\sigma \cdot B}{m}}}{2m} + \frac{V_S^{(r \cdot \nabla)^2 \frac{\sigma \cdot B}{m}}}{16m} (\mathbf{r} \cdot \nabla_R)^2 \right) \left(\sigma^{(1)} + \sigma^{(2)} \right) \cdot \langle \gamma | ee_Q \mathbf{B}^{em} |0\rangle \delta^3(\mathbf{r}' - \mathbf{r}) \delta^3(\mathbf{R}' - \mathbf{R}). \quad (16)$$

Since corrections to the state $|\underline{0}; \mathbf{r}, \mathbf{R}\rangle$ involving derivatives or spins are $1/m$ suppressed (see Eq. (3)), $\sigma \cdot ee_Q \mathbf{B}^{em}$ effectively behaves as the identity operator. As a consequence, the electromagnetic matrix element decouples in the left-hand side. From the normalization condition (6) it follows that

$$V_S^{\frac{\sigma \cdot B}{m}} = V_S^{(r \cdot \nabla)^2 \frac{\sigma \cdot B}{m}} = c_F^{em}. \quad (17)$$

This is a rather remarkable result that holds to all orders in the strong-coupling constant and non-perturbatively. It excludes that the $1/m$ magnetic coupling of the quarkonium field is affected by any soft contribution. A fortiori, it excludes large anomalous non-perturbative corrections to this coupling. Similar arguments lead to the following exact results at order $1/m^2$:

$$V_S^{\frac{\sigma \cdot (r \times r \times B)}{m^2}} = \frac{r^2}{2} V_S^{(0)\prime}, \quad V_S^{\frac{\sigma \cdot B}{m^2}} = 0, \quad V_S^{\frac{\sigma \cdot \nabla_r \times E}{m^2}} = V_S^{\frac{\sigma \cdot \nabla_r \times r \cdot \nabla E}{m^2}} = 2c_F^{em} - 1, \quad (18)$$

where $V_S^{(0)}$ is the static part of the V_S potential. The first equality follows from the fact that Poincaré invariance protects the spin-orbit coupling [17, 18]. The second one remarkably states that to all orders in the strong-coupling constant and non-perturbatively the existence of an effective scalar interaction, which has been often advocated in phenomenological models, is excluded. The third one that those matching coefficients, like the one in Eq. (17), get only hard contributions.

The matching of the $1/m^3$ terms is more complicated. One reason is that at this order kinetic energy and spin-dependent corrections affect the state $|\underline{0}; \mathbf{r}, \mathbf{R}\rangle$ and $\sigma \cdot ee_Q \mathbf{B}^{em}$ does not behave anymore like the identity operator.

Once the matching has been completed, the transition width is given by:

$$\Gamma_{H \to H'\gamma} = \int \frac{d^3 P'}{(2\pi)^3} \frac{d^3 k}{(2\pi)^3} (2\pi)^4 \delta^4(P_H - k - P') \overline{|\mathscr{A}[H(0) \to H'(\mathbf{P}')\gamma(\mathbf{k})]|^2}, \quad (19)$$

where

$$\mathscr{A}[H(0) \to H'(-\mathbf{k})\gamma(\mathbf{k})] \, \delta^3(\mathbf{P}' + \mathbf{k}) = \langle H'(\mathbf{P}') \gamma(\mathbf{k}) | - \int d^3 R \, \mathscr{L}_{\gamma \text{pNRQCD}} | H(0) \rangle. \quad (20)$$

The overline stands for the sum over the final-state polarizations and the average over the initial state ones. $P_H = (M_H, \mathbf{0})$ stands for the four-momentum of the initial-state quarkonium of mass M_H. The state $|H(\mathbf{P})\rangle$ is the state (11) boosted by $-\mathbf{P}/M_H$. The Lorentz-boost transformations may be read from [17, 18].

CONCLUSIONS

We have discussed in an unified framework inclusive and electromagnetic decays, and radiative transitions of heavy quarkonium in a regime where the typical momentum transfer is of order Λ_{QCD}. Noteworthy, also in this situation suitable effective field theories may be constructed, systematic expansions exploited and exact results derived.

It seems rather unlikely that the non-perturbative matching, once completed at order $1/m^3$, will support the formulas traditionally and universally used so far to describe radiative transitions at relative order v^2 and derived from phenomenological assumptions [19, 20]. This may possibly shade some light, for instance, on the radiative transition data for the Υ system recently collected at CLEO [21], whose understanding is problematic in many phenomenological models.

ACKNOWLEDGMENTS

The author acknowledges the financial support obtained inside the Italian MIUR program "incentivazione alla mobilità di studiosi stranieri e italiani residenti all'estero" and the Marie Curie Reintegration Grant contract MERG-CT-2004-510967.

REFERENCES

1. N. Brambilla et al., arXiv:hep-ph/0412158.
2. N. Brambilla, A. Pineda, J. Soto and A. Vairo, arXiv:hep-ph/0410047.
3. W. E. Caswell and G. P. Lepage, Phys. Lett. B **167**, 437 (1986).
4. G. T. Bodwin, E. Braaten and G. P. Lepage, Phys. Rev. D **51**, 1125 (1995) [Erratum-ibid. D **55**, 5853 (1997)] [hep-ph/9407339].
5. A. Pineda and J. Soto, Nucl. Phys. Proc. Suppl. **64**, 428 (1998) [arXiv:hep-ph/9707481].
6. N. Brambilla, A. Pineda, J. Soto and A. Vairo, Nucl. Phys. B **566**, 275 (2000) [arXiv:hep-ph/9907240].
7. N. Brambilla, A. Pineda, J. Soto and A. Vairo, Phys. Rev. D **63**, 014023 (2001) [arXiv:hep-ph/0002250].
8. A. Pineda and A. Vairo, Phys. Rev. D **63**, 054007 (2001) [Erratum-ibid. D **64**, 039902 (2001)] [arXiv:hep-ph/0009145].
9. N. Brambilla, A. Pineda, J. Soto and A. Vairo, Phys. Lett. B **580**, 60 (2004) [arXiv:hep-ph/0307159].
10. N. Brambilla, Y. Jia and A. Vairo, IFUM-841-FT (in preparation).
11. A. Czarnecki and A. G. Grozin, Phys. Lett. B **405**, 142 (1997) [arXiv:hep-ph/9701415].
12. A. Vairo, Mod. Phys. Lett. A **19**, 253 (2004) [arXiv:hep-ph/0311303].
13. N. Brambilla, D. Eiras, A. Pineda, J. Soto and A. Vairo, Phys. Rev. Lett. **88**, 012003 (2002) [arXiv:hep-ph/0109130].
14. N. Brambilla, D. Eiras, A. Pineda, J. Soto and A. Vairo, Phys. Rev. D **67**, 034018 (2003) [arXiv:hep-ph/0208019].
15. A. Vairo, Nucl. Phys. Proc. Suppl. **115**, 166 (2003) [arXiv:hep-ph/0205128].
16. A. V. Manohar, Phys. Rev. D **56**, 230 (1997) [hep-ph/9701294].
17. N. Brambilla, D. Gromes and A. Vairo, Phys. Lett. B **576**, 314 (2003) [hep-ph/0306107].
18. N. Brambilla, D. Gromes and A. Vairo, Phys. Rev. D **64**, 076010 (2001) [arXiv:hep-ph/0104068].
19. H. Grotch and K. J. Sebastian, Phys. Rev. D **25**, 2944 (1982).
20. H. Grotch, D. A. Owen and K. J. Sebastian, Phys. Rev. D **30**, 1924 (1984).
21. M. Artuso et al. [CLEO Collaboration], Phys. Rev. Lett. **94**, 032001 (2005) [hep-ex/0411068].

Model-independent Study on Magnetic Dipole Transition in Heavy Quarkonium

Yu Jia

INFN and Dipartimento di Fisica dell' Universita di Milano, via Celoria 16, 20133 Milano, Italy

Abstract.
Some new results on $M1$ transition in heavy quarkonium from (potential) NRQCD are briefly reported. This model-independent approach not only permits a systematic and lucid way to investigate the relativistic corrections, it also clarifies some inconsistent treatment in previous potential model approach. The impact of our formalism on $J/\psi \to \eta_c \gamma$, $\Upsilon(\Upsilon') \to \eta_b \gamma$ and $h_c \to \chi_{c0} \gamma$ are also discussed.

Keywords: $M1$ transition, heavy quarkonium, potential NRQCD
PACS: 13.20.Gd, 14.40.Gx, 12.39.Pn

Radiative transitions in heavy quarkonium are of considerable experimental and theoretical interest [1]. On the theory side, it provides us with further insight on the dynamics of quarkonium in addition to the knowledge we have gleaned from the spectra of $c\bar{c}$ and $b\bar{b}$ families.

Being an old subject, radiative transitions have been extensively studied in phenomenological models, notably the potential model approach. It is certainly desirable to study them from a model-independent perspective [2]. In this talk I will report such a study based on the effective-field-theory (EFT) approach [3]. Since $M1$ transition is theoretically cleaner and more interesting than $E1$ transition, I will focus on the former case, in spite of the fact that the latter is observed more copiously in nature.

In the nonrelativistic limit, the formula for the M1 transition between two S-wave onia has a particularly simple form:

$$\Gamma[n\,^3S_1 \to n'\,^1S_0 + \gamma] = \frac{4\alpha_{\rm em} e_Q^2}{3m^2}(1+\kappa_Q)^2 k_\gamma^3 \left|\int dr\, r^2 R_{n'0} R_{n0}\right|^2, \qquad (1)$$

where e_Q and κ_Q are the electric charge and anomalous magnetic moment of the heavy quark, and $R_{nl}(r)$ is the radial Schrödinger wave functions. Since the leading dipole operator $c_F e e_Q/2m\, \sigma \cdot \mathbf{B}^{\rm em}$ (where $c_F = 1+\kappa_Q$) only flips the spin and doesn't act on the spatial degrees of freedom, orthogonality of radial wave functions guarantees that: in the *allowed* transition ($n = n'$), the overlap integral equals 1; in the *hindered* transition ($n \neq n'$), the overlap integral vanishes.

It is well known that (1) overpredicts the observed $J/\psi \to \eta_c \gamma$ transition rate by a factor of $2 \sim 3$, with a normal input of m_c and κ_c from perturbative one-loop matching. This clearly indicates large relativistic corrections to (1), as is usually confronted in charmonium system. It has also been proposed that nonperturbative fluctuations may generate a large negative κ_c, so the discrepancy can be reduced.

The $\mathcal{O}(v^2)$ corrections to (1) has been available for a long time from potential model [1, 2]:

$$\Gamma[nS \to n'S+\gamma] = \frac{1}{2J_n+1}\frac{4\alpha_{em}e_Q^2}{m^2}k_\gamma^3\left|\sum_i I_i\right|^2, \qquad (2)$$

where $2J_n+1$ counts polarizations of the parent onium, and

$$I_1 = \left\langle n'0 \left| (1+\kappa_Q)\left(1-\frac{k_\gamma^2 r^2}{24}\right) + (1+2\kappa_Q)\frac{k_\gamma}{4m} \right| n0 \right\rangle, \qquad (3)$$

$$I_2 = -\left\langle n'0 \left| (1+\kappa_Q)\frac{\mathbf{p}^2}{2m^2} + \frac{\mathbf{p}^2}{3m^2} \right| n0 \right\rangle,$$

$$I_3 = \left\langle n'0 \left| \frac{\kappa_Q}{6m} rV_0' \right| n0 \right\rangle,$$

$$I_4 = \pm\frac{4}{E_{n0}^{(0)}-E_{n'0}^{(0)}} \left\langle n'0 \left| (1+\kappa_Q)\frac{V_{ss}}{m^2} \right| n0 \right\rangle,$$

$$I_5 = \left\langle n' \left| -\frac{\eta}{m} rV_S \right| n \right\rangle,$$

where V_0 stands for the static potential, the "+/−" sign in I_4 is associated with $^3S_1 \to {}^1S_0$ and $^1S_0 \to {}^3S_1$, respectively. Notice that I_4 accounts for the first-order correction to the wave function due to spin-spin potential (other higher dimensional potentials cease to contribute in S-wave transition), thus is only present in hindered transition.

I_5 is a prediction specific to popular assumption in potential models, where one usually decomposes the confining potential into a Lorentz scalar and a vector part, $V_{\rm conf} = \eta V_S + (1-\eta)V_V$. This term constitutes the major uncertainty in the potential model predictions, where contradictory claims often appear in the literature [2].

Presence of a hierarchy of scales in quarkonium, $m, mv, mv^2, \Lambda_{QCD}$, makes the EFT approach an ideal tool to analyze this transition process. One first descends from QCD to NRQCD by integrating out hard modes ($\sim m$) [4], then descends from NRQCD to potential NRQCD (pNRQCD) by further integrating out soft modes ($\sim mv$) [5]. As a result, the inter-quark potentials appear as Wilson coefficients in pNRQCD, and the only dynamical degrees of freedom of pNRQCD are ultrasoft modes ($\sim mv^2$).

A primary task of the EFT approach is to validate/invalidate (2). The real strength of EFT is, however, that it can further answer the following questions that are beyond the scope of potential models:

- Is it possible that a large nonperturbative correction to κ_Q arises when one descends from NRQCD to pNRQCD?
- Is it possible to reproduce I_5 in pNRQCD? If so, how to interpret it?
- Potential model focus exclusively on the $Q\bar{Q}$ Fock-state (with an exception of coupled-channel effects which may not be relevant as long as one excludes those states close to the open-flavor threshold). pNRQCD allows one to include ultrasoft gluons as dynamical degrees of freedom. Therefore, one naturally asks for the

possibility of large nonperturbative contribution arising from higher-Fock states $|Q\bar{Q}g\rangle$. This effect is usually referred to as *color-octet effect*.

In fact, the answer to these three questions are all `negative` [3], on which we will elaborate shortly. It turns out that the pNRQCD formalism is able to justify (2) except I_5. In the so called *weak coupling regime* (when $mv \gg \Lambda_{\rm QCD}$), this formula is complete; in the so-called *strong coupling regime* (when $mv \sim \Lambda_{\rm QCD}$), however, (2) is incomplete and further terms are needed.

Before going on to explanations, it is useful to first sketch the derivation of $\mathcal{O}(v^2)$ corrections in the framework of pNRQCD. One obvious source is from the contribution of higher dimensional $M1$ operators. These operators can be identified most conveniently by promoting the color gauge group of NRQCD Lagrangian to a larger gauge group $SU_c(3) \times U_{\rm em}(1)$. Explicitly, the relevant dipole operators up to $\mathcal{O}(1/m^3)$ read [6]:

$$\begin{aligned}
\mathcal{L}_{\rm NR} &= \psi^\dagger \left(iD_0 + \frac{\mathbf{D}^2}{2m} \right) \psi + \frac{c_F e e_Q}{2m} \psi^\dagger \sigma \cdot \mathbf{B}^{\rm em} \psi + \frac{i c_S e e_Q}{8m^2} \psi^\dagger \sigma \cdot [\nabla \times, \mathbf{E}^{\rm em}] \psi \\
&+ \frac{c_{W_1} e e_Q}{8m^3} \psi^\dagger \{\nabla^2, \sigma \cdot \mathbf{B}^{\rm em}\} \psi - \frac{c_{W_2} e e_Q}{4m^3} \psi^\dagger \nabla_i \sigma \cdot \mathbf{B}^{\rm em} \nabla_i \psi \\
&- \frac{c_{p'p} e e_Q}{8m^3} \left[\nabla \psi^\dagger \cdot \sigma \mathbf{B}^{\rm em} \cdot \nabla \psi + \nabla \psi^\dagger \cdot \mathbf{B}^{\rm em} \sigma \cdot \nabla \psi \right] + (\psi \to i\sigma^2 \chi^*), \quad (4)
\end{aligned}$$

where $D_\mu \equiv \partial_\mu + igT^a A^a_\mu + iee_Q A^{\rm em}_\mu$. Various Wilson coefficients can be computed through perturbative matching at the hard scale. For example, at one loop accuracy, $\kappa_Q = \alpha_s/2\pi$, is about a few percent for charm and bottom. It is well known that some of the Wilson coefficients are related with each other because of reparameterization invariance (RPI), notably $c_S = 2c_F - 1$, $c_{W2} = c_{W_1} - 1$ and $c_{p'p} = c_F - 1$ [6]. When inherited into pNRQCD, one make replacement $\nabla \to i\mathbf{p}$ in these operators, and these RPI relations can be utilized to condense the expressions.

There are other sources of $\mathcal{O}(v^2)$ corrections. One interesting contribution, first pointed out by Grotch and Sebastian [2], is the Lorentz boost effect due to the final-state recoil. Since the wave function of a moving S-wave state has a nonvanishing overlap with a P-wave, spin-flipped state at rest, the $M1$ transition can be effectively realized by a "E1" transition from the parent to this small component. Some subtlety arises in this effect, namely the recoil correction depends on which "E1" operator one uses, *i.e.*, $2ee_Q/m\mathbf{p} \cdot \mathbf{A}^{\rm em}$ or $ee_Q \mathbf{r} \cdot \mathbf{E}^{\rm em}$, which are connected by a redefinition of pNRQCD field. A solution to this problem is as following. The matching from NRQCD generates a new $M1$ operator induced by the spin-orbit potential, with a coefficient proportional to $V_{LS}^{\rm CM} = -V_0'/8r$ (Gromes relation). The contribution of this operator also depends on the field redefinition. However, the sum of these two corrections is convention-independent, thus comprises a meaningful relativistic effect.

We are now in a position in discussing general matching of $M1$ operators from NRQCD to pNRQCD. The operators in (4) are just an example of trivial matching. In general, new Wilson coefficients depending on inter-quark separation, r, will arise.

Obviously the correction to c_F can be interpreted as the $\mathcal{O}(1/m^0)$ matching coefficient of the leading $M1$ operator. Dimensional considerations require that this correction is function of $\log(r)$. We emphasize in passing that it may be inappropriate to attribute

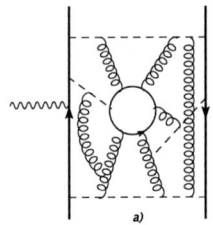 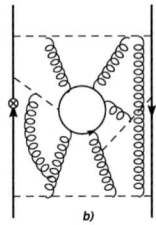

FIGURE 1. Typical NRQCD diagrams responsible for matching of the dipole operator at $\mathcal{O}(1/m^0)$. The left diagram represents the vertex correction, and the right one contributes to the wave-function renormalization factor $Z^s(r)$ due to soft modes, where the cross-cap implies insertion of a unit operator.

this new coefficient to the magnetic moment of an individual quark, since it really arises from entangled contributions from both quarks.

The simple answer, *no* corrections at all, is based on the heavy quark spin symmetry and that the $\sigma \cdot \mathbf{B}^{em}$ operator is a unit operator in spatial Hilbert space. These facts are independent of whether the matching is performed perturbatively as in weak coupling regime, or nonperturbatively, as in strong coupling regime. In a typical NRQCD diagram as shown in Fig. 1a), the external ultrasoft photon is attached to one of the quark lines, decorated by arbitrarily complicated loops made up of soft gluons and light quarks. Gluons attached to heavy quark lines must be longitudinal to avoid the $1/m$ suppression, and transverse gluons can only end on the internal gluons or light quark loops. For any vertex-correction diagram like Fig. 1a), there is one corresponding diagram in which the photon vertex is replaced by an insertion of unit operator, Fig. 1b), which contributes to the wave function renormalization $Z^s(r)$. Because all the propagators and vertices in Fig. 1 are spin-independent except the $M1$ vertex, and the $M1$ operator is a unit operator in coordinate (momentum) space, one finds Fig. 1a) exactly cancels with the lowest-order dipole operator multiplied by $Z^s(r)$ extracted from Fig. 1b), therefore there is no net contribution to the matching.

For the matching at $\mathcal{O}(1/m)$, a transverse gluon is allowed to end on quark line with a spin-independent $\mathbf{p} \cdot \mathbf{A}$ vertex (ending on a $\sigma \cdot \mathbf{B}$ vertex doesn't contribute). Repeating the argument as before, one doesn't obtain any new $M1$ operator besides the LS-potential induced one as discussed before, which comes from a different source. In particular there is no room for the operator of the form $V_S/m^2 \sigma \cdot \mathbf{B}^{em}$. Therefore one is forced to regard I_5 as an artifact from potential model, which cannot bear any physical significance.

However, the matching at $\mathcal{O}(1/m^2)$ become nontrivial, since several spin-dependent $\mathcal{O}(1/m)$ vertices will contribute and different topologies of diagrams will arise, our previous argument no longer applies. We will not dwell on the explicit expressions for these operators. However, it should be kept in mind that they can be neglected in the weak coupling regime because of suppressions by higher powers of $\alpha_s \sim v$. In the strong coupling regime, Unfortunately, since $\alpha_s \sim 1$, these operators might be as important as other $\mathcal{O}(v^2)$ corrections. These corrections involve some nonperturbative functions such as vacuum gluonic correlators, so that predictive power is unavoidably damaged.

The color-octet effect to radiative transitions was first envisaged by Voloshin long

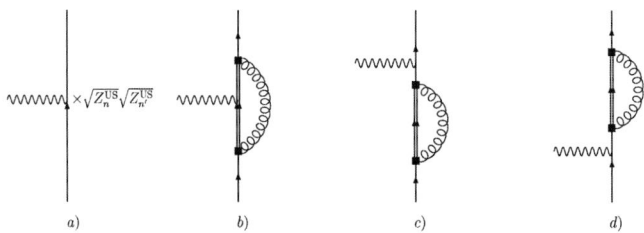

FIGURE 2. pNRQCD diagrams for color-octet contribution to radiative transition. Single and double lines correspond to the singlet and octet fields. The singlet-octet-gluon vertex is of chromo-E1 type.

time ago, but without a detailed study [7]. pNRQCD allows a systematic treatment of this effect. The corresponding diagrams are shown in Fig. 2. It is not surprising that the color-octet effect is nearly absent in $M1$ transition, because of the exactly same reason as before. It is interesting to note this bears some resemblance as absence of leading nonperturbative correction to $\bar{B} \to D^*$ form factor at the zero recoil, known as Luke's theorem [8]. Here we only give a heuristic argument based on the stationary perturbation theory in quantum mechanics. Treating chromo-E1 interaction as perturbation up to second order, one can schematically express a "true" quarkonium state as superpositions of the following Fock components (stripping off spin d.o.g.):

$$|N\rangle = \sqrt{Z_n^{\text{us}}}|Q\bar{Q}_1(n)\rangle + |Q\bar{Q}_8 g\rangle + \sum_{m \neq n}|Q\bar{Q}_1(m)\rangle \cdots, \qquad (5)$$

where $|Q\bar{Q}_1(n)\rangle$ is the unperturbed state, and Z_n^{us} is the wave function renormalization factor due to ultrasoft gluons. Spin-independence of chromo-E1 interaction, plus $M1$ operator being a unit operator in coordinate space, imply that the full $M1$ transition amplitude is nothing but the inner product $\langle N'|N\rangle$. Since orthogonality condition is preserved in perturbed states, it is equal to $\langle n'|n\rangle$, the color-octet contribution thus vanishes. We emphasize that sizable color-octet effect might arise in $E1$ transition, since the $E1$ operator is no longer a unit operator in coordinate space.

Finally, we turn to the phenomenological implication of (2), with the understanding that the unphysical I_5 term has been dropped. Thus far, the only observed $M1$ transitions are $J/\psi(\psi') \to \eta_c \gamma$, and upper bounds on $\Upsilon'(\Upsilon'') \to \eta_b \gamma$ have recently been set by CLEO [10]. To exploit this formula, however, one should first judge whether the problem at hand belongs to weak or strong coupling regime. (2) is a formula incorporating full $\mathcal{O}(v^2)$ corrections for the transition in weak coupling regime, but an incomplete one in the strong coupling regime, where $\mathcal{O}(1/m^2)$ nonperturbative matching from NRQCD sector is still missing.

Empirically, $\Upsilon(1S)$ and η_b are believed to lie in the weak coupling regime, whereas J/ψ, η_c fit in this regime to a less extent. Υ' and Υ'' are usually regarded as being in strong coupling regime, whereas ψ' and η_c' are too close to the open-flavor threshold and they cannot be correctly described by current formulation of pNRQCD, therefore we will exclude $\psi' \to \eta_c \gamma$ in our analysis.

To be concrete, let us first discuss the allowed transition $J/\psi \to \eta_c \gamma$. We assume that charmonium ground states are in the weak coupling regime, and their dynamics is largely

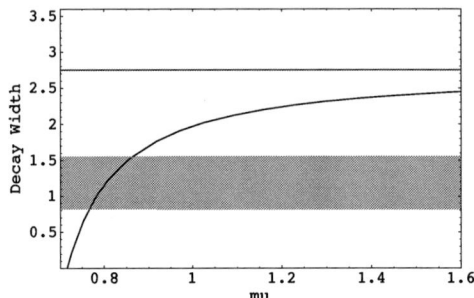

FIGURE 3. $\Gamma[J/\psi \to \eta_c \gamma]$ (in keV) vs. μ (in GeV). Dashed line is the prediction in NR limit, solid line includes $\mathcal{O}(v^2)$ corrections, and the band represents the measured width with error taken from [9].

governed by the perturbative potential, *i.e.*, $V_0 \approx -C_F \alpha(\mu)/r$, where the natural choice of μ is around the typical three-momentum scale. To make an unambiguous prediction, one needs also to specify the charm mass appearing in (2). Taking m_c to be naively half of J/ψ mass is somewhat sloppy, because this procedure may induce an error of order v^2. A more consistent way is to choose a threshold mass such as $1S$ mass, which is defined implicitly through $\hat{m} = m - \langle 10 | p^2/2m | 10 \rangle \approx m(1 - C_F^2 \alpha_s(\mu)^2/8)$, where $\hat{m}$ is half of the center-of-gravity ground state mass.

Many of the terms in (2) are practically negligible. Since κ_Q retains its NRQCD value and is only a few per cent for charm and bottom, we can simply put it to be zero in those $\mathcal{O}(v^2)$ terms (as a result, I_3 can be dropped in both allowed and hindered transitions). In fact, only I_2 in (2) accounts for genuine $\mathcal{O}(v^2)$ correction for allowed transition, and we end up with a simple expression

$$\Gamma[J/\psi \to \eta_c \gamma] \approx \frac{4\alpha_{em} e_c^2}{3\hat{m}_c^2} k_\gamma^3 \left(1 + 2\kappa_c - \frac{2C_F^2 \alpha_s^2(\mu)}{3}\right). \quad (6)$$

Fig. 3 shows a comparison between this formula and the data. It seems that (6) is compatible within the error to the data when μ is about 0.8 GeV. Note this value is consistent with the empirical mv value for J/ψ. Therefore, our reasonable success in describing $J/\psi \to \eta_c \gamma$ may be viewed as *a posteriori* support for the weak-coupling regime assignment. However, rather sharp μ dependence may suggest that J/ψ is in the border between weak and strong coupling regime. One can employ (6) to $\Upsilon \to \eta_b \gamma$ with more confidence, and finds a smaller $\mathcal{O}(v^2)$ correction with flatter scale dependence. Unfortunately, the narrow width of $2 \sim 3$ eV makes this transition unlikely to be observed.

As for hindered transitions, *e.g.* $\Upsilon'(\Upsilon'') \to \eta_b \gamma$, the experimental upper bounds are already rather tight and many model predictions have been ruled out [10]. Unfortunately, we would not expect (2) to reliably describe these transitions if, as usually assumed, excited bottomonia are truly in the strong coupling regime. Nevertheless, it may not be a too optimistic idea to treat, at least Υ', as being in weak coupling regime. Proceeding along this line, we find that the I_4 term is dominating over I_1 and I_2, and the latter two nearly cancel each other. As a result, the predicted width is about an order-of-magnitude larger than the experimental upper bound! Nevertheless, one should be alert there are lots of uncertainty associated with I_4. When higher-order corrections are included, one may wish this alarming discrepancy will decrease. In any rate, it seems to be fair to say that the weak coupling regime assignment of Υ' is problematic and a strong coupling regime analysis is awaited.

We end with a brief discuss of the P-wave $M1$ transition, which has received less attention in the past. It has now phenomenological interest since the recent discovery of h_c state [11]. It is straightforward to apply pNRQCD formalism to derive the $\mathcal{O}(v^2)$ corrections. For simplicity, we just quote the formula for the allowed transition:

$$\Gamma[n\,^3P_J \to n\,^1P_0 + \gamma] = \frac{4\alpha_{em}e_Q^2}{3m^2}k_\gamma^3 \left|1 + \kappa_Q - c_J \left\langle n1 \left| \frac{\mathbf{p}^2}{m^2} \right| n1 \right\rangle \right|^2, \quad (7)$$

where $c_J = 1/2, 1, 4/5$ for $J = 0, 1, 2$. For the $n\,^1P_1 \to n\,^3P_J\gamma$ transition, one simply multiplies the right side of (7) by a statistical factor of $(2J+1)/3$. Note this formula is incomplete since P-wave onia necessarily live in the strong coupling regime.

The fine splittings between h_c and χ_{c0}, χ_{c1} and χ_{c2} are about 110, -13, -32 MeV, respectively. It seems that only $h_c \to \chi_{c0}\gamma$ has a serious chance to be observed, with a width comparable to $\Gamma[J/\psi \to \eta_c\gamma]$. Future experimental input will enable us to infer the size of relativistic corrections, thus enriching our understanding of P-wave onia.

ACKNOWLEDGMENTS

I thank N. Brambilla and A. Vairo for collaboration on this work, also for their comments on the manuscript. This research is supported in part by Marie Curie ERG grant under contract MERG-CT-2004-510967 and by INFN.

REFERENCES

1. For a recent overview on radiative transitions in heavy quarkonia, see E. Eichten's review in Chap. 4 of CERN Yellow Report "Heavy quarkonium physics" [N. Brambilla *et al.*, hep-ph/0412158].
2. Here we give an incomplete list of references on $M1$ transition: G. Feinberg and J. Sucher, Phys. Rev. Lett. **35**, 1740 (1975); J. Sucher, Rept. Prog. Phys. **41**, 1781 (1978); J. S. Kang and J. Sucher, Phys. Rev. D **18**, 2698 (1978); R. McClary and N. Byers, *ibid*. **28**, 1692 (1983); V. Zambetakis and N. Byers, *ibid*. **28**, 2908 (1983); H. Grotch and K. J. Sebastian, *ibid*. **25**, 2944 (1982); H. Grotch, D. A. Owen and K. J. Sebastian, *ibid*. **30**, 1924 (1984); X. Zhang, K. J. Sebastian and H. Grotch, *ibid*. **44** (1991) 1606; T. A. Lahde, Nucl. Phys. A **714**, 183 (2003); D. Ebert, R. N. Faustov and V. O. Galkin, Phys. Rev. D **67**, 014027 (2003).
3. N. Brambilla, Y. Jia and A. Vairo, preprint IFUM-841-FT.
4. G. Bodwin, E. Braaten and G. Lepage, Phys. Rev. D **51**, 1125 (1995) [E-ibid. D **55**, 5853 (1997)].
5. A. Pineda and J. Soto, Nucl. Phys. Proc. Suppl. **64**, 428 (1998); N. Brambilla, A. Pineda, J. Soto and A. Vairo, Nucl. Phys. B **566**, 275 (2000). For a latest review on pNRQCD, see N. Brambilla, A. Pineda, J. Soto and A. Vairo, hep-ph/0410047.
6. A. V. Manohar, Phys. Rev. D **56**, 230 (1997); T. Kinoshita and M. Nio, Phys. Rev. D **53** (1996) 4909.
7. M. B. Voloshin, Nucl. Phys. B **154**, 365 (1979); Sov. J. Nucl. Phys. **36**, 143 (1982).
8. M. E. Luke, Phys. Lett. B **252**, 447 (1990).
9. S. Eidelman *et al.* [Particle Data Group Collaboration], Phys. Lett. B **592**, 1 (2004).
10. M. Artuso *et al.* [CLEO Collaboration], Phys. Rev. Lett. **94**, 032001 (2005).
11. J. L. Rosner *et al.* [CLEO Collaboration], Phys. Rev. Lett. **95** (2005) 102003.

Quark-antiquark bound state equation in the Wilson loop approach with minimal surfaces

F. Jugeau* and H. Sazdjian[†]

*Instituto de Fisica Corpuscular, IFIC,
Edificio Institutos de Investigacion, Apt. de Correus 22085, E-46071 València, Spain
E-mail: frederic.jugeau@ific.uv.es

[†]Institut de Physique Nucléaire, Groupe de Physique Théorique,
Université Paris XI, F-91406 Orsay Cedex, France
E-mail: sazdjian@ipno.in2p3.fr

Abstract.
The quark-antiquark gauge invariant Green function is studied through its dependence on Wilson loops. The latter are saturated, in the large-N_c limit and for large contours, by minimal surfaces. A covariant bound state equation is derived which in the center-of-mass frame and at equal-times takes the form of a Breit–Salpeter type equation. The large-distance interaction potentials reduce in the static case to a confining linear vector potential. In general, the interaction potentials involve contributions having the structure of flux tube like terms.

Keywords: QCD, Confinement, Wilson loop, Minimal surfaces, Bound states, Quarkonium.
PACS: 03.65.Pm, 11.10.St, 12.38.Aw, 12.38.Lg, 12.38.Ki.

The Wilson loop [1] is defined as the trace in color space of the path-ordered phase factor of the gluon field on a closed contour C:

$$\Phi(C) = \frac{1}{N_c} \text{tr}_c P e^{-ig \oint_C dx^\mu A_\mu(x)}. \tag{1}$$

Its vacuum expectation value, denoted $W(C)$,

$$W(C) = \langle \Phi(C) \rangle, \tag{2}$$

is a functional of the contour C. Loop equations were obtained and studied by Polyakov [2] and Makeenko and Migdal [3, 4, 5]. The Wilson loop essentially satisfies two types of equation, which are equivalent to the QCD equations of motion: The Bianchi identity and the loop equations (or Makeenko–Migdal equations). A third property, factorization, is obtained in the large-N_c limit [6] for two disjoint contours: $W(C_1, C_2) = W(C_1)W(C_2)$.

Further simplification is obtained in the large-N_c limit of the theory. In that limit, for large contours, i.e., at large distances, nonperturbative asymptotic solutions to the Wilson loops are represented by the minimal surfaces having as supports the loop contours [3, 7]. Therefore, if one is interested only in the large-distance properties of the theory, saturation of the Wilson loop averages by minimal surfaces provides a correct

---

[1] Talk presented by H. Sazdjian.

CP806, *QCD@Work 2005: International Workshop on Quantum Chromodynamics*
edited by P. Colangelo, F. De Fazio, E. Nappi, and G. Nardulli
© 2006 American Institute of Physics 0-7354-0302-3/06/$23.00

description of the theory in this regime. In that case, the Wilson loop average can be represented by the following functional of the contour C:

$$W(C) = e^{-i\sigma A(C)}, \qquad (3)$$

where σ is the string tension and $A(C)$ the minimal area with contour C.

Minimal surfaces also appear as natural solutions to the Wilson loop averages in two-dimensional gauge theories [8].

To deal with the quarkonium bound state problem, one starts with the two-particle gauge invariant Green function for quarks q_1 and q_2 with different flavors and with masses m_1 and m_2:

$$G(x_1,x_2;x_1',x_2') \equiv \langle \overline{\psi}_2(x_2) U(x_2,x_1) \psi_1(x_1) \overline{\psi}_1(x_1') U(x_1',x_2') \psi_2(x_2') \rangle_{A,q_1,q_2}. \qquad (4)$$

Here, $U(x_2,x_1)$ is the path-ordered phase factor,

$$U(x_2,x_1) = P e^{-ig \int_{x_1}^{x_2} dz^\mu A_\mu(z)}, \qquad (5)$$

taken along the straight-line $x_1 x_2$ (and similarly for $U(x_1',x_2')$). Integrating in the large-N_c limit with respect to the quark fields, one obtains:

$$G(x_1,x_2;x_1',x_2') = -\langle \text{tr}_c U(x_2,x_1) S_1(A;x_1,x_1') U(x_1',x_2') S_2(A;x_2',x_2) \rangle_A, \qquad (6)$$

where $S_1(A)$ and $S_2(A)$ are the quark and antiquark propagators in the presence of the external gluon field and tr_c designates the trace with respect to the color group. The quark propagator $S(A)$ satisfies the equation

$$\left(i\gamma.\partial_x - m - g\gamma.A(x) \right) S(A;x,x') = i\delta^4(x-x'). \qquad (7)$$

At this stage, the Green function G can schematically be represented as in Fig. 1.

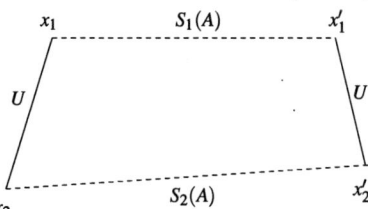

FIGURE 1. Schematic representation of the two-particle Green function.

In order to make the Wilson loop structure of G apparent, we adopt for the quark propagator in the external gluon field a representation based on an explicit use of the phase factor along straight lines [7]. Introducing the gauge covariant composite object $\widetilde{S}(A;x,x')$, made of a free fermion propagator $S_0(x-x')$ (without color group content) multiplied by the path-ordered phase factor $U(x,x')$ [Eq. (5)] taken along the straight segment $x'x$,

$$\widetilde{S}(A;x,x') \equiv S_0(x-x') U(x,x'), \qquad (8)$$

one shows that the quark propagator $S(A;x,x')$ in the external gluon field satisfies the following functional integral equation in terms of $\widetilde{S}$:

$$S(A;x,x') = \widetilde{S}(A;x,x') - \int d^4x'' S(A;x,x'') \gamma^\alpha \int_0^1 d\lambda\, (1-\lambda) \frac{\delta}{\delta x^\alpha(\lambda)} \widetilde{S}(A;x'',x'), \quad (9)$$

where the segment $x''x'$ has been parametrized with the parameter λ as $x(\lambda) = (1-\lambda)x'' + \lambda x'$ and where the operator $\delta/\delta x^\alpha(\lambda)$ acts on the factor U of $\widetilde{S}$, along the internal part of the segment $x''x'$, with x' held fixed. That equation is diagrammatically represented in Fig. 2.

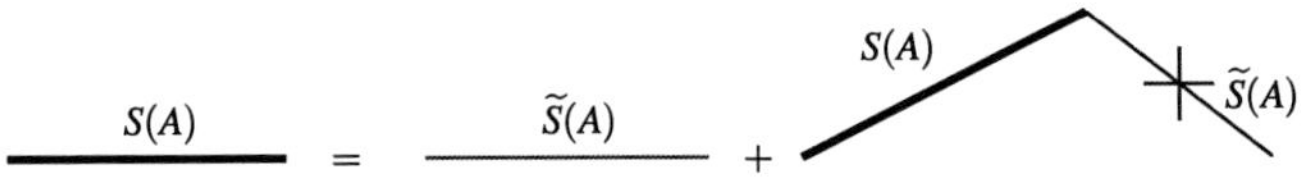

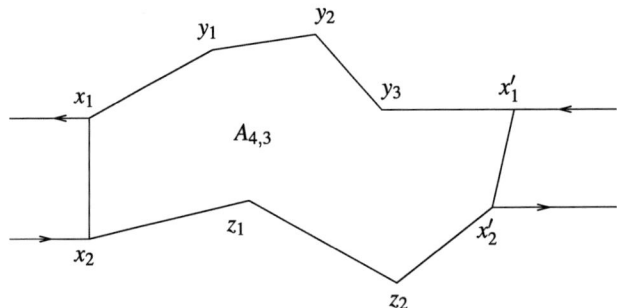

FIGURE 2. Diagrammatic representation of the integral equation satisfied by the quark propagator in the external gluon field. The cross represents the action of the functional derivative $\delta/\delta x(\lambda)$.

A similar equation in which the roles of x and x' are interchanged also holds. Those equations lead to iteration series for S in which the gauge covariance property is maintained at each order of the iteration.

Use of the above representations for the quark propagators in Eq. (6) leads for the two-particle Green function to a series expansion where each term contains a Wilson loop along a skew-polygon:

$$G = \sum_{i,j=1}^{\infty} G_{i,j}, \quad (10)$$

where $G_{i,j}$ represents the contribution of the term of the series having $(i-1)$ points of integration between x_1 and x'_1 (i segments) and $(j-1)$ points of integration between x_2 and x'_2 (j segments). We designate by $C_{i,j}$ the contour associated with the term $G_{i,j}$. A typical configuration for the contour of $G_{4,3}$ is represented in Fig. 3.

FIGURE 3. Contour $C_{4,3}$ associated with the term $G_{4,3}$. $A_{4,3}$ is the minimal surface with contour $C_{4,3}$.

Each segment of the quark lines supports a free quark propagator and except for the first segments (or the last ones, depending on the representation that is used) the Wilson loop is submitted to one functional derivative on each such segment. One then uses for the averages of the Wilson loops appearing in the above series the representation with minimal surfaces [Eq. (3)].

The Green function G satisfies the following equation with respect to the Dirac operator of particle 1 acting on x_1:

$$(i\gamma.\partial_{(x_1)} - m_1) G(x_1,x_2;x'_1,x'_2) = -i \langle \text{tr}_c U(x_2,x_1) \delta^4(x_1 - x'_1) U(x'_1,x'_2) S_2(x'_2,x_2) \rangle_A$$

$$-i\gamma^\alpha \langle \text{tr}_c \int_0^1 d\sigma (1-\sigma) \frac{\delta U(x_2,x_1)}{\delta x^\alpha(\sigma)} S_1(x_1,x'_1) U(x'_1,x'_2) S_2(x'_2,x_2) \rangle_A, \quad (11)$$

where the segment $x_1 x_2$ has been parametrized with the parameter σ as $x(\sigma) = (1-\sigma)x_1 + \sigma x_2$; furthermore, the operator $\delta/\delta x^\alpha$ does not act on the explicit boundary point x_1 of the segment, this contribution having been cancelled by the contribution of the gluon field A coming from the quark propagator S_1. A similar equation also holds with the Dirac operator of particle 2 acting on x_2. Representation (9) for the quark propagator can then be used in the above equation and its partner satisfied by the two-particle Green function G. One obtains two compatible equations for G where the right-hand sides involve the series of the terms $G_{i,j}$ of Eq. (10) and their functional derivative along the segment $x_1 x_2$. In order to obtain bound state equations, it is necessary to reconstruct in the right-hand sides the bound state poles contained in G [9]. In x-space, bound states are reached by taking the large separation time limit between the pair of points (x_1,x_2) and (x'_1,x'_2) [10]. To produce a bound state pole, it is necessary that there be a coherent sum of contributions coming from each $G_{i,j}$, since the latter, taken individually, do not have poles. Each $G_{i,j}$ involves a corresponding minimal surface $A_{i,j}$ on the contour of which act various functional derivatives. Those can be classified according to their possible irreducibility properties. Reducible contributions are those which are parts of the definition of the series of G. It does not seem possible to sum all these terms to reproduce exactly G with some kernel acting on it in the right-hand sides. However, for large separation time limits one can isolate terms that contribute to the pole terms. One notices that the derivative along the segment $x_1 x_2$ acts on areas $A_{i,j}$ with contour $C_{i,j}$ which are different from one term of the series to the other (the number of segments being different). To have a coherent sum of those contributions it is necessary to expand each such derivative term around the derivative of the lowest-order contour $C_{1,1}$, represented in Fig. 4.

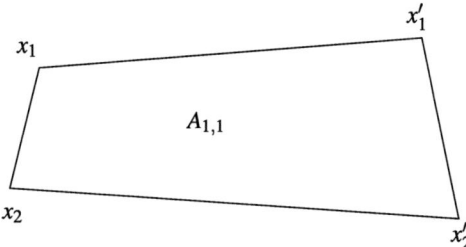

FIGURE 4. The lowest-order contour $C_{1,1}$ and its minimal surface $A_{1,1}$.

It is that term that can be factorized and can lead through the summation of the factored series to the reappearance of the Green function G and to its poles. The remaining terms do not lead to pole terms. Similarly, two derivative contributions should be expanded around the lowest-order contribution coming from the contours $C_{2,1}$ or $C_{1,2}$, and so forth.

In general, the derivative of the areas along $x_1 x_2$ depends among others on the slope of the areas in the orthogonal direction to $x_1 x_2$. One then associates that slope with the quark momenta. Taking then the large separation time limit and equal times in the center-of-mass frame, one ends up with a covariant three-dimensional equation, having the structure of a Breit–Salpeter type equation [11, 12] and where the interaction kernels or potentials are given by various functional derivatives involving at least one derivative along the segment $x_1 x_2$. Keeping for the potentials the terms containing one functional derivative of the area $A_{1,1}$ [Fig. 4], the equation takes the form [7]

$$\left[P_0 - (h_{10} + h_{20}) - \gamma_{10} \gamma_1^\mu A_{1\mu} - \gamma_{20} \gamma_2^\mu A_{2\mu} \right] \psi(\mathbf{x}) = 0, \tag{12}$$

where ψ is a 4×4 matrix wave function of the relative coordinate $x = x_2 - x_1$ considered at equal times, P_0 the center-of-mass total energy and h_{10} and h_{20} the quark and antiquark Dirac hamiltonians; the Dirac matrices of the quark (with index 1) act on ψ from the left, while the Dirac matrices of the antiquark (with index 2) act on ψ from the right. The potentials A_1 and A_2 are defined through the equations

$$A_{1\mu} = \sigma \int_0^1 d\sigma' (1 - \sigma') \frac{\delta A_{1,1}}{\delta x^\mu(\sigma')}, \quad A_{2\mu} = \sigma \int_0^1 d\sigma' \sigma' \frac{\delta A_{1,1}}{\delta x^\mu(\sigma')}, \tag{13}$$

$x(\sigma')$ belonging to the segment $x_1 x_2$.

The time components of A_1 and A_2 add up in the wave equation. For their sum, one has the expression (in the c.m. frame)

$$A_{10} + A_{20} = \sigma r \frac{E_1 E_2}{E_1 + E_2} \left\{ \left(\frac{E_1}{E_1 + E_2} \varepsilon(p_{10}) + \frac{E_2}{E_1 + E_2} \varepsilon(p_{20}) \right) \right.$$
$$\times \sqrt{\frac{r^2}{\mathbf{L}^2}} \left(\arcsin\left(\frac{1}{E_2} \sqrt{\frac{\mathbf{L}^2}{r^2}} \right) + \arcsin\left(\frac{1}{E_1} \sqrt{\frac{\mathbf{L}^2}{r^2}} \right) \right)$$
$$+ (\varepsilon(p_{10}) - \varepsilon(p_{20})) \left(\frac{E_1 E_2}{E_1 + E_2} \right) \left(\frac{r^2}{\mathbf{L}^2} \right) \left(\sqrt{1 - \frac{\mathbf{L}^2}{r^2 E_2^2}} - \sqrt{1 - \frac{\mathbf{L}^2}{r^2 E_1^2}} \right) \right\}. \tag{14}$$

Here, $r = \sqrt{\mathbf{x}^2}$, $E_a = \sqrt{m_a^2 + \mathbf{p}^2}$, $a = 1, 2$, with m_a the quark masses, $\mathbf{p}$ the c.m. momentum, $\mathbf{p} = (\mathbf{p}_2 - \mathbf{p}_1)/2$, $\mathbf{L}$ the c.m. orbital angular momentum, and $\varepsilon(p_{10})$ and $\varepsilon(p_{20})$ the energy sign operators of the free quark and the antiquark, respectively:

$$\varepsilon(p_{a0}) = \frac{h_{a0}}{E_a}, \quad a = 1, 2. \tag{15}$$

The space components of A_1 and A_2 are orthogonal to $\mathbf{x}$. The expression of $\mathbf{A}_1$ is (in the c.m. frame):

$$\mathbf{A}_1 = -\sigma r \frac{E_1 E_2}{E_1 + E_2} \left\{ \frac{r^2}{2\mathbf{L}^2} \frac{E_1 E_2}{E_1 + E_2} \mathbf{p}^t \right.$$

$$\times \sqrt{\frac{r^2}{\mathbf{L}^2}} \left(\arcsin\left(\frac{1}{E_2}\sqrt{\frac{\mathbf{L}^2}{r^2}}\right) + \arcsin\left(\frac{1}{E_1}\sqrt{\frac{\mathbf{L}^2}{r^2}}\right) \right)$$

$$+ \frac{1}{E_2}\mathbf{p}^t\left(\frac{E_1 E_2}{E_1+E_2}\right)\left(\frac{r^2}{\mathbf{L}^2}\right)\left(\sqrt{1-\frac{\mathbf{L}^2}{r^2 E_2^2}} - \sqrt{1-\frac{\mathbf{L}^2}{r^2 E_1^2}}\right)$$

$$-\frac{1}{2}\mathbf{p}^t\left(\frac{r^2}{\mathbf{L}^2}\right)\left(\frac{E_1}{E_1+E_2}\sqrt{1-\frac{\mathbf{L}^2}{r^2 E_2^2}} + \frac{E_2}{E_1+E_2}\sqrt{1-\frac{\mathbf{L}^2}{r^2 E_1^2}}\right)\Bigg\}. \quad (16)$$

Here, $\mathbf{p}^t$ is the transverse part of $\mathbf{p}$ with respect to $\mathbf{x}$:

$$\mathbf{p}^t = \mathbf{p} - \mathbf{x}\frac{1}{\mathbf{x}^2}\mathbf{x}\cdot\mathbf{p}. \quad (17)$$

The expression of $\mathbf{A}_2$ is obtained from that of $\mathbf{A}_1$ by an interchange in the latter of the indices 1 and 2 and a change of sign of $\mathbf{p}^t$.

For sectors of quantum numbers where $\mathbf{L}^2 = 0$, the expressions of the potentials become:

$$A_{10} + A_{20} = \frac{1}{2}(\varepsilon(p_{10}) + \varepsilon(p_{20}))\sigma r, \quad (18)$$

$$\mathbf{A}_1 = -\frac{1}{E_1 E_2}\left(\frac{1}{3}(E_1+E_2) - \frac{1}{2}E_1\right)\mathbf{p}^t\sigma r,$$

$$\mathbf{A}_2 = +\frac{1}{E_1 E_2}\left(\frac{1}{3}(E_1+E_2) - \frac{1}{2}E_2\right)\mathbf{p}^t\sigma r. \quad (19)$$

The potentials are generally momentum dependent operators and necessitate an appropriate ordering of terms.

From the structure of the wave equation (12) and the expressions of the potentials, one deduces that the interaction is confining and of the vector type. However, compared to the conventional timelike vector potential, it has additional pieces of terms contributing to the orbital angular momentum dependent parts. A closer analysis of those terms shows that they can be interpreted as being originated from the moments of inertia of the segment $x_1 x_2$ carrying a constant linear energy density equal to the string tension. The interaction potentials are therefore provided by the energy-momentum vector of the segment joining the quark to the antiquark, in similarity with the color flux tube picture of confinement. An analogous equation had also been proposed by Olsson et al. on the basis of a model where the quarks are attached at the ends of a straight string or a color flux tube [13, 14]. A similar conclusion had also been reached by Brambilla, Prosperi et al. on the basis of the analysis of the relativistic corrections to the nonrelativistic limit of the Wilson loop [15, 16, 17].

For heavy quarks, one can expand equation (12) around the nonrelativistic limit and obtain the hamiltonian to order $1/c^2$ [7].

The relativistic corrections to the interquark potential arising from the Wilson loop were analyzed and evaluated in the literature by Eichten and Feinberg [18], Gromes [19], Brambilla, Prosperi et al. [15, 16, 17], Brambilla, Pineda, Soto and Vairo [20].

The Wilson loop approach was also used for the study of quarkonium systems by Dosch, Simonov et al. with the use of the stochastic vacuum model [21].

In conclusion, the saturation of the Wilson loop averages in the large-N_c limit by minimal surfaces provides a systematic tool for investigating the large-distance dynamics of quark-antiquark bound state systems.

ACKNOWLEDGMENTS

Institut de Physique Nucléaire of Orsay is Unité Mixte de Recherche 8608. This work was supported in part for H.S. by the EU RTN network EURIDICE under contract No. CT2002-0311.

REFERENCES

1. K. G. Wilson, *Phys. Rev. D* **10**, 2445 (1974).
2. A. M. Polyakov, *Nucl. Phys.* **B164**, 171 (1979).
3. Yu. M. Makeenko and A. A. Migdal, *Phys. Lett.* **88B**, 135 (1979); *ibid.* **97B**, 253 (1980); *Nucl. Phys.* **B188**, 269 (1981).
4. A. A. Migdal, *Phys. Rep.* **102**, 199 (1983).
5. Yu. Makeenko, *Large N gauge theories*, hep-th/0001047.
6. G. 't Hooft, *Nucl. Phys.* **B72**, 461 (1974).
7. F. Jugeau and H. Sazdjian, *Nucl. Phys. B* **670**, 221 (2003).
8. V. A. Kazakov and I. K. Kostov, *Nucl. Phys.* **B176**, 199 (1980); V. A. Kazakov, *Nucl. Phys.* **B179**, 283 (1981).
9. E. E. Salpeter and H. A. Bethe, *Phys. Rev.* **84**, 1232 (1951).
10. M. Gell-Mann and F. Low, *Phys. Rev.* **84**, 350 (1951).
11. G. Breit, *Phys. Rev.* **34**, 553 (1929); *ibid.* **36**, 383 (1930); *ibid.* **39**, 616 (1932).
12. E. E. Salpeter, *Phys. Rev.* **87**, 328 (1952).
13. M. G. Olsson and K. Williams, *Phys. Rev. D* **48**, 417 (1993).
14. D. LaCourse and M. G. Olsson, *Phys. Rev. D* **39**, 2571 (1989); C. Olson, M. G. Olsson and K. Williams, *Phys. Rev. D* **45**, 4307 (1992); C. Olson and M. G. Olsson, *Phys. Rev. D* **49**, 4675 (1994).
15. A. Barchielli, E. Montaldi and G. M. Prosperi, *Nucl. Phys.* **B296**, 625 (1988); *ibid.* **B303**, 752 (1988) (E); A. Barchielli, N. Brambilla and G. M. Prosperi, *Nuovo Cimento* **103 A**, 59 (1990).
16. N. Brambilla, P. Consoli and G. M. Prosperi, *Phys. Rev. D* **50**, 5878 (1994); N. Brambilla, E. Montaldi and G. M. Prosperi, *Phys. Rev. D* **54**, 3506 (1996).
17. M. Baldicchi and G. M. Prosperi, *Phys. Lett. B* **436**, 145 (1998); *Phys. Rev. D* **62**, 114024 (2000); *ibid.* **66**, 074008 (2002).
18. E. Eichten and F. Feinberg, *Phys. Rev. D* **23**, 2724 (1981).
19. D. Gromes, *Z. Phys. C* **22**, 265 (1984); *ibid.* **26**, 401 (1984).
20. A. Pineda and A. Vairo, *Phys. Rev. D* **63**, 054007 (2001); *ibid.* **64**, 039902 (2001) (E); N. Brambilla, A. Pineda, J. Soto and A. Vairo, *Phys. Rev. D* **63**, 014023 (2001).
21. H. G. Dosch, *Phys. Lett. B* **190**, 177 (1987); Yu. A Simonov, *Nucl. Phys.* **B307**, 512 (1988); H. G. Dosch and Yu. A. Simonov, *Phys. Lett. B* **205**, 339 (1988); A. Di Giacomo, H. G. Dosch, V. I. Shevchenko and Yu. A. Simonov, *Phys. Rep.* **372**, 319 (2002).

QCD at High Temperature : Results from Lattice Simulations with an Imaginary μ

Massimo D'Elia*, Francesco Di Renzo† and Maria Paola Lombardo**

*University and I.N.F.N., Genova
†University and I.N.F.N., Parma
**I.N.F.N., L.N.F.

Abstract.
We summarize our results on the phase diagram of QCD in the temperature, imaginary chemical potential plane with emphasis on the high temperature regime. For $T \geq 1.5 T_c$ the results are compatible with a free field behavior, while for $T \simeq 1.1 T_c$ this is not the case, suggesting that the system remains strong interacting in this region.

Keywords: Field theory thermodynamics; QCD; Critical Phenomena, Lattice Field Theory
PACS: PACS Nos.: 12.38.Aw 12.38.Gc 12.28.Mh. 11.15.Ha 11.10.Wx

INTRODUCTION

The historical developments of the phase diagram of QCD is characterized by an increasing complication: according to early views based on a straightforward application of asymptotic freedom , the phase diagram was sharply divided into an hadronic phase and a quark gluon plasma phase. In the late 90's it was appreciated that the high density region is much more complicated than previously thought [1]. In the last couple of years, it was the turn of the region above T_c to become more rich: the survival of bound states above the phase transition brought up the idea of a more complicated, highly non–perturbative phase, whose precise nature has not been clarified yet[2]. The properties of the high temperature phase are especially interesting in view of the ongoing ultrarelativistic heavy ions collisions experiments with at RHIC, which explore temperatures close to T_c, and of the future experiments at LHC, which will approach the perturbative, free gas limit of QCD.

Up to which extent the properties of the matter produced in these ultrarelativistic heavy ion collisions can be predicted by the basic theory of strong interactions, Quantum Chromo Dynamics? In this note we review our results on this point,

QuantumChromoDynamics (QCD) basic degrees of freedom are quarks and gluons. The Lagrangian built by use of these fundamental fields enjoys local and (approximate) global symmetries. The realization of the global chiral symmetry depends on the thermodynamic conditions of the system : it is spontaneously broken, with the accompanying phenomena of a Goldstone mode and a mass gap, in ordinary conditions, and it gets restored at high temperature. At the same time, confinement, which is realized in the normal phase, disappears at high temperature: all in all, in our low temperature world quarks are confined within hadrons, there is one light preudoscalar meson, the Goldstone boson, the pion, and there are massive mesons and baryons. In the high temperature phase - the Quark Gluon Plasma - quarks and gluons are no longer confined, and the mass spectrum reflects the symmetries of the Lagrangian.

In a standard nuclear physics approach these features of the two phases are imposed by fiat, and the phase transition is obtained by equating the free energies of the quark-gluon gas on one side, and the hadron gas on the other side of the transition. At a variance with this, an approach based on QCD derives the different degrees of freedom of the two phases, as well as the phase transition line, from the same Lagrangian. These calculations, being completely non–perturbative, require a specific technique, Lattice QCD.

LATTICE QCD THERMODYNAMICS

Without entering into the details of this approach [3], let me just remind that the QCD equations are put on a 'grid' which should be fine enough to resolve details, and large enough to accommodate hadrons within: obviously, this would call for grids with a large number of points. On the other hand, the calculations complexity grows fast with the number of nodes in the grid, and the actual choices rely on a compromise between physics requirements and computer capabilities[11].

In practical numerical approaches, the lattice discretization is combined with a statistical techniques for the computation of the physical observables. This requires a positive 'measure', which is given by the exponential of the Action. A notorious problem plagues these calculations at finite baryon density, as the Action itself becomes complex, with a non–positive definite real part: because of this, for many years QCD at nonzero baryon density was not progressing at all.

Luckily, in the last four years a few lattice techniques – imaginary chemical potential, Taylor expansion, multiparameter reweighting – proven successful for $\mu_B/T < 1$. [4, 5, 6, 7, 8, 9, 10]. It has to be stressed, however, that these techniques are just dodges and workaround, and do not provide a real solution to the 'sign problem'. Moreover, due to the computer limitations sketched above, which we hope will be soon overcome by the next generation of supercomputers [11], the results have not yet reached the continuum, infinite volume limit.

While waiting for final results in the scaling limit and with physical values of the parameters, it is very useful to contrast and compare current lattice results with model calculations and perturbative studies. The imaginary chemical potential approach[13, 14, 8, 4, 5, 15]to QCD thermodynamics seems to be ideally suited for the interpretation and comparison with analytic results. Results from an imaginary μ have been obtained for the critical line of the two, three and two plus one flavor model [8], as well as for four flavor [4]. Thermodynamics results – order parameter, pressure, number density – were obtained for the four flavor model [5], and are extended in this note, where we concentrate on the region $T > T_c$.

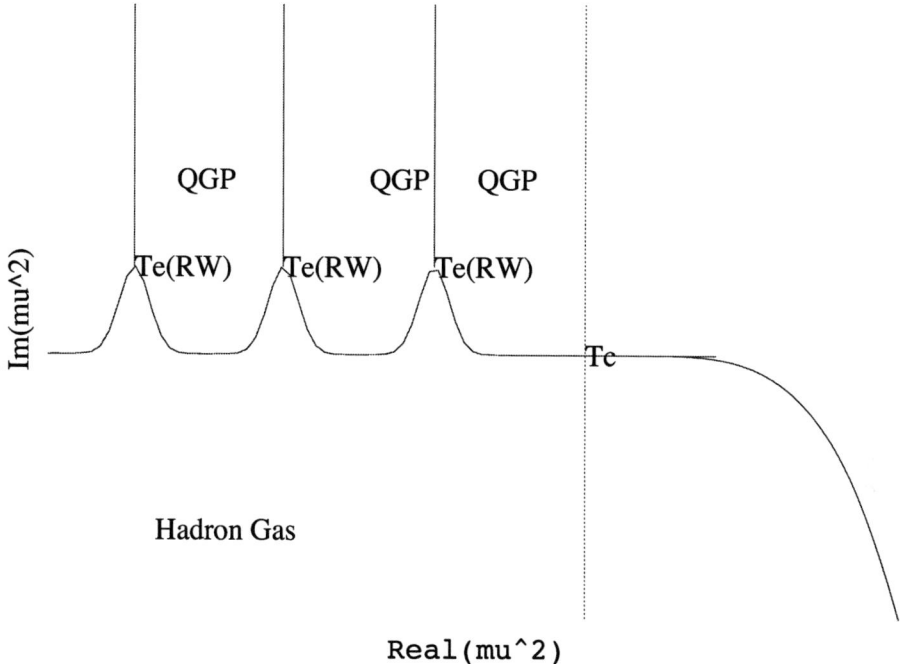

FIGURE 1. Sketch of the phase diagram in the μ^2, T plane: the solid line is the chiral transition, the dashed line is the Roberge Weiss transition. Simulations can be carried out at $\mu^2 \leq 0$ and results continued to the physical domain $\mu^2 \geq 0$. The derivative and reweighting methods have been used so far to extract informations from simulations performed at $\mu = 0$. The imaginary chemical potential approaches uses results on the left hand half plane. Different methods could be combined to improve the overall performance.

IMAGINARY CHEMICAL POTENTIAL

The imaginary chemical potential method uses information from all of the negative μ^2 half plane (Fig. 1) to explore the positive, physical relevant region.

The main physical idea behind any practical application is that at $\mu = 0$ fluctuations allow the exploration of $N_b \neq 0$ hence tell us about $\mu \neq 0$. Mutatis mutandis, this is the same condition for the reweighting methods to be effective: the physics of the simulation ensemble has to overlap with that of the target ensemble.

A practical way to use the results obtained at negative μ^2 relies on their analytical continuation in the real plane. For this to be effective[13] $\mathscr{L}(\mu, T)$ must be analytical, nontrivial, and fulfilling this rule of thumb:

$$\chi(T, \mu) = \partial \rho(\mu, T)/\partial \mu = \partial^2 logZ(\mu, T)/\partial \mu^2 > 0 \qquad (1)$$

This approach has been tested in the strong coupling limit [13] of QCD, in the dimensionally reduced model of high temperature QCD [14] and, more recently, in the two color model [15].

THE HOT PHASE AND THE APPROACH TO A FREE GAS

At high temperature, in the weak coupling regime, finite temperature perturbation theory might serve as a guidance, suggesting that the first few terms of the Taylor expansion might be adequate in a wider range of chemical potentials. So, at a variance with the expansion in the hadronic phase, where the natural parametrization is given by a Fourier analysis[4, 5], in this phase the natural parametrization for the grand partition function is a polynomial.

The leading order result for the pressure $P(T, \mu)$ in the massless limit is easily computed, given that at zero coupling the massless theory reduces to a non–interacting gas of quarks and gluons, yielding for the pressure

$$p(T,\mu) = \frac{\pi^2}{45}T^4\left(8+7N_c\frac{n_f}{4}\right) + \frac{n_f}{2}\mu^2 T^2 + \frac{n_f}{4\pi^2}\mu^4. \tag{2}$$

Obviously, when analytically continued to the negative μ^2 side, this gives

$$p(T,\mu_I) = \frac{\pi^2}{45}T^4\left(8+7N_c\frac{n_f}{4}\right) - \frac{n_f}{2}\mu_I^2 T^2 + \frac{n_f}{4\pi^2}\mu_I^4. \tag{3}$$

Because of the Roberge Weiss [12] periodicity this polynomial behavior should be cut at the Roberge Weiss transition $\mu_I = \pi T/3$: this is consistent with the Roberge Weiss critical line being strongly first order at high temperature. We discuss first the results of the fits of the number density to polynomial form; then we contrast these results with a free field behavior.

The considerations above suggests a natural ansatz for the behavior of the number density in this phase as a simple polynomial with only odd powers. We performed then fits to

$$n(T,\mu_I) = a(T)\mu_I - b(T)\mu_I^3 \tag{4}$$

whose obvious analytic continuation is

$$n(T,\mu) = a(T)\mu + b(T)\mu^3. \tag{5}$$

Note again that $a(T) = \chi_q(T, \mu = 0)$.

In ref. [5] we contrasted the results for the particle number at $T = 1.5T_c$, $T = 2.5T_c$, $T = 3.5T_c$ with a free field behaviour.

Some deviations are apparent, whose origin we would like to understand. It would be however arduous, given the strong lattice artifacts, to try to make contact with a rigorous perturbative analysis carried out in the continuum [19, 20, 21]. Rather then attempting that, we parametrize the deviation from a free field behavior as [16, 18]

$$\Delta P(T,\mu) = f(T,\mu)P^L_{free}(T,\mu) \tag{6}$$

where $P^L_{free}(T,\mu)$ is the lattice free result for the pressure. For instance, in the discussion of Ref. [18]

$$f(T,\mu) = 2(1 - 2\alpha_s/\pi) \tag{7}$$

and the crucial point was that α_s is μ dependent.

We can search for such a non trivial prefactor $f(T,\mu)$ by taking the ratio between the numerical data and the lattice free field result $n^L_{free}(\mu_I)$ at imaginary chemical potential:

$$R(T,\mu_I) = \frac{n(T,\mu_I)}{n^L_{free}(\mu_I)} \tag{8}$$

A non-trivial (i.e. not a constant) $R(T,\mu_I)$ would indicate a non-trivial $f(T,\mu)$.

We found that $R(T,\mu_i)$ is constant within errors, so that our data do not permit to distinguish a non trivial factor within the error bars: rather, the results for $T \geq 1.5T_c$ seem consistent with a free lattice gas, with an fixed effective number of flavors $N^{eff}_f(T)/4 = R(T)$: $N^{eff}_f = 0.92 \times 4$ for $T = 3.5T_c$, and $N^{eff}_f = 0.89 \times 4$ for $T = 1.5T_c$.

One last remark concerns the mass dependence of the results, which, as in the broken phase, can be computed from the derivative of the chiral condensate. In the chiral limit this gives $\frac{\partial n}{\partial m} = 0$, since the chiral condensate is identically zero. We have verified that $\frac{\partial n}{\partial m}$ remains very small compared to n itself: in a nutshell, in the quark gluon plasma phase $<\bar{\psi}\psi>$ is very small (zero in the chiral limit), while the number density grows larger, and this implies that the mass sensitivity is greatly reduced with respect to that in the broken phase.

The discussions presented above bring very naturally to the consideration of a dynamical region which is comprised between the deconfinement transition, and the endpoint of the Roberge Weiss transition.

In this dynamical region the analytic continuation is valid till $\mu = \infty$ along the real axis, since there are no singularities for real values of the chemical potential. The interval accessible to the simulations at imaginary μ is small, as simulations in this area hits the chiral critical line for $\mu^2 < 0$.

This region is of special interest and it is here that we are concentrating our efforts: In Fig. 2 we show our new results obtained at $T/T_c = 1.095$, indicating a non trivial deviation from a free field behaviour.

Let us make some general consideration about the thermodynamic behavior in this region by considering the critical line at imaginary chemical potential. Let us consider first the case of a second order transition: the analytic continuation of the polynomial predicted by perturbation theory for positive μ^2 would hardly reproduce the correct critical behavior at the second order phase transition for $\mu^2 < 0$. In fact, for a second order chiral transition at negative μ^2, $\Delta P(T,\mu^2) \propto (\mu^2 - \mu^2_c)^\chi$, where χ is a generic exponent. As the window between the critical line and the $\mu = 0$ axis is anyway small, such behavior - possibly with subcritical corrections - should persist in the proximity of the real axis. For generic values of the exponent a second order chiral transition seems incompatible with a free field behavior. The same discussion can be repeated for a first order transition of finite strength, by trading the critical point μ_c with the spinodal point μ^*. So deviations from free field are to be expected in this intermediate regime.

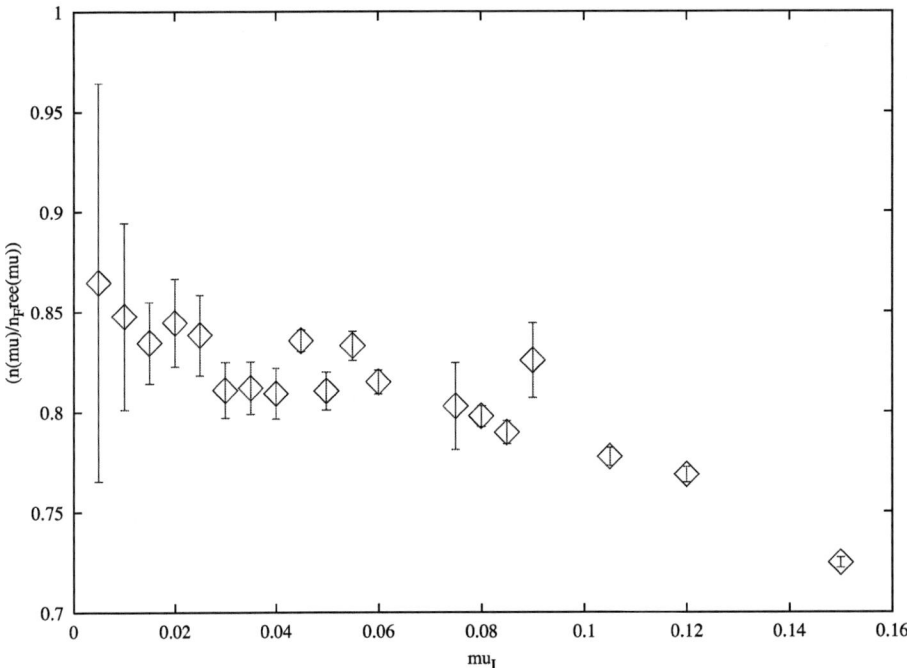

FIGURE 2. The ratio of the lattice results to the free fields results $R(T,\mu_I)$ at $T \simeq 1.1T_c$: an horizontal line would indicate a nearly–free behaviour which is clearly incompatible with the data. At this temperature the chiral transition takes place at imaginary chemical potential, i.e. at a negative μ^2 (Fig.1)

A more detailed discussion of these results, and their interrelation (or lack thereof) with a strongly interactive quark gluon plasma will be given elsewhere [22].

FUTURE DIRECTIONS

The approach to a free gas of quarks and gluons is a fascinating subject: we are moving from a world of colorless hadrons to a world of colored particles - quarks, gluons, and perhaps many more.

Three different, independent methods which afford a quantitative approach to this problem have been proposed and exploited in the past few years, producing several interesting and coherent results. In particular we have focussed on the properties of the hot phase right above the critical temperature, where we have observed a clear non–perturbative behaviour for the thermodyamical observables, showing that the system is still very far from a free gas of quark gluons.

These results, however, still need improvement: in particular, small quark masses, and a controlled approach to the continuum limit. All this requires a large amount of computer resources, and there is hope that the new dedicated supercomputers for lattice

QCD - QCDOC and apeNEXT - will produce significant advances in this field.

High quality numerical results togheter with a careful consideration of phenomenological models and critical behaviour in the negative μ^2 half–plane should produce a coherent and complete description of the high temperature phase of the strong interactions, which will hopefully confront soon ongoing and future experiments.

ACKNOWLEDGMENTS

The new calculations reported here were performed on the APEmille computers of the MI11 *Iniziativa Specifica*, and we wish to thank our colleagues in Milano and in Parma for their kind help.

REFERENCES

1. M. Alford, this volume.
2. E. V. Shuryak and I. Zahed, Phys. Rev. C **70**, 021901 (2004); F. Karsch, S. Ejiri and K. Redlich, arXiv:hep-ph/0510126.
3. see e.g. J. Smit, Cambridge Lect. Notes Phys. **15** (2002) 1.
4. M. D'Elia and M. P. Lombardo, Phys. Rev. D **67** (2003) 014505.
5. M. D'Elia and M. P. Lombardo, Phys. Rev. D **70** (2004) 074509.
6. Z. Fodor and S. D. Katz, Phys. Lett. B **534** (2002) 84; JHEP **0404** (2004) 50;
   Z. Fodor, S. D. Katz and K. K. Szabo, Phys. Lett. B **568** (2003) 73;
   F. Csikor *et al.* JHEP **0405** (2004) 046;
   S. D. Katz, Nucl. Phys. Proc. Suppl. **129** (2004) 60.
7. Ph. de Forcrand *et al.*, Nucl. Phys. Proc. Suppl. **119** (2003) 541.
8. Ph. de Forcrand and O. Philipsen, Nucl. Phys. B **642** (2002) 290;
   Nucl. Phys. B **673** (2003) 170.
9. C. R. Allton *et al.*, Phys. Rev. D **66** (2002) 067801;
   Phys. Rev. D **68**, (2003) 014081.
10. R. Gavai, S. Gupta and R. Roy, Prog. Theor. Phys. Suppl.**153** (2004) 270.
11. T. Wettig, talk at Lattice2005, Dublin, to appear in the Proceedings.
12. A. Roberge and N. Weiss, Nucl. Phys. B **275**, 734 (1986).
13. M. P. Lombardo, Nucl. Phys. Proc. Suppl. **83**, 375 (2000).
14. A. Hart, M. Laine and O. Philipsen, Phys. Lett. B **505**, 141 (2001).
15. P. Giudice and A. Papa, Phys. Rev. D **69**, 094509 (2004).
16. K. K. Szabo and A. I. Toth, JHEP **0306**, 008 (2003).
17. F. Csikor, G. I. Egri, Z. Fodor, S. D. Katz, K. K. Szabo and A. I. Toth, Prog. Theor. Phys. Suppl. **153**, 93 (2004).
18. J. Letessier and J. Rafelski, Phys. Rev. C **67**, 031902 (2003).
19. A. Vuorinen, arXiv:hep-ph/0402242.
20. A. Vuorinen, Phys. Rev. D **68**, 054017 (2003).
21. A. Ipp, A. Rebhan and A. Vuorinen, Phys. Rev. D **69**, 077901 (2004).
22. M. D'Elia, F. Di Renzo and M.P. Lombardo, work in progress.

Jet quenching: RHIC results and phenomenology

David d'Enterria

Nevis Laboratories, Columbia University
Irvington, NY 10533, and New York, NY 10027, USA

Abstract. I review the main experimental results on jet physics in high-energy nucleus-nucleus collisions as studied via inclusive leading hadron spectra and di-hadron correlations at high transverse momentum. In central Au+Au at RHIC ($\sqrt{s_{NN}}$ = 200 GeV), the observed large suppression of high-p_T hadron spectra as well as the strongly modified azimuthal dijet correlations compared to baseline p+p results in free space, provide crucial information on the thermodynamical and transport properties of QCD matter.

Keywords: Jet quenching, relativistic nucleus-nucleus, QGP, QCD
PACS: 12.38.Mh,13.87.Fh,24.85.+p,25.75.-q

INTRODUCTION

The research program of high-energy heavy-ion physics is centered on the study of the collective properties of extended quark-gluon systems. By colliding two heavy nuclei at relativistic energies one expects to form a hot and dense deconfined medium whose collective (color) dynamics can be quantitatively described by QCD thermodynamics calculations on the lattice [1]. In this context, the main goal of the RHIC experiments is the production and study under laboratory conditions of the Quark Gluon Plasma (QGP) predicted to be formed when strongly interacting matter attains energy densities above $\varepsilon \approx 1$ GeV/fm^3. The production of such an extremely hot and dense partonic system should manifest itself in a variety of experimental signatures [2]. One of the first proposed QGP "smoking guns" was "jet quenching" [3] i.e. the disappearance of the collimated spray of hadrons resulting from the fragmentation of a hard scattered parton due to its "absorption" in the dense medium produced in the reaction. Extensive theoretical work on high-energy parton propagation in a QCD environment [4, 5, 6, 7] has shown that the main mechanism of attenuation is of radiative nature: the traversing parton loses energy mainly by multiple gluon emission ("gluonstrahlung"). Such a medium-induced *non-Abelian* energy loss should result in several observable experimental consequences:

(i) **depleted** production of **high p_T leading hadrons** (dN/dp_T) [4],

(ii) **unbalanced** back-to-back **di-jet azimuthal correlations** $(dN_{pair}/d\phi)$ [8], and

(iii) **modified energy flow** and **particle multiplicity** within the final jets [9, 10].

By quantitatively comparing the jet structure modifications in A+A relative to baseline p+p collisions in free space, one can have experimental access to the properties of the produced QCD matter. In this overview, we discuss several significant experimental measurements from Au+Au reactions at RHIC which have been phenomenologically linked to key thermodynamical and transport properties which can, in some cases, be directly computed in lattice QCD. E.g., if the observed high-p_T leading hadron suppression is due to medium-induced gluon radiation off hard scattered partons, then

- the initial **gluon density** dN^g/dy of the expanding plasma (with transverse area $A_\perp$ and length L) can be estimated from the measured energy loss via [6]:

$$\Delta E \propto \alpha_S^3 C_R \frac{1}{A_\perp} \frac{dN^g}{dy} L, \qquad (1)$$

- and also the **transport coefficient** $\langle \hat{q} \rangle$, characterizing the squared average momentum transfer from the medium to the hard parton per unit distance, can be derived from the average energy loss according to [5, 7]:

$$\langle \Delta E \rangle \propto \alpha_S C_R \langle \hat{q} \rangle L^2. \qquad (2)$$

Likewise, it has been argued that a fast parton propagating (and loosing energy) through the medium can generate a wake of lower energy gluons with Mach- [11, 12] or Čerenkov-like [12, 13] conical angular patterns.

- In the first case, the **speed of sound** of the traversed matter, $c_s^2 = \partial P / \partial \varepsilon$, can be determined from the characteristic angle of the emitted secondaries with respect to the (quenched) jet axis [11, 12]:

$$\cos(\theta_M) = c_s, \text{ where } \theta_M \text{ is the Mach shock wave angle.} \qquad (3)$$

- In the second scenario, the **gluon dielectric constant** in the medium ε or, equivalently, its index of refraction $n = \sqrt{\varepsilon}$, can be estimated from the distinctive angular pattern of emission of the (soft) radiated gluons [12, 13]:

$$\cos(\theta_c) \approx \frac{1}{\sqrt{\varepsilon}} \approx \frac{1}{n} \qquad (4)$$

HIGH P_T LEADING HADRON SUPPRESSION

The standard method to quantify the (initial- and final-state) medium effects on the production yields of a given hard probe in a nucleus-nucleus reaction is given by the *nuclear modification factor*:

$$R_{AA}(p_T, y; b) = \frac{\text{"hot/dense QCD medium"}}{\text{"QCD vacuum"}} = \frac{d^2 N_{AA}/dy dp_T}{\langle T_{AA}(b) \rangle \cdot d^2 \sigma_{pp}/dy dp_T}, \qquad (5)$$

which measures the deviation of A+A at impact parameter b from an incoherent superposition of nucleon-nucleon collisions ($T_{AA}(b)$ is the corresponding Glauber nuclear overlap function at b). Among the most exciting results from the first 5 years of operation at RHIC is the large high p_T hadron suppression ($R_{AA} \ll 1$) observed in central Au+Au reactions at $\sqrt{s_{NN}} = 200$ GeV, expected in jet quenching scenarios. Most of the empirical properties of the suppression factor are in quantitative agreement with the predictions of non-Abelian parton energy loss models:

(1) **Magnitude** of the suppression: The experimental $R_{AA} \approx 0.2$ value at top RHIC energies can be well reproduced assuming the formation of a very dense system

with initial gluon rapidity density $dN^g/dy \approx 1000$ [14] (Eq. 1) or transport coefficient $\langle \hat{q} \rangle \approx 14$ GeV^2/fm [15] (Eq. 2), both consistent with the total charged hadron multiplicities measured in the reaction: $dN/d\eta \approx 3/2 \cdot dN_{ch}/d\eta \approx 1000$ [16].

(2) **Universal** (light) hadron suppression: Above $p_T \approx 5$ GeV/c, π^0 [17], η [18], and inclusive charged hadrons [19, 20] (dominated by $\pi^\pm$ [20]) show all a common factor of $\sim$5 suppression relative to the $R_{AA} = 1$ perturbative expectation which holds for hard probes, such as direct photons, insensitive to final-state interactions [21] (Fig. 1, left). Such a "universal" hadron deficit is consistent with in-medium *partonic* energy loss of the parent quark or gluon prior to vacuum fragmentation.

(3) Flat **transverse momentum** dependence: Above $p_T \approx 5$ GeV/c, $R_{AA}(p_T)$ remains constant up to the highest transverse momenta measured so far ($p_T \approx 14$ GeV/c for π^0 [18], Fig. 1 left). Such p_T-independence of the quenching factor is also well accommodated by parton energy loss models [14, 15, 22].

(4) **Centrality** dependence: The amount of suppression in Au+Au reactions decreases with impact parameter as expected (from Eqs. 1,2) for the different parton production points and, hence, the different densities and lengths encountered by the traversing parton through the medium [15, 23].

(5) **Center-of-mass energy** dependence: The amount of quenching rises in the range $\sqrt{s_{NN}} \approx 20 - 200$ GeV as expected due to the growing initial parton densities and the increasingly longer duration of the QGP phase [16] (Fig. 1, right).

(6) **Non-Abelian nature** of the energy loss: At $y = 0$, high-p_T hadroproduction is dominated by quark (gluon) scattering at large (small) fractional momentum $x_T = 2p_T/\sqrt{s_{NN}}$. In the range $\sqrt{s_{NN}} \approx 20 - 200$ GeV and for a *fixed* (high) p_T value, the suppression factor increases as expected in the canonical non-Abelian scenario where there is an increasingly large relative fraction of hard scattered gluons radiating with a $C_A/C_F = 9/4$ larger probability than quarks (Fig. 2, left).

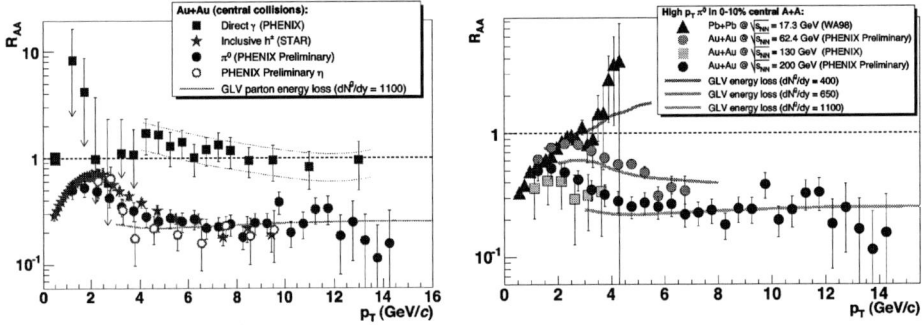

FIGURE 1. Left: $R_{AA}(p_T)$ measured in central Au+Au at 200 GeV for: π^0 and η mesons [18], charged hadrons [19], and direct photons [21] compared to theoretical predictions for parton energy loss in a dense medium with $dN^g/dy = 1100$ [14]. Right: Compilation of all measured $R_{AA}(p_T)$ for high p_T neutral pions in central A+A collisions in the range $\sqrt{s_{NN}} \approx 20 - 200$ GeV [16], compared to GLV parton energy loss calculations [14] for different initial gluon densities ($dN^g/dy = 400$, 650 and 1100).

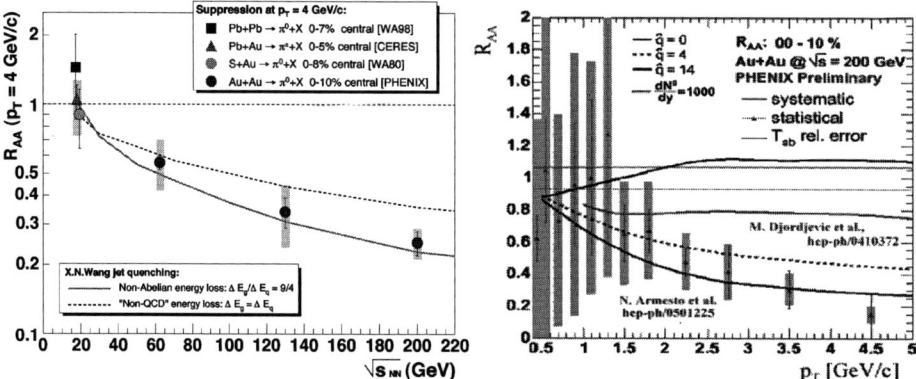

FIGURE 2. Left: Excitation function of $R_{AA}(p_T = 4$ GeV/$c)$ for π^0 [16] with two different implementations of partonic energy loss [24]: (i) canonical non-Abelian (gluons loose $C_A/C_F = 9/4$ more energy than quarks; solid line) and (ii) *ad hoc* "non-QCD" (q,g radiate with equal probability; dashed line) prescriptions. Right: R_{AA} for "non-photonic" $e^\pm$ measured in central Au+Au at $\sqrt{s_{NN}} = 200$ GeV [30] compared to theoretical predictions of heavy-quark energy loss [31, 32].

Though most of the experimental results on high p_T hadron suppression in A+A reactions are in agreement with the non-Abelian energy loss "paradigm", it is worth to stress that there are a few aspects of the data which are less well reproduced:

(1) The pronounced **path-length** L dependence of the energy loss, – as determined by the π^0 suppression factor along different azimuthal angles ϕ with respect to the Au+Au reaction plane [25] –, does not support the theoretical $\propto L$ or L^2 behaviour given by Eqs. 1, 2. Such a failure of the jet-quenching models points to an extra source of azimuthal anisotropy that enhances the in-plane π^0 production even in a kinematic domain beyond $p_T \approx 4.5$ GeV/c where non-perturbative effects should play a minor role [16].

(2) The **unsuppressed baryon** ($p,\bar{p}$ [26] and $\Lambda, \bar{\Lambda}$ [27]) spectra within $p_T \approx 2 - 5$ GeV/c have been explained in terms of an extra mechanism of baryon production based on in-medium quark coalescence [28] which compensates for the energy loss suffered by the fragmenting parent partons. Though such a mechanism successfully describes many aspects of the data, it cannot explain (in its simplest "thermal recombination" form) the similar ("jet-like") azimuthally-correlated hadron yields measured for trigger baryons and mesons at intermediate p_T [29].

(3) The **large** amount of **heavy-quark quenching** indicated by the suppressed high-p_T spectra of electrons from semi-leptonic D and B meson decays measured by PHENIX in central Au+Au [30] (Fig. 2, right) is in apparent conflict with the robust $\Delta E_Q < \Delta E_q < \Delta E_g$ prediction of parton energy loss models. State-of-the-art theoretical predictions [31, 32] require much larger gluon densities to reproduce the high p_T open charm/bottom results at RHIC, than they needed to describe the quenched light hadron spectra.

MODIFIED HIGH P_T DI-HADRON ϕ, η CORRELATIONS

Full jet reconstruction in A+A collisions with standard jet algorithms [33] is unpractical at RHIC energies due to the overwhelming background of soft particles in the underlying event. Thus, beyond the leading hadron spectra measurements discussed in the previous section, more detailed studies of the modifications of the jet structure in a dense QCD environment have been addressed via the study of high-p_T two-particle ϕ, η correlations. Jet-like correlations are measured on a statistical basis by selecting high-p_T *trigger* particles and measuring the azimuthal ($\Delta\phi = \phi - \phi_{trig}$) and rapidity ($\Delta\eta = \eta - \eta_{trig}$) distributions of *associated* hadrons ($p_{T,assoc} < p_{T,trig}$) relative to the trigger:

$$C(\Delta\phi, \Delta\eta) = \frac{1}{N_{trig}} \frac{d^2 N_{pair}}{d\Delta\phi d\Delta\eta}. \tag{6}$$

Combinatorial background contributions, corrections for finite pair acceptance, and the superimposed effects of *collective* azimuthal modulations (elliptic flow) are then taken care of with different techniques [34, 35, 36]. If no initial- or final- state interactions affect the parton-parton scattering process, a dijet signal should appear to first order as two distinct back-to-back Gaussian peaks at $\Delta\phi \approx 0$, $\Delta\eta \approx 0$ (near-side) and at $\Delta\phi \approx \pi$ (away-side). At variance with this standard dijet topology in the QCD vacuum, the di-hadron correlations in Au+Au reactions at RHIC show several striking features:

(1) The gradual **disappearance** of the **away-side azimuthal peak** with centrality (observed at $\Delta\phi \approx \pi$ in the $dN_{pair}/d\Delta\phi$ distributions for hadrons with $2 < p_{T,assoc} < 4 < p_{T,trig} < 6$ GeV/c), consistent with strong suppression of the leading fragments of the recoiling jet traversing the medium [34].

(2) The **broadening** of the **nearside pseudo-rapidity** correlations $dN_{pair}/d\Delta\eta$ ("stretching" of the jet cone along η), reminiscent of the coupling of the induced radiation with the longitudinal expansion of the system [35].

(3) The vanishing away-side peak, observed in the $dN_{pair}/d\Delta\phi$ distribution for recoiling hadrons with $p_{T,assoc} = 2 - 4$ GeV/c, is accompanied with an *enhanced* production of *lower* p_T hadrons ($p_{T,assoc} = 1 - 2.5$ GeV/c [36] or 0.15 – 4 GeV/c [35]) with a characteristic **"double-peak"** structure at $\Delta\phi \approx \pi \pm 1.3$ or $\pi \pm 1.1$ (Fig. 3).

Figure 3 shows the double-peak structure appearing in the away-side azimuthal correlations of central Au+Au (top left-plot, and star symbols in the right-plot), compared to the standard back-to-back dijet topology seen in peripheral Au+Au (bottom, left) and in d+Au and p+p collisions (right). Such a non-Gaussian "volcano"-like shape seen in the away-side hemisphere has attracted much theoretical attention because it suggests conical patterns induced by Mach-shock [11, 12] or Čerenkov-like [12, 13] emissions.

In the "Mach cone" scenario [11, 12], the local maxima (red arrows in the plots) in central Au+Au at an angle $\Delta\phi \approx \pi \pm 1.2$ relative to the high-p_T trigger are caused by the Mach shock of the supersonic recoiling (quenched) parton through the medium. The resulting preferential emission of secondary partons from the plasma at an angle $\theta_M \approx 1.2$, yields (Eq. 3) a value of the speed sound $c_s \approx 0.36$, close to that of an ideal QGP ($c_s = 1/\sqrt{3}$). In the Čerenkov gluonstrahlung picture [12], – developed at a more quantitative level [13] after this conference –, it is argued that the combination of the

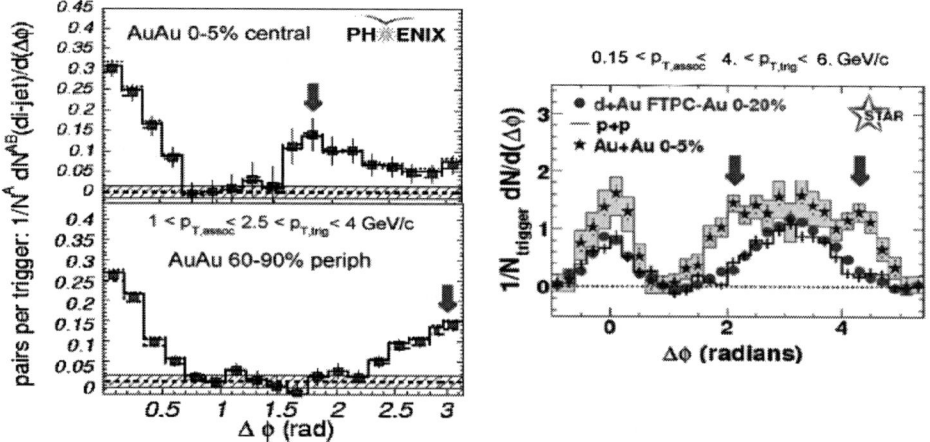

FIGURE 3. Azimuthal dihadron distributions normalized per trigger particle, $1/N_{trigg}\,dN_{pair}/d\Delta\phi$ (the arrows indicate the local maxima in the away-side hemisphere) measured at RHIC. Left: PHENIX results in central (top) and peripheral (bottom) Au+Au [36]. Right: STAR results in central Au+Au, d+Au and p+p collisions [35]. (Note that each experiment has different p_T values for the associated hadrons).

Landau-Pomeranchuk-Migdal interference characteristic of gluon bremsstrahlung and a medium with a *large* dielectric constant ($n \approx 2.75$ is needed to reproduce the location of the experimental peaks using Eq. 4) should also result in the two-peak shape observed in the data. At variance with the cone angle of the "sonic boom" mechanism (which is constant in the fluid but effectively *increases* with $p_{T,assoc}$ at the spectra level [11]), the Čerenkov angle *decreases* with the momentum of the radiated gluon. Such a trend is, however, seemingly in disagreement with the fact that PHENIX (STAR) measures a *larger* (lower) $\theta_c \approx 1.3$ (1.1) for higher (lower) average values of $p_{T,assoc}$.

SUMMARY

Experimental results on single inclusive spectra and dihadron correlations measured at high transverse momentum in Au+Au at RHIC collider energies ($\sqrt{s_{NN}} = 200$ GeV) have been reviewed as a means to learn about jet production and fragmentation in hot and dense QCD matter. The analysis of jet structure modifications in A+A collisions provides quantitative information on the thermodynamical and transport properties of the strongly interacting medium produced in the reactions. Two notable experimental results have been discussed: (i) the observed factor ~ 5 suppression of high p_T leading hadrons in central Au+Au relative to p+p collisions in free space; and (ii) the conical-like shape of the azimuthal distributions of secondary hadrons emitted in the away-side hemisphere of a high-p_T trigger hadron. Most of the properties of the observed high p_T suppression (such as its magnitude, light flavor "universality", p_T, reaction centrality, and $\sqrt{s_{NN}}$ dependences) are in quantitative agreement with predictions of non-Abelian energy loss models. The confrontation of these models to the data permits to derive the initial gluon density $dN^g/dy \approx 1000$ and transport coefficient $\langle \hat{q} \rangle \approx 14$ GeV^2/fm of the produced medium. The second striking observation of a softer and broadened angular distribution of secondary hadrons peaking at a finite angle away from the (quenched) jet axis has been attributed to Mach conical flow caused by the propagation of a supersonic parton through the dense system. If such a phenomenon is confirmed, the speed of sound of the medium could be extracted. The same angular pattern could also be the result of Čerenkov gluon radiation and provide, in that case, information on the gluon dielectric constant in hot and dense QCD matter.

REFERENCES

1. F. Karsch, Lect. Notes Phys. **583**, 209 (2002).
2. J. W. Harris and B. Muller, Ann. Rev. Nucl. Part. Sci. **46**, 71 (1996).
3. J. D. Bjorken, FERMILAB-PUB-82-059-THY.
4. M. Gyulassy, M. Plümer, Phys. Lett. **B243**, 432 (1990); X.N. Wang, M. Gyulassy, Phys. Rev. Lett. **68**, 1480 (1992).
5. R. Baier, Y. L. Dokshitzer, A. H. Mueller, S. Peigné and D. Schiff, Nucl. Phys. **B484**, 265 (1997); R. Baier, D. Schiff, B.G. Zakharov, Ann. Rev. Nucl. Part. Sci. **50**, 37 (2000).
6. M. Gyulassy, P. Levai and I. Vitev, Phys. Rev. Lett. **85**, 5535 (2000); Nucl. Phys. **B594**, 371 (2001).
7. U. A. Wiedemann, Nucl. Phys. **B588**, 30 (2000); C. A. Salgado and U. A. Wiedemann, Phys. Rev. **D68**, 014008 (2003).
8. D. A. Appel, Phys. Rev. D **33**, 717 (1986); J. P. Blaizot, L. D. McLerran, Phys. Rev. D **34**, 2739 (1986).
9. C. A. Salgado and U. A. Wiedemann, Phys. Rev. Lett. **93**, 042301 (2004).
10. N. Borghini and U. A. Wiedemann, hep-ph/0506218.
11. H. Stoecker, Nucl. Phys. **A750**, 121 (2005); J. Casalderrey, E. Shuryak, D. Teaney, hep-ph/0411315.
12. J. Ruppert and B. Muller, Phys. Lett. **B618**, 123 (2005).
13. A. Majumder and X. N. Wang, nucl-th/0507062; V. Koch, A. Majumder and X. N. Wang, nucl-th/0507063; I. M. Dremin, hep-ph/0507167.
14. I. Vitev and M. Gyulassy, Phys. Rev. Lett. **89**, 252301 (2002); I. Vitev, J. Phys. G **30**, S791 (2004).
15. A. Dainese, C. Loizides and G. Paic, Eur. Phys. J. **C38**, 461 (2005).
16. D. d'Enterria, Eur. Phys. J. **C43**, 295 (2005).
17. S.S. Adler *et al.* [PHENIX Collaboration], Phys. Rev. Lett. **91**, 072301 (2003).
18. H. Busching [PHENIX Collaboration], Eur. Phys. J. **C43**, 303 (2005).
19. J. Adams *et al.* [STAR Collaboration], Phys. Rev. Lett. **91**, 172302 (2003).
20. S. S. Adler *et al.* [PHENIX Collaboration], Phys. Rev. **C69**, 034910 (2004).
21. S.S. Adler *et al.* [PHENIX Collaboration], nucl-ex/0503003.
22. S. Jeon and G. D. Moore, Phys. Rev. C **71**, 034901 (2005).
23. X.N. Wang, Phys. Rev. **C70**, 031901 (2004).
24. Q. Wang and X.N. Wang, Phys. Rev. **C71**, 014903 (2005).
25. B. Cole [PHENIX Collaboration], Eur. Phys. J. **C43**, 271 (2005).
26. S. S. Adler *et al.* [PHENIX Collaboration], Phys. Rev. Lett. **91**, 172301 (2003).
27. J. Adams *et al.* [STAR Collaboration], Phys. Rev. Lett. **92**, 052302 (2004).
28. R. C. Hwa, C. B. Yang, Phys. Rev. **C67**, 034902 (2003); R. J. Fries, B. Muller, C. Nonaka, S. A. Bass, Phys. Rev. **C68**, 044902 (2003); V. Greco, C. M. Ko, P. Levai, Phys. Rev. Lett. **90**, 202302 (2003).
29. S. S. Adler *et al.* [PHENIX Collaboration], Phys. Rev. **C71**, 051902 (2005).
30. B. Jacak [PHENIX Collaboration], Proceeds. ICPAQGP 2005, India, Feb 2005; nucl-ex/0508036.
31. M. Djordjevic, M. Gyulassy and S. Wicks, Phys. Rev. Lett. **94**, 112301 (2005).
32. N. Armesto, A. Dainese, C. A. Salgado and U. A. Wiedemann, Phys. Rev. **D71**, 054027 (2005).
33. See e.g G. C. Blazey *et al.*, hep-ex/0005012.
34. C. Adler *et al.* [STAR Collaboration], Phys. Rev. Lett. **90**, 082302 (2003).
35. J. Adams *et al.* [STAR Collaboration], Phys. Rev. Lett. **95**, 152301 (2005); F. Wang [STAR Collaboration] in RIKEN/BNL Workshop on Jet Correlations at RHIC, March 2005.
36. S. S. Adler *et al.* [PHENIX Collaboration], nucl-ex/0507004.

Hadronic Modes in the Quark-Gluon Plasma

M. Mannarelli* and R. Rapp[†]

*Center for Theoretical Physics, Laboratory for Nuclear Science and Department of Physics
Massachusetts Institute of Technology, Cambridge, MA 02139
[†]Cyclotron Institute and Physics Department, Texas A&M University, College Station, Texas 77843-3366

Abstract. We study the properties of quark-antiquark interactions in the Quark-Gluon Plasma at moderate temperatures, $T \simeq 1\text{-}2\ T_c$, within a Brueckner-type many-body approach. The quark-antiquark T-matrix, including both color-singlet and -octet channels, and corresponding quark self-energies are evaluated self-consistently. We find that light mesonic states persist as resonances in the Quark-Gluon Plasma up to temperatures $T \simeq 1.75 T_c$ and fermionic quasiparticles acquire masses of up to $\sim$150 MeV and widths of around $\sim$200 MeV.

Keywords: Quark gluon plasma, quasi-particle plasma
PACS: 25.75.-q, 25.75.Dw, 25.75.Nq

INTRODUCTION

Since the pioneering works of Hagedorn [1] and Cabibbo and Parisi [2] our understanding of the possible phases of Quantum Chromodynamics (QCD) has much improved (see, e.g., Ref. [3] for a recent overview). It is now widely accepted that at sufficiently high temperatures, due to asymptotic freedom, matter consists of a weakly interacting gas of quark and gluons, the so-called Quark-Gluon Plasma (QGP).

However, it is less clear what the nature of matter is, and what the relevant degrees of freedom are, at comparatively moderate temperatures, $T \simeq 1\text{-}2\ T_c$ ($T_c \simeq 170$ MeV: critical temperature). On the one hand, the experimental investigations pursued at the Relativistic Heavy-Ion Collider (RHIC) suggest the formation of a of a "strongly interacting QGP" (sQGP), to reconcile the phenomenological success of hydrodynamic approaches with the inherent short thermalization times of $\sim$0.5 fm/c (see Refs. [4, 5, 6, 7] for recent theoretical overviews). On the other hand, lattice computations of QCD (lQCD) [8, 9] indicate the formation of mesonic bound (and/or resonance) states in the QGP, which are also supported by applications of lQCD-based heavy-quark potentials [10] in a Klein-Gordon equation [11].

An important question in this context is how the mesonic states in the QGP affect the scattering properties of quarks and gluons. In the present talk we report results of Ref. [12] regarding our investigation of the properties of the light pseudoscalar mesonic nonet. Implementing a lQCD potential into a Brueckner-type many-body scheme [13] we calculate quark-antiquark scattering matrices to address both quasiparticle and mesonic properties at temperatures above the phase transition.

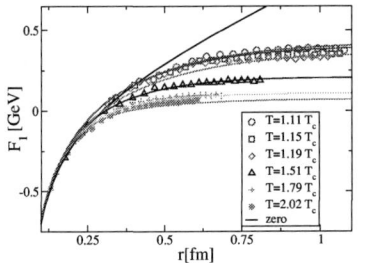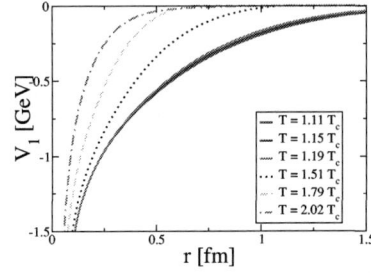

FIGURE 1. Left panel: lattice QCD results for the color-singlet free energy from unquenched simulations (symbols) for 6 different values of the temperature compared to our fit function, Eq. (1), represented by the various curves. Right panel: corresponding potentials in the color-singlet channel obtained with Eq. (4) for the same values of temperature.

QUARK-ANTIQUARK POTENTIAL FROM LATTICE QCD

The main input required for a two-body scattering equation is a driving potential. To obtain the latter for interactions of a q-$\bar{q}$ pair we take recourse to lQCD calculations of the static free energy for a (heavy) Q-$\bar{Q}$ pair. For the temperature range $T = 1.1$-$2\,T_c$, it turns out that unquenched singlet free energy [10] can be reasonably well reproduced by a parametrization of the following form

$$F_1(r,T) = -\frac{\alpha}{r}e^{-a\mu(r,T)r} + \frac{\sigma}{\mu(r,T)}\left(1 - e^{-\mu(r,T)r}\right), \tag{1}$$

where

$$\mu(r,T) = \frac{\sigma}{b}e^{-0.3/r} \tag{2}$$

is a "screening mass", a and b are two fitting functions (see [12] for details), $\alpha = 0.4$ and $\sigma = 1.2\,\text{GeV}^2$, left panel of Fig. 1.

The internal energy is obtained by subtracting the entropy contribution to the free energy according to

$$E_1 = F_1 - T\frac{dF_1}{dT}, \tag{3}$$

and the potential is finally calculated subtracting the nonzero asymptotic value of the internal energy (which is interpreted as an in-medium mass term). Therefore, the potential in the color-singlet channel takes the form

$$V_1(r,T) = E_1(r,T) - E_1(\infty,T). \tag{4}$$

The singlet potential is shown in the right panel of Fig. 1 for the same values of temperature as the singlet free-energy (left panel). Toward a more complete description of the q-$\bar{q}$ interactions in the QGP we also consider the (repulsive) contributions from

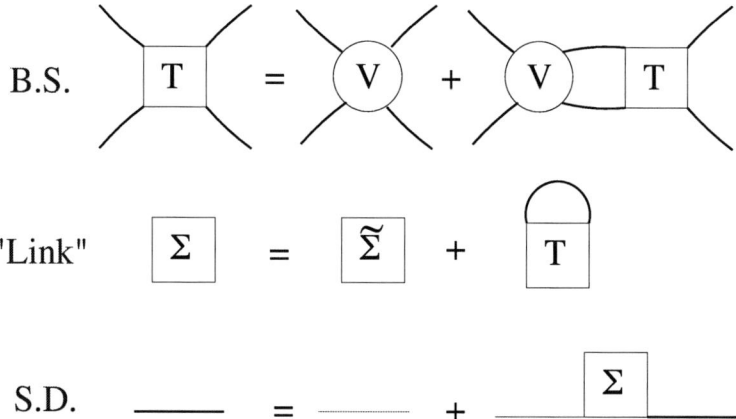

FIGURE 2. Schematic representation of the self-consistency problem. Upper panel (B.S.): Bethe-Salpeter Equation (6) for the T-matrix, Eq. (6). Middle panel ("Link"): quark self-energy Σ, Eq. (7). Lower panel (S.D.): Schwinger-Dyson equation for the self-energy, Eq. (8). Thick (blue) lines: full quark propagators, thin (red) line: bare quark propagators.

the color-octet channel assuming that the octet potential follows the leading-order result of perturbation theory,

$$F_8 = -\frac{1}{8}F_1. \tag{5}$$

Finally, relativistic corrections are included via velocity-velocity interactions [14].

BRUECKNER APPROACH

To evaluate quark-antiquark interactions in the QGP we employ the T-matrix approach, as is well known from the nuclear many-body problem. In relativistic field theory, the starting point is a 4-dimensional Bethe-Salpeter (BS) equation,

$$T = K + \int KSST, \tag{6}$$

where K denotes the interaction kernel and S is the single-particle propagator. In the following, the kernel is approximated with the potential in the pertinent color channel derived in the previous section, and the medium effects in the quark propagator are encoded in a self-energy which we decompose according to

$$\Sigma = \tilde{\Sigma} + \int TS. \tag{7}$$

The first term on the right-hand side, $\tilde{\Sigma}$, represents a "gluon-induced" contribution due to interactions of quarks/antiquarks with surrounding thermal gluons and will be treated as a mass term m in the quark dispersion law [12]. The full quark propagator obeys a

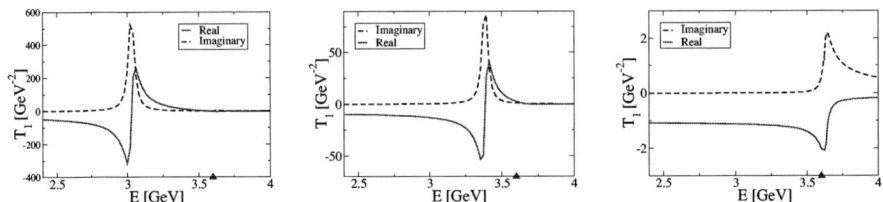

FIGURE 3. Real (full red line) and imaginary part (absolute value, dashed blue line) of the T-matrix in the color-singlet channel for charmonium (with a charm-quark mass of $m = 1.8$ GeV) at $T = 1.2T_c$, $T = 1.5T_c$, $T = 2T_c$ (left, middle and right panel, respectively) as a function of CM energy E. The red triangles on the x-axis correspond to the threshold energy, $E_{thr} = 3.6$ GeV.

Schwinger-Dyson equation,

$$S = S_0 + S_0 \Sigma S, \qquad (8)$$

where S_0 is the free quark propagator. The three equations (6), (7) and (8) constitute a self-consistency problem which is diagrammatically illustrated in Fig. 2.

NUMERICAL RESULTS

After employing an appropriate non-relativistic reduction scheme in line with the potential approximation we solve the set of equations (6), (7) and (8) numerically. Self-consistency is achieved by iteration, starting with the calculation of the T-matrix using a constant self-energy in the first step. The self-energy is then calculated from (7) and used to determine the on-shell propagator (8). The propagator is then re-inserted into the T-matrix equation and the procedure is iterated until T-matrix and self-energy converge (typically within less than 10 iteration steps).

Scattering amplitude

In order to check the reliability of our procedure we first apply our approach to the c-$\bar{c}$ (charmonium) sector by using a quark mass of $m = 1.8$ GeV and a fixed value for the self-energy. For numerical purposes we set $\Sigma_I = -10$ MeV and $\Sigma_R = 0$. The results are displayed in Fig. 3 for three different temperatures, $1.2T_c$ (left panel), $1.5T_c$ (middle panel) and $2T_c$ (right panel). As the temperature increases, the charmonium state moves up in energy reaching the threshold, $E_{thr} = 3.6$ GeV (corresponding to the red triangle on the x-axis), at $T \simeq 2T_c$, after which the resonance peak essentially dissolves. This behavior is in reasonable (qualitative) agreement with previous calculations within lQCD [15] and effective models [16, 17].

Turning to the light-quark sector, the self-consistent results for real and imaginary part of the on-shell T-matrix with a gluon-induced mass-term of $m = 0.1$ GeV are summarized in Figs. 4 and 5 for temperatures $T = 1.2T_c$, $T = 1.5T_c$ and $T = 1.75T_c$.

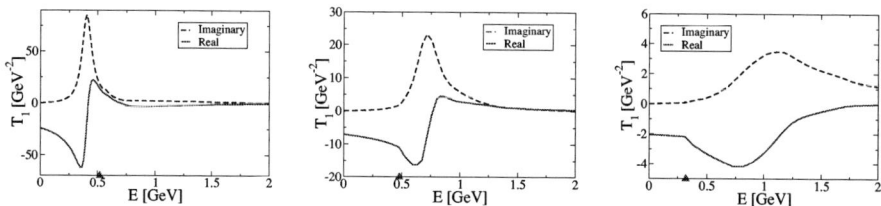

FIGURE 4. Real (full red line) and (absolute value of the) imaginary part (dashed blue line) of the light-quark (on-shell) T-matrix in the color-singlet channel at temperatures $T = 1.2T_c$, $T = 1.5T_c$ and $T = 1.75T_c$ (left, middle and right panel, respectively) as a function of the $q\bar{q}$ CM energy E, with a "gluon-induced" quark-mass term $m = 0.1$ GeV. The red triangles on the x-axis correspond to the threshold energies.

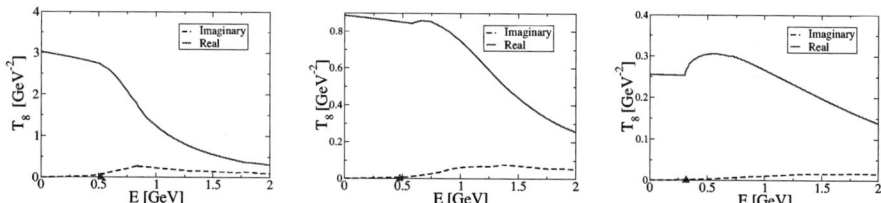

FIGURE 5. Light-quark T-matrix in the color-octet channel vs. $q\bar{q}$ CM energy at $T = 1.2T_c$, $T = 1.5T_c$ and $T = 1.75T_c$ (left, middle and right panel, respectively) with $m = 0.1$ GeV. Solid (red) line: real part; dashed (blue) line: imaginary part (absolute value). The red triangles on the x-axis correspond to the threshold energies.

At $T = 1.2T_c$ the color-singlet T-matrix exhibits a relatively narrow bound state located significantly below the q-$\bar{q}$ threshold energy of $E_{thr} \simeq 0.52$ GeV. As the temperature is increased to $1.5T_c$, the state moves to higher CM energy above the threshold ($E_{thr} \simeq 0.48$ GeV) and broadens. Also note that the peak value is substantially reduced as compared to the $1.2\,T_c$ case. We attribute this behavior to the decrease in the potential, see right panel of Fig. 1, reflecting an overall reduction in interaction strength. The trends in suppression, broadening and upward energy-shift continue at $T = 1.75T_c$ where the resonance has now essentially melted as indicated by a width of almost 1 GeV. These results are again in reasonable agreement with the computations of mesonic spectral functions in (quenched) lQCD [8, 9].

The T-matrix in the color-octet channel is displayed in Fig. 5 for the same set of temperatures. As to be expected for a purely repulsive potential, we find a smooth dependence of both real and imaginary part with CM energy.

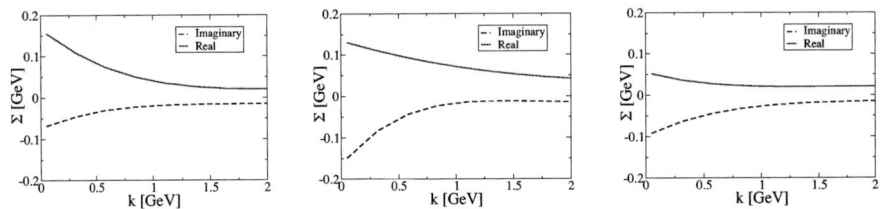

FIGURE 6. Real (solid line, red) and imaginary (dashed line, blue) parts of the on-shell quark self-energy as a function of 3-momentum at temperatures $T = 1.2T_c$, $T = 1.5T_c$ and $T = 1.75T_c$ (left, middle and right panel, respectively) with $m = 0.1$ GeV.

Self-Energy

In Fig. 6 the on-shell self-energy is displayed for the same selection of temperatures as in the previous section. Both real and imaginary parts are smooth functions of the quark 3-momentum with maximal values at $k = 0$. Note that the real part is positive, implying that the repulsive contribution from the octet channels overcomes the attractive singlet channels. This is due to the fact that the octet contribution to the self-energy enters with a degeneracy factor of 8 times larger than the singlet one. The imaginary part (width), on the other hand, chiefly arises due to resonant scattering in the singlet channel.

More quantitatively, in the temperature regime 1.2-1.5T_c, the nonperturbative contribution to the thermal quark mass reaches values of around 150 MeV at small momenta, decreasing to ~50 MeV at 1.75 T_c. With the underlying "gluon-induced" mass term of $m = 100$ MeV, the total thermal mass, $m + \Sigma_R$, amounts to 150-250 MeV.

An important aspect of our results is the rather large imaginary part of the quark self-energy, translating into widths of about 200 MeV at low momenta for temperatures around 1.5 T_c. As mentioned above, the width is almost entirely generated by the resonant scattering in the singlet channel; this is nicely illustrated by the significant increase in ImΣ when going from 1.2 to 1.5 T_c (left and middle panel in Fig. 6), during which the state in the T-matrix moves from below to above threshold (left and middle panel in Fig. 4), i.e. converts from bound state to resonance[1]. The magnitude of the quark widths is quite comparable to the thermal masses, qualitatively supporting the notion that the QGP could be in a liquid-like regime [18].

CONCLUSIONS

In this talk we have presented a self-consistent many-body (Brueckner) approach to assess nonperturbative properties of (anti-) quarks and mesonic composites in a Quark-Gluon Plasma at temperatures $T \simeq 1.2$-$2T_c$. Our key ingredient to describe the q-$\bar{q}$ inter-

---

[1] We recall that bound states are not accessible in on-shell 2 → 2 scattering; even if a resonance is close to threshold it does not contribute effectively to rescattering processes if the average thermal energy of particles from the heat bath is appreciable.

action in the QGP was a driving kernel (potential) extracted from unquenched finite-T lattice QCD calculations for the free energy of a heavy-quark pair, supplemented with corrections for relativistic motion. Our main objective was to go beyond earlier applications to bound states by solving the scattering problem thereby accounting for absorptive effects (finite imaginary parts). One of our main new findings is that the lQCD potentials (dynamically) generate S-wave resonance states *above* the q-$\bar{q}$ threshold up to temperatures of $\sim 1.75\ T_c$. These resonances play a key role in inducing large quark scattering rates as indicated by single-particle widths of $\Gamma \simeq 200$ MeV at temperatures around $1.5\ T_c$. At the same time, significant (positive) real parts arise from repulsive interactions in the color-octet channel entailing thermal masses of up to $\sim$150 MeV. Further consequences regarding the transport properties of the "sQGP" and RHIC phenomenology remain to be worked out.

ACKNOWLEDGMENTS

One of us (MM) is grateful to the organizers of this conference for the invitation, and acknowledges partial support by U.S. Department of Energy under cooperative research agreement #DE-FC02-94ER40818. One of us (RR) is supported in part by a U.S. National Science Foundation CAREER award under grant PHY-0449489.

REFERENCES

1. R. Hagedorn, *Nuovo Cim. Suppl.* **3**, 147–186 (1965).
2. N. Cabibbo, and G. Parisi, *Phys. Lett.* **B59**, 67 (1975).
3. T. Schaefer, e-print arXiv: hep-ph/0509068 (2005).
4. R. Rapp, *J. Phys.* **G30**, S951–S962 (2004).
5. M. Gyulassy, and L. McLerran, *Nucl. Phys.* **A750**, 30–63 (2005).
6. E. V. Shuryak, *Nucl. Phys.* **A750**, 64–83 (2005).
7. B. Muller, e-print arXiv: nucl-th/0508062 (2005).
8. M. Asakawa, T. Hatsuda, and Y. Nakahara, *Nucl. Phys.* **A715**, 863–866 (2003).
9. F. Karsch, and E. Laermann, e-print arXiv: hep-lat/0305025 (2003).
10. O. Kaczmarek, F. Karsch, P. Petreczky, and F. Zantow, *Nucl. Phys. Proc. Suppl.* **129**, 560–562 (2004).
11. E. V. Shuryak, and I. Zahed, *Phys. Rev.* **D70**, 054507 (2004).
12. M. Mannarelli, and R. Rapp, e-print arXiv: hep-ph/0505080 (2005).
13. R. Brockmann, and R. Machleidt, e-print arXiv: nucl-th/9612004 (1996).
14. G. E. Brown, *Philos. Mag.* **43**, 467 (1952).
15. M. Asakawa, and T. Hatsuda, *Phys. Rev. Lett.* **92**, 012001 (2004).
16. C.-Y. Wong, e-print arXiv: hep-ph/0408020(2004).
17. A. Mocsy, and P. Petreczky, e-print arXiv: hep-ph/0411262 (2004).
18. M. H. Thoma, *J. Phys.* **G31**, L7 (2005).

Production of multiply heavy flavoured baryons from Quark Gluon Plasma

F. Becattini

Universita di Firenze and INFN Sezione di Firenze, Via G. Sansone 1, I-50019, Sesto F.no (Firenze), Italy

Abstract. We show that in heavy ion collisions at LHC there could be a measurable production of baryons containing two or three heavy quarks from statistical coalescence. This production mechanism is peculiar of Quark Gluon Plasma and the predicted rates, in heavy ion collisions at LHC energy, exceed those from a purely hadronic scenario, particularly for Ξ_{bc} and Ω_{ccc}. Thus, besides the interest in the discovery of these new states, enhanced ratios of these baryons over singly heavy flavoured hadrons, like B or D, in heavy ion collisions with respect to pp at the same energy, would be a clear indication of kinetical equilibration of heavy quarks in the Quark Gluon Plasma.

INTRODUCTION

Charm and bottom quarks are expected to be abundantly produced in hadronic collisions at very high energies. In heavy ion collisions (HIC), multiple pair production is expected to occur, with average multiplicities which, at the LHC energy of 5.5 TeV, attain $\mathscr{O}(10)$ for bottom and $\mathscr{O}(100)$ for charm in central collisions. These quarks, produced in the early stage of the collision off hard scatterings, lose energy in the Quark Gluon Plasma (QGP) and, if the lifetime of the source is long enough, may reach thermal [1] (not chemical, as their reannihilation rate is very low) equilibrium within the medium. There are by now clues of such a possibility from measurements of elliptic flow at RHIC [2], which indicate that charm quarks participate in the collective hydrodynamical expansion. At the hadronization point, heavy quarks will coalesce into hadrons. If coalescence process occurs statistically at the hadronization temperature, there is a finite chance that two or even three of them coalesce into the same particle, thus giving rise to multiply heavy flavoured hadrons, particularly baryons. This phenomenon is likely to occur only if heavy quarks get very close to thermal equilibrium reshuffling over a large region because high momentum quarks, most likely, will hadronize into different particles unless two or three of them emerge very close in momentum from the hard process. The latter production mechanism, where multiply heavy flavoured hadrons arise from *correlated* quarks, predicts multiplicities which, at some large energy, are exceeded by those predicted by coalescence of *uncorrelated* quarks. The ultimate reason of this effect is that the average multiplicity of heavy quark pairs increases faster than soft hadrons multiplicity as a function of centre-of-mass energy. This is in turn related to the volume of the system at freeze-out. Consequently, the system gets denser in heavy quarks at chemical freeze-out as energy increases and so does the chance of formation of multiply heavy flavoured hadrons. Therefore, an enhanced production rate of these objects relative to singly heavy flavoured hadrons (like B's or D's) is distinctive of HIC and can be used as a probe of thermalization of heavy quarks within the QGP, thence as a signal of QGP itself. This idea was advocated in refs. [3, 4] where the authors envisaged an enhancement of B_c and J/ψ mesons production from the QGP with respect to hadronic collisions at RHIC. In this paper, we amend and reinforce this picture by observing that the main advantage of baryons (and especially Ξ_{bc} and Ω_{ccc} whose signal should be detectable at LHC) over quarkonia and B_c is to enhance the difference from the "background" of coherent production, i.e. direct production of these states from early hard scatterings, possibly followed by melting in the plasma. Furthermore, in view of this relatively large production, HIC become a suitable place to discover many of these yet unobserved states.

STATISTICAL COALESCENCE

The basic idea of the statistical coalescence model (SCM) has been introduced in ref. [4] following a work on statistical production of J/ψ [5], and used thereafter by many authors [6]. In practice, the SCM is the statistical hadronization model supplemented with the constraint of a fixed number of heavy quarks and antiquarks. Let us now proceed to calculate of the mean multiplicities $\langle n_j \rangle$ of heavy flavoured particles in the framework of this model. In HIC at high energy, electric charge, strangeness and baryon number conservation can be treated grand-canonically whereas charm and beauty conservation cannot because the multiplicities of heavy flavoured hadrons are not large. Thus, besides the number N_c of $c + \bar{c}$ quarks and N_b of $b + \bar{b}$ quarks, also the net charm C and net beauty B should be fixed. In fact, instead of C, B, N_c, N_b, any combination of these integers can be used to constrain the partition function. Indeed, it is advantegeous to use the numbers v_c of c, $v_{\bar{c}}$ of $\bar{c}$, v_b of b and $v_{\bar{b}}$ of $\bar{b}$ quarks. The relevant partition function reads:

$$Z(v_c, v_{\bar{c}}, v_b, v_{\bar{b}}) = Z_l \left[\prod_{f=c,\bar{c},b,\bar{b}} \int_{-\pi}^{\pi} \frac{d\phi_f}{2\pi} e^{iv_f \phi_f} \right] \exp \left[\sum_j z_j \lambda_j e^{-iv_{cj}\phi_c - iv_{\bar{c}j}\phi_{\bar{c}} - iv_{bj}\phi_b - iv_{\bar{b}j}\phi_{\bar{b}}} \right] \quad (1)$$

where Z_l is the grand-canonical partition function including all light-flavoured species; z_j are the one-particle partition functions:

$$z_j = \frac{g_j V}{2\pi^2} m^2 T K_2\left(\frac{m}{T}\right) \simeq (\text{for } m \gg T) \, g_j V \left(\frac{mT}{2\pi}\right)^{3/2} e^{-m/T} \quad (2)$$

λ_j are the fugacities (with regard to electric, baryonic and strangeness charge) and v_{cj}, $v_{\bar{c}j}$, v_{bj}, $v_{\bar{b}j}$ are the number of c, $\bar{c}$, b, $\bar{b}$ quarks respectively of the of the jth hadronic species; the factor g_j in Eq. (2) is its spin degeneracy and T is the temperature.

If we now denote by a_{fn} the sum of $z_j \lambda_j$ for hadrons with n units of open flavour f, $a_{\bar{f}n}$ for those with n units of open anti-flavour $\bar{f}$, and a_{f0} for $f\bar{f}$ states, it can be proved [7] that the the primary average multiplicity of heavy flavoured hadrons for fixed number of c, c̄, b, b̄ quarks has a particularly simple expression:

$$\langle n_j \rangle = z_j \lambda_j \prod_{f=c,\bar{c},b,\bar{b}, \nu_{fj}>0} \frac{\nu_f(\nu_f-1)\ldots(\nu_f-\nu_{fj}+1)}{a_{f1}^{\nu_{fj}}} \quad (3)$$

provided that the ratios:

$$R_f \equiv \frac{a_{f2}}{a_{f1}^2} \nu_f(\nu_f-1) \approx \frac{\nu_f(\nu_f-1)e^{2m_{u,d}/T}}{g_{\text{eff}} V [\frac{(m_f+m_u)^2 T}{2\pi m_f}]^{3/2}} \quad (4)$$

and

$$R_{cb} = \frac{\nu_c \nu_{\bar{b}} e^{2m_{u,d}/T}}{g_{\text{eff}} V [\frac{m_c m_b T}{2\pi(m_c+m_b)}]^{3/2}} \quad (5)$$

are $\ll 1$. In the Eqs. (4,5) $g_{\text{eff}}(T,\lambda)$ is an effective degeneracy parameter including the spin degeneracy and the different states with the same numbers of heavy quarks, weighted by the ratio $z_j\lambda_j/z_1\lambda_1$, z_1 being the one-particle partition function of the lowest lying state. In Eq. (4) we tacitly assumed that g_{eff} is the same for hadrons with one or two heavy quarks, which approximately holds according to our numerical check. From known states with one heavy quark, either c or b, one expects $g_{\text{eff}} \sim 10$ at $T = 165$ MeV and $\lambda_j = 1$. Taking constituent quark masses $m_c = 1.54$ GeV, $m_b = 4.95$ GeV, $m_{u,d} = 0.33$ GeV $m_s = 0.51$ GeV and $T = 165$ MeV, i.e. the fitted chemical freeze-out temperature at very large energy [8, 9], it turns out that $R_c \sim \nu_c(\nu_c-1)/0.34\,\text{fm}^{-3}V$ and $R_b \sim \nu_b(\nu_b-1)/1.4\,\text{fm}^{-3}V$. Therefore, with the large volumes involved in HIC, R_c, R_b, as well as R_{cb} are likely to be $\ll 1$ unless ν_c or ν_b are consistently large.

The (3) is to be further averaged over the multiplicity distribution p_{ν_c} of cc̄ and p_{ν_b} of bb̄ pairs created in a single collision. If they are independently produced, this is a Poisson distribution and, for open flavoured hadrons, the sum (3) yields its factorial moments, i.e.:

$$\langle\langle n_j \rangle\rangle = z_j \lambda_j \prod_{f=c,\bar{c},b,\bar{b}} \left(\frac{\langle \nu_f \rangle}{a_{f1}}\right)^{\nu_{fj}} \equiv z_j \lambda_j \prod_{f=c,\bar{c},b,\bar{b}} \eta_f^{\nu_{fj}} \quad (6)$$

whereas for quarkonia it is more complicated since $\nu_c = \nu_{\bar{c}}$ and $\nu_b = \nu_{\bar{b}}$. Eq. (6) is our final formula. As has been mentioned, it is an approximated expression valid if $R_c, R_b, R_{cb} \ll 1$. However, it can be shown [10] that it still holds under the weaker condition $R_f/\nu_f \ll 1$ if ν_f is large. Altogether, the Eq. (6) implies that the contribution of hadrons carrying more than one heavy flavoured quark in the balance equations $\sum_j \langle n_j \rangle \nu_{fj} = \langle \nu_f \rangle$ is neglected. It is interesting to note that the enhancement factors $\eta_f = \langle \nu_f \rangle/a_{f1}$ are proportional to the *density* of heavy quarks at the hadronization temperature, as $a_{f1} \propto V$. Therefore, the ratio between multiply and singly heavy flavoured hadrons, proportional to $(\langle \nu_f \rangle/V)^{\nu_{fj}-1}$, increases with centre-of-mass energy because the volume (or the charged multiplicity) increases much slower than $\sigma_{c\bar{c}}$ and $\sigma_{b\bar{b}}$ do.

PRODUCTION RATES AT RHIC AND LHC

The formula (6) can now be applied to estimate the average multiplicity of multiply heavy flavoured hadrons in HIC at RHIC and LHC. For sake of simplicity, we will confine ourselves to full phase space integrated quantities, disregarding spectra. To get started, we need the cross sections $\sigma_{c\bar{c}}$ and $\sigma_{b\bar{b}}$ in pp collisions at relevant energies. There is a large uncertainty on these values; recent calculations indicate $\sigma_{c\bar{c}} = 110 - 656$ μb and $\sigma_{b\bar{b}} = 1.2 - 2.86$ μb at $\sqrt{s} = 200$ GeV [11] and $\sigma_{c\bar{c}} = 3.4 - 9.2$ mb and $\sigma_{b\bar{b}} = 88 - 260$ μb at $\sqrt{s} = 5.5$ TeV [12]. The production of heavy quark pairs is a hard process and should scale like the number of collisions N_{coll} in the Glauber model. Specifically, if σ_{inel} is the total inelastic NN cross section, the average multiplicity of $c\bar{c}$ pairs is $\langle v_f \rangle = \langle v_{\bar{f}} \rangle = N_{\text{coll}} \sigma_{f\bar{f}}/\sigma_{\text{inel}}$. At RHIC, in Au-Au collisions at $\sqrt{s_{NN}} = 200$ GeV, $\sigma_{\text{inel}} \simeq 42$ mb and for a 5.5% centrality selected sample, corresponding to an impact parameter range 0-3.5 fm, $N_{\text{coll}} = 1080$ [13]. Thus, the average multiplicity of $c\bar{c}$ pairs ranges from 2.8 to 17, whereas for $b\bar{b}$ pairs from 0.03 to 0.07. On the other hand, at LHC, in Pb-Pb collisions at $\sqrt{s_{NN}} = 5.5$ TeV, $\sigma_{\text{inel}} \simeq 60$ mb and for a 5.1% centrality selected sample, corresponding to the same impact parameter above, $N_{\text{coll}} = 1670$ [13]. In this latter case, the average multiplicity of $c\bar{c}$ pairs ranges from 95 to 256 and from 2.4 to 7.2 for $b\bar{b}$ pairs. It should be noted that these estimates do not take into account possible structure function saturation effects, which are predicted to reduce heavy quark cross section at LHC [14]. Since there is not a clearcut evidence of this phenomenon as yet, we stick to the traditional picture of N_{coll} scaling, though this possibility is worthy of consideration in the future.

The other key ingredient in our calculation is the volume V. In order to extrapolate from SPS to RHIC, we take advantage of the fact that V is proportional to the average multiplicity of pions. In Pb-Pb at SPS at $\sqrt{s_{NN}} = 17.2$ GeV, $V \simeq 3.5 \, 10^3$ fm^3 in full phase space [9], and $\langle \pi^+ + \pi^- \rangle \simeq 1258$ [15]. At RHIC, at $\sqrt{s_{NN}} = 200$ GeV in full phase space $\langle \pi^+ + \pi^- \rangle \simeq 3343$ [16] leading to $V \approx 10^4$ fm^3. This extrapolation assumes very little variation of temperature and baryon-chemical potentials from SPS to RHIC, which approximately holds [9]. In order to extrapolate from RHIC to LHC, we pragmatically use the saturation model (which proved to be successful in extrapolating multiplicity from SPS to RHIC) which predicts an increase in $\langle n_{\text{ch}} \rangle$ by a factor $\simeq 4.5$ from $\sqrt{s_{NN}} = 200$ GeV to 5.5 TeV [17]. This means that $V_{LHC} \approx 4.5 \, 10^4$ fm^3 with a fair uncertainty up to a factor 2. By using the above values for V, and, conservatively, the upper estimates for $\langle v_c \rangle$ and $\langle v_b \rangle$ we obtain, from Eqs. (4) and (5) $R_c = \mathcal{O}(10^{-2}), R_b = \mathcal{O}(10^{-7}), R_{cb} = \mathcal{O}(10^{-4})$ at RHIC and $R_c = \mathcal{O}(1), R_b = \mathcal{O}(10^{-4}), R_{cb} = \mathcal{O}(10^{-1})$ at LHC, by using as input parameters $m_c, m_b, m_{u,d}$ and T the same quoted below Eq. (4). Therefore, both at RHIC and LHC energies the formula (6) should be fairly accurate, because either the R's are $\ll 1$ or, like in the charm sector at LHC, $\langle v_c \rangle \gg 1$ and $R_c/\langle v_c \rangle \ll 1$.

We can now perform our predictions. To estimate the η_f's, we use the approximation $a_{f1} \simeq g_{\text{eff}} V [(m_f + m_u)T/2\pi]^{3/2} \exp[(-m_f + m_u)/T]$ with input parameters like for the ratios R_f. We then obtain $\eta_c \approx 1.7 - 10$ and $\eta_b \approx (3.5 - 8.2) \cdot 10^6$ at RHIC and $\eta_c \approx 12 - 34$ and $\eta_b \approx (0.63 - 1.9) \cdot 10^8$ at LHC. With these numbers, assuming the mass of multiply heavy flavoured hadrons to be the sum of its quarks constituent masses, using Eq. (6) with $\lambda_j = 1$ (i.e. taking vanishing chemical potentials) and appropriate spins, we get average *primary* yields of doubly charmed baryons (like Ξ_{cc} and Ω_{cc}) between $0.7 \cdot 10^{-4}$ and $7 \cdot 10^{-3}$ in central collisions at RHIC and between 0.019 and 0.38 at LHC. For mixed charmed-beautiful hadrons (like Ξ_{bc}, Ω_{bc} and B_c meson), the yields should range between $4 \cdot 10^{-7}$ and $6 \cdot 10^{-5}$ at RHIC and $3 \cdot 10^{-4}$ and 0.022 at LHC. For doubly beautiful baryons (like Ξ_{bb} and Ω_{bb}), our estimates range between $2 \cdot 10^{-9}$ and $3 \cdot 10^{-8}$ at RHIC and between $2.6 \cdot 10^{-6}$ and $7 \cdot 10^{-5}$ at LHC. For the Ω_{ccc} baryon, the predicted yields are affected by a large uncertainty due to the cubic dependence on η_c; they range between $7 \cdot 10^{-7}$ and 10^{-4} at RHIC and between 10^{-3} and 0.03 at LHC. For charmed baryon yields at LHC, the predictions of the Eq. (6) turn out to be in good agreement with a preliminary calculation with the exact formula. All of the previous yields are enhanced by the feeding from heavier states, even by factor of about 4-5; another factor ≈ 2 comes from antiparticle yields. These factors should roughly compensate for the limited rapidity window accessible to experiments. Therefore, while at RHIC only doubly charmed hadrons seem to be within reach, at LHC, with a statistics of 10^7 central events/year, doubly and triply charmed, charmed-beautiful, and perhaps doubly beautiful hadrons could in principle be observed.

HADRONIC VS HEAVY ION COLLISIONS

We can now compare the above yields with the predictions by production models based on QCD hard scattering. At LHC, a model where heavy diquarks produced in $\mathscr{O}(\alpha_S^4)$ diagrams are assumed to fully hadronize into ccq-baryons, yields an upper bound on inclusive production of $(10^{-4} - 10^{-3})\langle v_c \rangle$ at $\sqrt{s} = 14$ TeV [18] in pp, to be compared with $g_{\text{eff}}(0.8-2) \cdot 10^{-3}\langle v_c \rangle$ from coalescence in HIC at $\sqrt{s_{NN}} = 5.5$ TeV. A larger difference is found in the Ξ_{bc} sector, where in p$\bar{\text{p}}$ at $\sqrt{s} = 1.8$ TeV the 1S-wave cross-section is predicted to be around 1 nb [19], implying a ratio $\langle \Xi_{bc} \rangle / \langle v_b \rangle \sim 10^{-5}$ to be compared with $g_{\text{eff}}^{1S}(3-9) \cdot 10^{-4}$ from coalescence, i.e. at least one order of magnitude larger. Since the production process is $\mathscr{O}(\alpha_S^6)$, the difference is even larger for Ω_{ccc}, for which a recent calculation [20] predicts a ratio $\langle \Omega_{ccc} \rangle / \langle v_c \rangle = \mathscr{O}(10^{-7})$ in pp at $\sqrt{s} = 14$ TeV; this is between 2-3 orders of magnitude lower than our estimated ratio from coalescence at $\sqrt{s_{NN}} = 5.5$ TeV i.e. $(0.1-1) \cdot 10^{-4}$. For charmonia and B_c, the two mechanisms give closer predictions. For $\langle J/\psi \rangle / \langle v_c \rangle$, the difference is estimated to be a factor about 2.5 at LHC [12], whilst for B_c cross sections calculations [21] at $\sqrt{s} = 14$ TeV imply a ratio $\langle B_c \rangle / \langle v_b \rangle = \mathscr{O}(10^{-3})$, around the same as from coalescence at $\sqrt{s_{NN}} = 5.5$ TeV. Also, it should be pointed out that, unlike J/ψ and B_c, multiply heavy flavoured hadrons have not been measured in hadronic collisions, so the predictions of the models based on hard scattering are still to be checked.

The predominant uncertainty on the previous estimates is that on heavy quark cross section. Other relevant uncertainties are the those on masses, effective degeneracy, extrapolated temperature and charged multiplicities, a modulation of the production as a function of rapidity as well as the replacement of the approximated formula (6) with the exact one. Yet, all these effects, which will be discussed in detail in a forthcoming paper [10], cannot alter the ratios of multiply to singly flavoured hadron yields by one order of magnitude. So the conclusion remains that if statistical coalescence scheme applies, a large enhancement in the measurement of $\langle \Xi_{bc} \rangle / \langle B \rangle$, which becomes dramatic for

$\langle\Omega_{ccc}\rangle/\langle D\rangle$, could be found in heavy ion collisions with respect to pp at the LHC energy. This could be a clear indication of QGP formation.

ACKNOWLEDGMENTS

The author would like to thank the organizer of this conference for an enjoyable and stimulating atmosphere. Useful discussions with F. Antinori and D. Treleani are gratefully acknowledged.

REFERENCES

1. H. van Hees and R. Rapp, Phys. Rev. C **71**, 034907 (2005).
2. S. S. Adler *et al.*, PHENIX Coll., Phys. Rev. C **72**, 024901 (2005).
3. M. Schroedter, R. Thews and J. Rafelski, Phys. Rev. C **62**, 024905 (2000)
4. P. Braun-Munzinger and J. Stachel, Phys. Lett. B **490**, 196 (2000).
5. M. Gazdzicki and M. Gorenstein, Phys. Rev. Lett. **83**, 4009 (1999).
6. R. Thews, M. Schroedter and J. Rafelski, Phys. Rev. C **63**, 054905 (2001); M. Gorenstein *et al.*, Phys. Lett. B **509**, 277 (2001); L. Grandchamp and R. Rapp, Phys. Lett. B **523**, 60 (2001); A. Andronic *et al.*, Phys. Lett. B **571**, 36 (2003); A. Kostyuk, nucl-th/0502005.
7. F. Becattini, Phys. Rev. Lett. **95**, 022301 (2005).
8. W. Broniowski and W. Florkowski, hep-ph/0202059.
9. F. Becattini *et al.*, Phys. Rev. C **69**, 024905 (2004)
10. F. Becattini, A. Bieniek, in preparation.
11. M. Cacciari, P. Nason and R. Vogt, hep-ph/0502203
12. M. Bedjidian *et al.*, hep-ph/0311048
13. D. Miskowiec, http://www-linux.gsi.de/ misko/overlap/
14. D. Kharzeev and K. Tuchin, Nucl. Phys. A **735**, 248 (2004)
15. S. Afanasiev *et al.*, Phys. Rev. C **66**, 054902 (2002)
16. I. Bearden *et al.*, nucl-ex/0403050
17. M. Nardi, J. Phys. Conf. **5**, 148 (2005)
18. A. Berezhnoi *et al.*, Phys. Rev. D **57**, 4385 (1998)
19. V. Kiselev and A. Likhoded, Phys. Usp. **45**, 455 (2002)
20. M. Nobary and R. Sepahvand, Phys. Rev. D **71**, 034024 (2005)
21. I. Gouz *et al.*, Phys. Atom. Nucl. **67**, 1559 (2004)

Latest results from the NA57 experiment

G. E. Bruno on behalf of the NA57 Collaboration [1]

Dipartimento IA di Fisica dell'Università e del Politecnico di Bari and INFN, Bari, Italy

Abstract. The NA57 experiment at the CERN SPS has measured the production of strange and multi-strange particles in Pb-Pb (and p-Be) collisions at mid-rapidity. The collective dynamics of the collision is studied in the transverse and longitudinal directions as a function of centrality and beam momentum and analysed on the basis of hydrodynamical models. Central-to-peripheral nuclear modification factors at 158 A GeV/c are presented and compared with other measurements and theory.

Keywords: QGP, Heavy Ions, Strange particles, collective dynamics
PACS: 12.38.Mh, 25.75.Nq, 25.75.Dw

INTRODUCTION

The measurement of strange particle production provides a powerful tool to study the dynamics of the reaction in heavy-ion collisions. In particular, an enhanced production of strange particles in nucleus–nucleus collisions with respect to proton–induced reactions was suggested long ago as a possible signature of the phase transition from colour confined hadronic matter to a Quark Gluon Plasma (QGP) [1]. Results on hyperon enhancements at 160 and 40 A GeV/c obtained by NA57 can be found in reference [2] and they will not be discussed in this paper.

Important insights into the reaction dynamics can been determined from the p_T distributions of strange particles: at low transverse momenta ($p_T \lesssim 2$ GeV/c) they allow to study the transverse expansion of the collision [3]; at higher values ($p_T \gtrsim 2$ GeV/c) the p_T dependence of the nuclear modification factors can be used to probe the properties of the medium produced after the collisions [4, 5].

Rapidity distributions provide a tool to study the longitudinal dynamics. Hydrodynamical properties of the expanding matter created in heavy ion reactions have been discussed by Landau [6] and Bjorken [7] using different initial conditions. In both scenarios, thermal equilibrium is quickly achieved and the subsequent isentropic expansion is governed by hydrodynamics.

COLLECTIVE DYNAMICS

The presence of strong collective dynamics in relativistic heavy-ion collisions has been widely proven and its features measured (see, e.g., ref. [8, 9] for reviews). While longitudinal 'flow' is not necessarily a signature for nuclear collectivity, because it may

---

[1] For the author list see http://wa97.web.cern.ch/WA97/NA57authors/index.html.

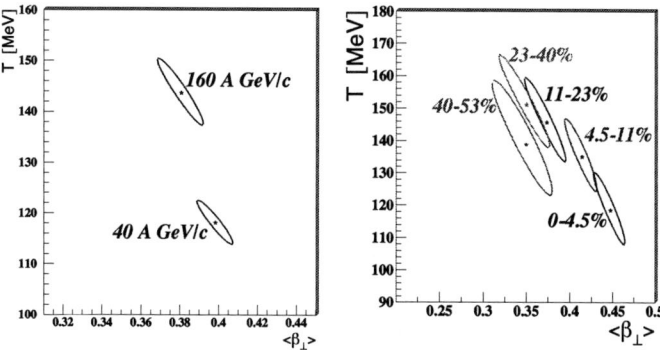

FIGURE 1. Contour plots in the $\langle \beta_\perp \rangle$–$T$ plane at the 1σ confidence level. Left: energy dependence for the most central 53% of the Pb-Pb interactions. Right: centrality dependence at 160 A GeV/c.

be due simply to incomplete stopping of the two colliding nuclei, transverse collective expansion flow can only be driven by the transverse pressure gradients between the hot center of the fireball and the surrounding vacuum. The relative extent of the longitudinal flow with respect to the transverse flow can be used to determine the amount of nuclear stopping.

Transverse dynamics

The *transverse* dynamics of the collisions has been studied at 160 and 40 A GeV/c respectively in ref. [10] and [11] from the analysis of the transverse momentum distributions of the strange particles in the framework of the blast-wave model [3]. Here we recall some results of this analysis. In the model the double differential cross-section for a particle species j assumes the form:

$$\frac{d^2 N_j}{m_T dm_T dy} = \mathscr{A}_j \int_0^{R_G} m_T K_1\left(\frac{m_T \cosh\rho}{T}\right) I_0\left(\frac{p_T \sinh\rho}{T}\right) r\, dr \qquad (1)$$

where $m_T = \sqrt{p_T^2 + m^2}$, $\rho(r) = \tanh^{-1}\beta_\perp(r)$, K_1 and I_0 are two modified Bessel functions and $\mathscr{A}_j$ is a normalization constant. The parameters of the model are the thermal freeze-out temperature T and the transverse flow velocity $\beta_\perp$, whose average has been computed assuming a linear transverse profile $\beta_\perp(r) = \beta_S \left(\frac{r}{R_G}\right)$ [10].

In fig. 1 we show the energy (left panel) and the centrality (right panel) dependences of the freeze-out parameters obtained from the combined fits of the strange particles m_T spectra to formula 1. With increasing centrality, the transverse flow velocity increases and the freeze-out temperature decreases; the temperature also decreases at lower energy.

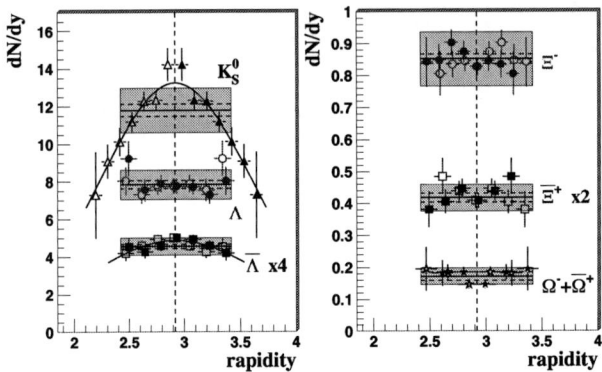

FIGURE 2. Rapidity distributions of strange particles in the most central 53% of Pb-Pb interactions at 158 A GeV/c. Closed symbols are measured data, open symbols are measured points reflected around mid-rapidity. The $\overline{\Lambda}$ and $\overline{\Xi}^+$ results have been scaled by factors 4 and 2, respectively, for display purposes. The superimposed boxes show the yields measured in one unit of rapidity with the dashed and full lines indicating the statistical and systematic errors, respectively.

Longitudinal dynamics

Results on the rapidity distributions of strange particles are discussed in detail in ref. [12]; here we outline the main results of the analysis.

The measured rapidity distributions for the centrality range corresponding to the most central 53% of the Pb-Pb inelastic cross-section (total sample) are shown in fig. 2 with closed symbols. The symmetry of the Pb-Pb colliding system allows us to reflect the rapidity distributions around mid-rapidity (open symbols in fig. 2). The rapidity distributions of Λ, Ξ^-, $\overline{\Xi}^+$ and $\Omega^- + \overline{\Omega}^+$ are compatible, within the error bars, with being flat within the NA57 acceptance window. For the K_S^0 and $\overline{\Lambda}$ spectra, instead, we observe a rapidity dependence. The rapidity distributions for these particles are well described by Gaussians centered at mid-rapidity.

The rapidity distributions can be used to extract information about the *longitudinal* dynamics. We use an approach outlined in ref. [3] (i.e., the same blast-wave model used for the study of the transverse dynamics) and [13], where, respectively, Bjorken [7] and Landau [6] hydrodynamics are folded with a thermal distribution of the fluid elements.

In fig. 3 (left panel) the measured rapidity distributions are compared with the expectation for a stationary thermal source and with a longitudinally boost-invariant superposition of multiple isotropic, locally-thermalized sources (i.e. Bjorken hydrodynamics). The average longitudinal flow velocity is evaluated from the fit to be $<\beta_L> = 0.42 \pm 0.03$ with $\chi^2/ndf = 28.2/32$. The freeze-out temperature has been fixed to the value $T_f = 144$ MeV obtained from the analysis of the transverse expansion. The average *transverse* flow velocity has been determined to be $<\beta_\perp> = 0.38 \pm 0.02$, i.e. slightly less than the *longitudinal* flow velocity obtained from the same data sample; this indicates substantial stopping of the incoming nuclei.

In the Landau model the width of the rapidity distribution is sensitive to the speed of

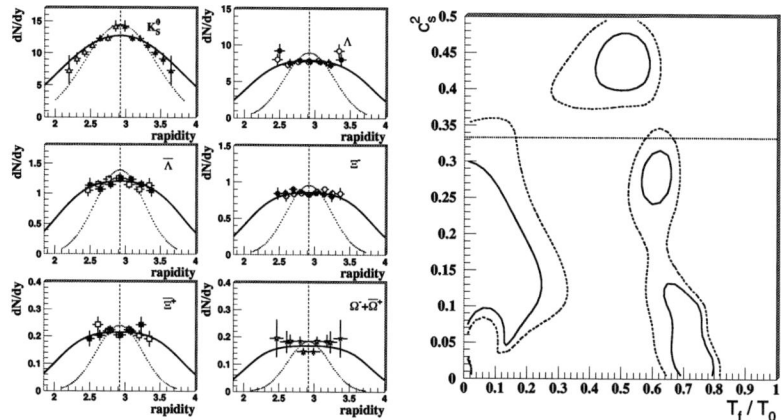

FIGURE 3. Left: Rapidity distributions of strange particles for the centrality range corresponding to the most central 53% of the inelastic Pb-Pb cross-section as compared to the thermal model calculation (dotted lines, in red) and a thermal model with Bjorken longitudinal flow (full lines, in black). Right: The square of the speed of sound in the medium (in unit of c^2) versus the ratio of the freeze-out temperature to the inital temperature. The 1σ (full curves) and the 3σ (dashed curves) confidence contours are shown. The dotted line at $c_s^2 = 1/3$ shows the ideal gas limit.

sound and to the ratio of the freeze-out temperature to the initial temperature. Landau hydrodynamics can also reproduce simultaneously the distributions for all the strange particles considered ($\chi^2/ndf \simeq 28/32$) but we are not able to put stringent constraints on both the speed of sound and the ratio T_f/T_0. The confidence level contours in the c_s^2 vs. $\frac{T_f}{T_0}$ parameter space are shown in the right-hand panel of fig. 3.

R_{CP} NUCLEAR MODIFICATION FACTORS

The main results on the nuclear modification factors at 160 A GeV/c (i.e. $\sqrt{s_{NN}}$=17.3 GeV) are discussed in this section. The complete sets of results and the details of the analysis can be found in ref. [14].

The central-to-peripheral nuclear modification factor is defined as

$$R_{CP}(p_T) = \frac{\langle N_{coll}\rangle_P}{\langle N_{coll}\rangle_C} \times \frac{d^2N_{AA}^C/dp_T dy}{d^2N_{AA}^P/dp_T dy}, \qquad (2)$$

where $\langle N_{coll}\rangle_C$ and $\langle N_{coll}\rangle_P$ are the average numbers of nucleon–nucleon (NN) collisions for *central* (C) and *peripheral* (P) classes of collisions. The factor would be equal to unity if the Pb-Pb collision were a mere superposition of N_{coll} independent nucleon–nucleon collisions. The collision centrality is determined using the charged particle multiplicity N_{ch} in the pseudorapidity range $2 < \eta < 4$, sampled by the microstrip silicon detectors (MSD) as described in [15]. In table 1 we define the centrality classes used in

TABLE 1. Average number of participants and of NN collisions with their systematic errors.

Class (% σ_{inel}^{Pb-Pb})	0–5.0%	10.0–20.0%	20.0–30.0%	30.0–40.0%	40.0–55.0%
$\langle N_{part} \rangle$	345.3 ± 1.7	214.7 ± 5.8	143.0 ± 6.6	92.6 ± 6.4	49.5 ± 5.0
$\langle N_{coll} \rangle$	779.2 ± 26.6	421.7 ± 26.1	247.7 ± 21.5	140.5 ± 16.2	63.8 ± 9.8

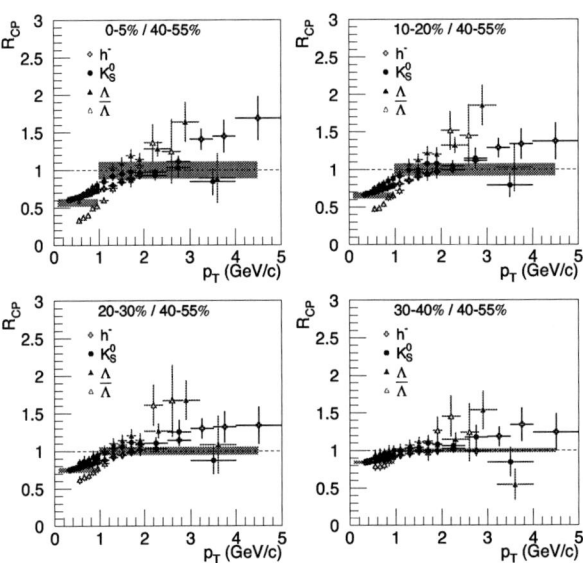

FIGURE 4. Centrality dependence of $R_{CP}(p_T)$ for h^-, K_S^0, Λ and $\overline{\Lambda}$ in Pb–Pb collisions at $\sqrt{s_{NN}} = 17.3$ GeV. Shaded bands centered at $R_{CP} = 1$ represent the systematic error due to the uncertainty in the ratio of the values of $\langle N_{coll} \rangle$ in each class; shaded bands at low p_T represent the values expected for scaling with the number of participants, together with their systematic error.

this analysis along with the corresponding values of $\langle N_{part} \rangle$ and $\langle N_{coll} \rangle$ with their systematic errors. We use class 40–55% as the reference peripheral class in the denominator of R_{CP}, see Eq. (2), and vary the 'central' class in the numerator from 0–5% to 30–40%; the results are shown in fig. 4. The shaded bands centered at $R_{CP} = 1$ represent the p_T-independent systematic error due to the uncertainty in the ratio $\langle N_{coll} \rangle_P / \langle N_{coll} \rangle_C$, while the shaded bands at low p_T represent the R_{CP} values corresponding to N_{part}-scaling, with the band indicating the systematic error due to the uncertainty in the ratio $\langle N_{part} \rangle_C / \langle N_{part} \rangle_P$. In fig. 5 we compare our results to R_{CP} measurements at the SPS and at RHIC. In the left-hand panel, the WA98 π^0 data [16] for the ratio 1–6%/22–43% in Pb–Pb collisions at $\sqrt{s_{NN}} = 17.3$ GeV are plotted together with the NA57 h^- and K_S^0 data for the same centrality classes. The K_S^0 R_{CP} is approximately constant at 0.9 for $p_T > 1$ GeV/c and is significantly larger than that measured by the WA98 Collaboration for π^0 ($R_{CP} \approx 0.6$), even when taking into account the normalization systematic errors, independent for the two experiments. The h^- data from NA57 are compatible, within the

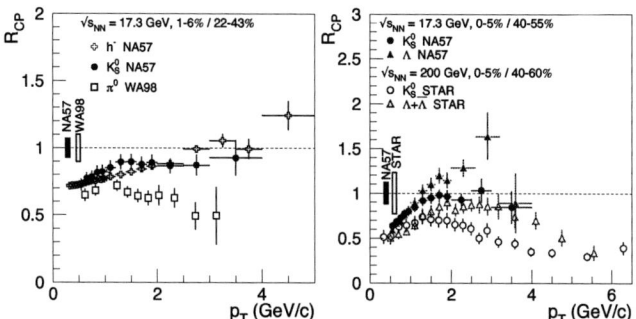

FIGURE 5. Left: $R_{CP}(p_T)$ for h^- and K_S^0 from NA57 and π^0 from WA98 [16] in Pb–Pb collisions at $\sqrt{s_{NN}} = 17.3$ GeV. Right: $R_{CP}(p_T)$ for K_S^0 and Λ in Pb–Pb at $\sqrt{s_{NN}} = 17.3$ GeV (NA57) and in Au–Au at $\sqrt{s_{NN}} = 200$ GeV (STAR) [17]; slightly different peripheral classes are employed for the comparison. The bars centered at $R_{CP} = 1$ represent the normalization errors; the point-by-point bars are the quadratic sum of statistical and systematic errors.

systematic errors, with the π^0 data from WA98 for $p_T \lesssim 1.5$ GeV/c, where the h^- sample is expected to be dominated by π^-. For higher p_T, h^- have a larger R_{CP} than π^0; this may be due to increasing contributions from K^- and $\bar{p}$ in the h^- sample. The comparison for K_S^0 and Λ at SPS and RHIC (STAR data for Au–Au at $\sqrt{s_{NN}} = 200$ GeV [17]) is presented in the right-hand panel of Fig. 4. In the p_T range covered by our data, up to 4 GeV/c, the *relative* pattern for K_S^0 and Λ is similar at the two energies, while absolute values are higher at SPS than at RHIC. At RHIC, the larger R_{CP} for Λ with respect to kaons [17] or, more generally, for baryons with respect to mesons [18], in the intermediate p_T range, 2–4 GeV/c, has been interpreted as due to parton coalescence in a high-density medium with partonic degrees of freedom. Our data show that a similar Λ–K pattern is present also at $\sqrt{s_{NN}} = 17.3$ GeV. We note that in our case such a pattern may also be explained in terms of larger Cronin effect for Λ with respect to kaons. We have compared our results with perturbative-QCD-based theoretichal predictions [19, 20] with and without in-medium energy loss, which include the initial-state partonic intrinsic p_T broadening tuned on the original Cronin effect data [21]. The calculations with in-medium energy loss describe the data much better [14].

CONCLUSION

We have measured the dN/dp_T and dN/dy distributions of high purity samples of K_S^0, Λ, Ξ and Ω particles produced at central rapidity in Pb-Pb collisions over a wide centrality range of collision (i.e. the most central 53% of the Pb–Pb inelastic cross-section).

The analysis of the transverse mass spectra of strange particles in Pb-Pb collisions at SPS energies suggests that after a central collision the system expands explosively and then it freezes-out when the temperature is of the order of 120 MeV, with an average transverse flow velocity of about one half of the speed of light. Similar transverse flow velocities are measured at 40 and 158 A GeV/c but the freeze-out temperature is lower at the lower energy. The results on the centrality dependence of the expansion dynamics indicate that with increasing centrality the transverse flow velocity increases and the freeze-out temperature decreases.

Boost-invariant Bjorken hydrodynamics can describe simultaneously the rapidity spectra of all the strange particles under study with $\chi^2/ndf \approx 1$, yielding an average longitudinal flow velocity $<\beta_L> = 0.42 \pm 0.03$, sligthly larger than the measured transverse flow. The *longitudinal* flow velocity being close to the transverse one is suggestive of large nuclear stopping. A fairly good description is also provided by Landau hydrodynamics, which allows us to put constraints in the parameter space of the speed of sound in the medium and the ratio of the freeze-out temperature to the initial temperature.

Central-to-peripheral nuclear modification factors for K_S^0, Λ, $\overline{\Lambda}$ and h^- in Pb–Pb collisions at top SPS energy have been measured as a function of p_T up to about 4 GeV/c. At low p_T, R_{CP} agrees with N_{part} scaling for all the particles under consideration, except the $\overline{\Lambda}$, for which the yields at low p_T are found to increase slower than the number of participants. For $p_T \gtrsim 1$ GeV/c, K_S^0, Λ and $\overline{\Lambda}$ show a pattern similar to that observed in Au–Au collisions at top RHIC energy, although the R_{CP} values are found to be larger at SPS. At RHIC, this pattern has been interpreted in the framework of models that combine parton energy loss with hadronization via coalescence, at intermediate p_T, and via fragmentation, at higher p_T. The measured K_S^0 0–5%/40–55% R_{CP} is not reproduced by a theoretical calculation that includes only initial-state nuclear effects. The data can be better described by including final-state parton energy loss as predicted for SPS energy on the basis of RHIC data.

REFERENCES

1. J. Rafelski and B. Müller, *Phys. Rev. Lett.* **48** 1066 (1982), ibidem **56** 2334 (1986).
2. F. Antinori et al., *Phys. Lett.* B **595** 68–74 (2004); G.E. Bruno et al., *J. Phys.* G **30** S1329–1332 (2004).
3. E. Schnedermann, J. Sollfrank and U. Heinz *Phys. Rev.* C **48** 2462 (1993), ibidem **50** 1675 (1994).
4. J. Adam et al., *Phys. Rev. Lett.* **91** 172302 (2003); S.S. Adler et al., *Phys. Rev.* C **69** 034910 (2004).
5. M. Gyulassy and X.N. Wang, *Phys. Rev. Lett.* **68** 1480 (1992); ibidem *Nucl. Phys.* B **420** 583 (1994);
   R. Baier, D. Schiff and B.G. Zakharov, *Annu. Rev. Nucl. Part. Sci.* **50** 37 (2000);
   C.A. Salgado and U.A. Wiedemann, *Phys. Rev.* D **68** 014008 (2003).
6. L.D. Landau, *Izv. Akad. Nauk.* SSSR **17** 51 (1953);
   S. Belenkij and L.D. Landau, *Usp. Fiz. Nauk.* **56** 309 (1955); ibidem *Nuovo Cimento* **3** 15 (1956).
7. J.D. Bjorken, *Phys. Rev.* D **27** 140 (1983).
8. W. Florkowski and W. Broniowski, *Acta. Phys. Pol.* B **35** 2895–2910 (2004).
9. G. Torrieri and J. Rafelski, *Phys. Rev.* C **68** 034912 (2003).
10. F. Antinori et. al., *J. Phys.* G **30** 823–840 (2004).
11. G.E. Bruno et. al., *J. Phys.* G **31** S127–S133 (2005).
12. F. Antinori et. al., *J. Phys.* G in press; *pre-print* nucl-ex/0509009 (2005).
13. B. Mohanty and J. Alam, *Phys. Rev.* C **68** 064903 (2003).
14. F. Antinori et al., *Phys. Lett.* B **623** 17–25 (2005).
15. F. Antinori et al., *Eur. Phys. J.* G **18** 57 (2000); F. Antinori et al., *J. Phys.* G **31** 321–335 (2005).
16. M.M. Aggarwal et al., *Eur. Phys. J.* C **23** 225 (2002).
17. J. Adams et al., *Phys. Rev. Lett.* **92** 0502002 (2004).
18. S.S. Adler et al., *Phys. Rev.* C **69** 034909 (2004).
19. X.N. Wang, *Phys. Rev.* C **61** 064910 (2000); *Phys. Rev. Lett.* **81** 2655 (1998); private communication.
20. A. Dainese et al., *Eur. Phys. J.* C **38** 461 (2005); private communication.
21. J. Cronin et al., *Phys. Rev.* D **11** 3105 (1975); D. Antreasyan et al., *Phys. Rev.* D **19** 764 (1979).

J/ψ production in indium-indium collisions at SPS energies

P. Pillot[*], R. Arnaldi[†], R. Averbeck[**], K. Banicz[‡,§], J. Castor[¶],
B. Chaurand[‖], C. Cicalo[††], A. Colla[†], P. Cortese[†], S. Damjanovic[§],
A. David[‡,‡‡], A. de Falco[††], A. Devaux[¶], A. Drees[**], L. Ducroux[*],
H. En'yo[§§], A. Ferretti[†], M. Floris[††], P. Force[¶], N. Guettet[‡,¶], A. Guichard[*],
H. Gulkanian[¶¶], J. Heuser[§§], M. Keil[‡,§], L. Kluberg[‡,‖], J. Lozano[‡‡],
C. Lourenço[‡], F. Manso[¶], A. Masoni[††], P. Martins[‡,‡‡], A. Neves[‡‡],
H. Ohnishi[§§], C. Oppedisano[†], P. Parracho[‡], G. Puddu[††], E. Radermacher[‡],
P. Ramalhete[‡,‡‡], P. Rosinsky[‡], E. Scomparin[†], J. Seixas[‡,‡‡], S. Serci[††],
R. Shahoyan[‡,‡‡], P. Sonderegger[‡‡], H.J. Specht[§], R. Tieulent[*], G. Usai[††],
R. Veenhof[‡,‡‡] and H.K. Wöhri[‡,‡‡]

[*]*IPN-Lyon, Univ. Claude Bernard Lyon-I and CNRS-IN2P3, Lyon, France*
[†]*Univ. di Torino and INFN, Italy*
[**]*SUNY Stony Brook, New York, USA*
[‡]*CERN, Geneva, Switzerland*
[§]*Univ. Heidelberg, Heidelberg, Germany*
[¶]*LPC, Univ. Blaise Pascal and CNRS-IN2P3, Clermont-Ferrand, France*
[‖]*LLR and CNRS-IN2P3, Palaiseau, France*
[††]*Univ. di Cagliari and INFN, Cagliari, Italy*
[‡‡]*IST-CFTP, Lisbon, Portugal*
[§§]*RIKEN, Wako, Saitama, Japan*
[¶¶]*YerPhI, Yerevan, Armenia*

Abstract.
The NA60 experiment collected data on dimuon production in indium-indium collisions at 158 GeV/c per incident nucleon, in year 2003, to contribute to the clarification of several questions raised by previous experiments studying high-energy heavy-ion physics at the CERN SPS in search of the quark gluon plasma. Among these previous results stands the observation, by NA50, that the production yield of J/ψ mesons is suppressed in central Pb-Pb collisions beyond the normal nuclear absorption defined by proton-nucleus data. By comparing the centrality dependence of the suppression pattern between different colliding systems, S-U, Pb-Pb and In-In, we should be able to identify the corresponding scaling variable, and the physics mechanism driving the suppression. In this paper, we will present the ratio of J/ψ and Drell-Yan production cross-sections in indium-indium collisions, in three centrality bins, and how these values compare to previous measurements. We will also present a study of the transverse momentum distributions of the J/ψ mesons, in seven centrality bins.

Keywords: Quark-gluon plasma, Resonance, J/psi, Dimuon, Transverse momentum
PACS: 25.75.Dw Particle and resonance production - 25.75.Nq Quark deconfinement, quark-gluon plasma production, and phase transitions - 13.20.Gd Decays of J/psi, Upsilon, and other quarkonia

PHYSICS MOTIVATIONS

Lattice QCD predicts that, above a critical temperature or energy density, strongly interacting matter undergoes a phase transition to a new state named Quark and Gluon Plasma, where quarks and gluons are no longer confined into hadrons and chiral symmetry is restored. Since 1986, many experiments have studied high energy nuclear collisions at the CERN SPS to search for this QCD phase transition through several signatures. The observations done so far have provided evidence that a new state of matter is produced in heavy ion collisions at SPS energies. However, several questions remain open or have been raised by these observations and require further work to be clarified.

In the following, we will concentrate in the J/ψ suppression, a signature of the QGP formation which has been proposed by Matsui and Satz in 1986 [1]. The NA3, NA38, NA50 and NA51 experiments studied the J/ψ production in different colliding systems. The results obtained in proton-nucleus collisions at 200, 400 and 450 GeV/c are well reproduced by taking only into account the "normal" nuclear absorption. The same behaviour seems to be observed in S-U data at 200 GeV/c while the J/ψ production in Pb-Pb collisions at 158 GeV/c, measured by the NA50 experiment as a function of the centrality of the collision, shows a pattern different from the expected one, derived from the p-A data. An "abnormal suppression" of the J/ψ is observed above a certain centrality threshold. This behaviour can be interpreted as the melting of the χ_c in hot and dense nuclear matter, leading to the suppression of the $\sim 30\%$ of the measured J/ψ which normaly come from the χ_c decay. However, these results are not enough to fully understand the J/ψ behaviour and several questions remain open. What is the physics mechanism behind the J/ψ suppression seen in Pb-Pb collisions? What is the real impact of the χ_c feed-down on the observed J/ψ suppression pattern? Is the normal nuclear absorption of the χ_c identical to the one of the J/ψ? Can we compare p-A data at 450 GeV/c with Pb-Pb data at 158 GeV/c?

The NA60 experiment has been explicitly designed to provide new and accurate measurements which should help to clarify the previous questions, among others. In particular, in order to disentangle between different models proposed to explain the anomalous suppression of the J/ψ, it is important to understand which is the variable driving this suppression. For instance, the normal nuclear absorption is governed by the average length of nuclear matter traversed by the charmonium state, L. If the J/ψ is suppressed in a geometrical phase transition, such as in the percolation model [2], the scaling variable should be the density of nucleons participating in the collision, while in case of the formation of a QGP (thermal phase transition) it would be the local energy density [3]. The correlations between different variables, as a function of the centrality of the collision, depend on the colliding system, which offers an experimental handle to determine the variable driving the J/ψ suppression. For instance, if the good variable is L, the results obtained with different colliding systems will overlap when plotted as a function of L, but not when plotted as a function of the number of participants (N_{part}). In 2003, NA60 collected data in indium-indium collisions, to be compared with the S-U and Pb-Pb results obtained by NA38/NA50.

EXPERIMENTAL APPARATUS AND DATA SELECTION

NA60 is first composed of a muon spectrometer inherited from the NA50 experiment. It is made of 8 multi-wire proportionnal chambers for the tracking and 4 scintillator hodoscopes providing the dimuon trigger. The muon tracks are bent by a toroidal magnet. In 2003, two different values of the magnet current have been used: 4000 A and 6500 A. The lower current gives a better acceptance for low mass dimuons while the larger current gives a better mass resolution at the J/ψ mass. The last scintillator hodoscope is located behind a 120 cm iron wall to ensure that only muons can give a trigger. The spectrometer is protected by a 5.5 m long hadron absorber.

However, the multiple scattering and energy loss fluctuations of the muons in the absorber degrade the mass resolution. In order to overcome this problem, NA60 placed a radiation-hard and high granularity silicon tracking detector upstream of the absorber, $\sim$ 7 cm downstream of the center of the target system. This vertex telescope was installed in a 2.5 T dipole magnet in order to measure the momentum of charged particles. The muon tracks reconstructed in the muon spectrometer can then be extrapolated to the target region and matched, in coordinate and momentum, with the charged particle tracks reconstructed in the telescope, thus improving significantly the dimuon mass resolution. Quantitatively, from the indium data, the width of the J/ψ mass peak improves from $\sim$ 105 MeV/c^2 to $\sim$ 70 MeV/c^2. Moreover, the back-tracking of the charged particles allows us to determine the position of the interaction vertex with a resolution better than 20 μm perpendicularly to the beam axis and better than 200 μm in the beam direction. Such a resolution permits to distinguish the 7 indium targets in-between the target box windows, as we can see on the left panel of Fig. 1. In addition, it becomes possible to determine the origin of the muons, allowing us to disentangle between prompt dimuons and offsetted dimuons (from D meson decays). The only draw-back of requiring the track matching is the reduction of the available statistics, since the matching efficiency in the J/ψ mass region is $\sim$ 70 % (regardless of the centrality of the collision). Details on the tracking, vertexing and muon track matching can be found in Ref. [4].

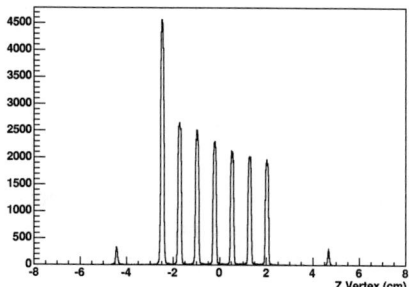 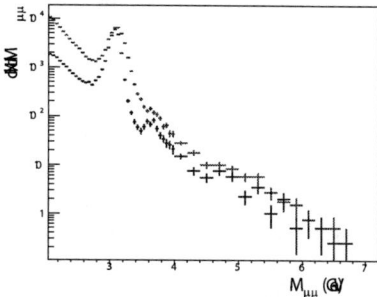

FIGURE 1. Left: Z-coordinate of the reconstructed vertices, showing the 7 indium targets surrounded by the two vacuum windows of the target box. To cover the entire beam profile, the first target has a bigger transverse size (12 mm diameter). **Right:** comparison between the dimuon mass spectra obtained from the 6500 A data sample, before (above) and after (below) applying the muon track matching. The ψ' peak becomes more clearly visible if the matching is required.

In 2003, the vertex telescope was made of eight small planes followed by four stations each composed by two large planes, providing 12 tracking points and covering the angular acceptance of the spectrometer ($3 < \eta_{lab} < 4$). The basic unit of these tracking planes is a single chip assembly — four chips for the small planes and eight chips for the large planes — made of one sensor chip bump-bonded to the radiation-tolerant ALICE1LHCb readout pixel chip. Each sensor is composed of 32×256 cells of 425×50 μm^2. A more detailed description of the vertex telescope can be found in Ref. [5].

The trajectory of the incoming ions was measured with ~ 20 μm resolution by a beam tracker made of four silicon microstrip detectors arranged in two double stations placed respectively 30 cm and 10 cm upstream of the center of the target system. Each detector plane was composed of 24 strips of 50 μm pitch operated at 130 K. Finally, a Zero Degree Calorimeter (ZDC) measured the energy released by the incoming nucleons which did not take part in the interaction, giving access to the centrality of the collision.

In the 5 weeks long run of year 2003, approximately 4×10^{12} ions were delivered on the seven 1.5 mm thick indium targets, giving rise to ~ 230 million dimuon triggers. The average beam intensity was 5×10^7 ions per 4.8 s long spill, every 16.8 s and the total interaction probability of ~ 20 %.

The dimuons are studied in the phase space window $2.92 < y_{lab} < 3.92$ and $-0.5 < cos\theta_{CS} < 0.5$, where y_{lab} and $cos\theta_{CS}$ are respectively the rapidity in the lab frame and the polar decay angle of the muons in the Collins-Soper reference system. In order to select only dimuons produced in In-In collisions and not in a collision induced by a beam fragment, we mainly use the vertexing information provided by the vertex telescope. Due to the good vertexing accuracy, we can easily select the events in which the first interaction took place in one of the seven indium targets. Moreover, as peviously mentioned, the muon track matching gives us access to the precise origin of the dimuons. Nevertheless, for the studies of J/ψ production, the use of the muon track matching is not mandatory. In this case, we can still reject muons coming from downstream of the target (e.g. due to collisions in the ZDC) by using the $p \cdot D_{targ}$ cut on the individual muons, where p is the muon momentum and D_{targ} is the transverse distance between the extrapolated muon track and the beam axis, at the target center. The invariant mass spectra of the opposite sign dimuons, obtained with and without using the muon track matching, are presented in the right part in Fig. 1.

J/ψ PRODUCTION IN INDIUM-INDIUM COLLISIONS

The opposite sign dimuon mass distribution above 2 GeV/c^2 contains the J/ψ and ψ′ resonances, sitting on a continuum composed of Drell-Yan (DY) dimuons, muon pairs from the simultaneous semi-muonic decay of D and $\overline{D}$ mesons and the combinatorial background from π and K decays. By fitting this distribution, we can extract the ratio between the J/ψ and the DY cross sections. The advantage of this ratio is that it is not sensitive to most experimental inefficiencies and to the integrated luminosity. Moreover, the DY process is a good reference to study J/ψ production, since the DY cross section is known to scale linearly with the product of the projectile and target mass numbers, and is insensitive to the nature of the medium formed in the collision.

The combinatorial background is estimated from the measured like sign pairs while the "signal" contributions to the opposite sign mass spectra, as well as their acceptances, are evaluated through a detailed Monte Carlo simulation. We used Pythia [6] as event generator with GRV94 LO parton distribution functions [7], and GEANT to reproduce the detector effects. The events are then reconstructed as the real data. The acceptances obtained from these simulations are between 12.4 % and 13.8 % for the J/ψ, and between 13.2 % and 14.1 % for the DY, depending on the magnetic field value and on whether we apply or not the muon track matching.

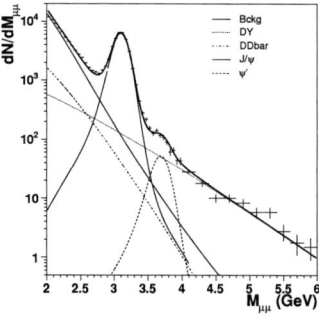 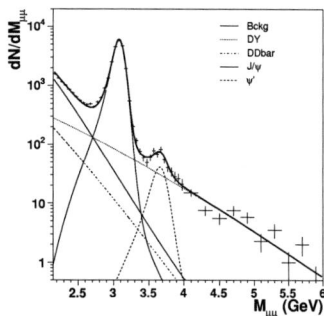

FIGURE 2. Fit to the dimuon mass spectra obtained without (left) and with (right) the muon track matching, for the 6500 A event sample.

The fit proceeds in three consecutive steps: first we determine the Drell-Yan yield from the mass region above 4.2 GeV/c^2, where it is the dominant contribution; then, keeping the Drell-Yan normalization fixed, we determine the charm normalization from the mass window 2.2 < M < 2.5 GeV/c^2. Finally, fixing the charm and Drell-Yan contributions, we extract the J/ψ and ψ' yields with a fit to the mass region 2.9 < M < 4.2 GeV/c^2. In this last step, also the exact position and width of the J/ψ peak are left as free parameters. The result of the fit to the invariant mass spectrum collected at 6500 A is shown Fig. 2, without (left) and with (right) the muon track matching. From this data set together with the one collected at 4000 A, and without the muon track matching, we obtain $\sim$ 520 DY events above 4.2 GeV/c^2 and $\sim$ 70 000 J/ψ counts.

To be able to directly compare our cross section ratio $B_{\mu\mu}\sigma(J/\psi)/\sigma(DY)$ with the results previously obtained by the NA38 and NA50 experiments, the DY cross section must be integrated in the mass range 2.9 < M < 4.5 GeV/c^2. By averaging the cross section ratio extracted from the 4000 A and the 6500 A samples, we obtained $B_{\mu\mu}\sigma(J/\psi)/\sigma(DY) = 20.2 \pm 0.9$ without matching and 20.8 ± 1.2 with matching. These results are almost insensitive to reasonable changes in the background normalization, different event selection criteria or fitting procedures.

As previously mentioned, the centrality of the collision can be estimated from the ZDC measurements, allowing us to study the J/ψ production within several centrality bins. To ensure the correctness of these measurements, additional selection cuts must be used, resulting in a $\sim$ 40 % loss of statistics. Consequently, only the analysis done without applying muon track matching provides reliable values of the J/ψ over DY cross section ratio and only three centrality bins can be considered.

For each bin, the variables related to the centrality of the collision, L and N_{part}, are calculated using the Glauber model, and the cross section ratio $B_{\mu\mu}\sigma(J/\psi)/\sigma(DY)$ is extracted using exactly the same procedure as the one presented above. To be able to compare our results to those obtained with other colliding systems, we must then apply an isospin correction which convert the In-In system to the p-p reference. In Fig. 3, the In-In measurements together with the p-p, p-nucleus, S-U and Pb-Pb measurements [8] are plotted as a function of L (left) and as a function of N_{part} (right). On the right panel, the results are divided by the normal nuclear absorption curve (derived from the p-A data), which is presented as a continuous line on the left panel.

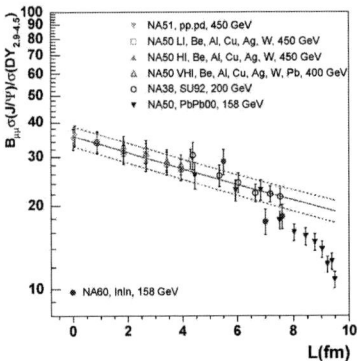
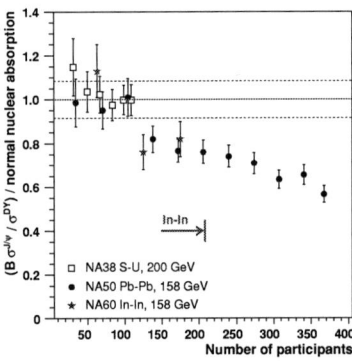

FIGURE 3. J/ψ over DY cross section ratio vs. L (left) and vs. N_{part} (right), for several collision systems, compared to (left) and divided by (right) the normal nuclear absorption curve.

These results show that the J/ψ is "anomalously" suppressed in In-In collisions, and that the In-In, Pb-Pb and S-U seems to better overlap when plotted as a function of N_{part}. These observations are limited by the large statistical errors on the $B_{\mu\mu}\sigma(J/\psi)/\sigma(DY)$ values due to the small yield of DY above 4.2 GeV/c^2. Work is in progress to extract the J/ψ suppression pattern in In-In collisions without referring to the DY yield. Studies of the J/ψ production as a function of energy density and other centrality variables are also under way.

TRANSVERSE MOMENTUM DISTRIBUTION

Besides the J/ψ / DY behaviour, also other preliminary information on the J/ψ transverse momentum has been extracted. For this study, the muon track maching was not applied and 7 centrality bins were performed. For each bin, the $\langle p_T^2 \rangle$ value is extracted from the acceptance corrected p_T distribution obtained by selecting opposite sign dimuons in the mass range 2.9–3.3 GeV/c^2 (mainly from J/ψ decays). The acceptance correction is evaluated through a Monte Carlo simulation in which the generated distributions in p_T, y_{lab} and $\cos\theta_{CS}$ reproduce the experimental ones, in order to account for the correlations between the kinematical variables. Results presented below are almost insensitive to reasonable changes in these generated distributions. More details on this analysis can be found on Ref. [9].

Fig. 4 shows our results as a function of L, together with the 1996 Pb-Pb data published by NA50 [10]. The plotted errors are only statistical. The dependence of $\langle p_T^2 \rangle$ as a function of L follows the previous observations, which have been related to initial state interactions through the gluon-nucleon (gN) scattering [11]: $\langle p_T^2 \rangle = \langle p_T^2 \rangle_{pp} + a_{gN} L$. The numerical value $a_{gN} = 0.081 \pm 0.003$ GeV^{2}c^{-2}fm^{-1} is obtained with a $\chi^2/ndf = 0.5$ when fitting In-In and Pb-Pb points together. The result of the fit is shown as a continuous line in Fig. 4.

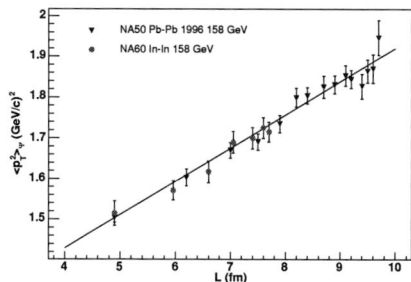

FIGURE 4. J/ψ $\langle p_T^2 \rangle$ measured in In-In and Pb-Pb collisions, plotted as a function of L and fitted with the initial state parton multiple scattering model.

SUMMARY

The analysis of J/ψ production from the indium-indium data collected in the year 2003 is well advanced. Our measurements of the cross section ratio $B_{\mu\mu}\sigma(J/\psi)/\sigma(DY)$ in three centrality bins seem to be in good agreement with previous results when plotted as a function of N_{part}. The analysis of the J/ψ suppression pattern as a function of energy density and other centrality variables, and without referring to the DY yield, is ongoing.

First results from the J/ψ transverse momentum analysis shows a L dependence of the $\langle p_T^2 \rangle$ compatible with previous measurements, reinforcing the interpretation in terms of initial state parton multiple scattering.

REFERENCES

1. T. Matsui and H. Satz, Phys. Lett. **B178** (1986) 416.
2. S. Digal et al., Eur. Phys. J. **C32** (2004) 547.
3. D. Kharzeev and H. Satz, "Color Deconfinement and Quarkonium Dissociation", in "Quark-gluon plasma" (Ed. R.C. Hwa), vol 2, 395; hep-ph/9505345.
4. R. Shahoyan et al. (NA60 Coll.), Eur. Phys. J. **C43** (2005) 209.
5. K.Banicz et al., Nucl. Instrum. Meth. **A539** (2005) 137.
6. T. Sjöstrand, Comp. Phys. Commun. **135** (2001) 238.
7. M. Glück et al., Z. Phys. **C67** (1995) 433.
8. B. Alessandro et al. (NA50 Coll.), Eur. Phys. J. **C39** (2005) 335.
9. P. Pillot, Ph.D. Thesis, Université Claude Bernard, Lyon, France, 2005.
10. M.C. Abreu et al. (NA50 Coll.), Phys. Lett. **B499** (2001) 85.
11. J. Hüfner et al., Phys. Lett. **B215** (1988) 218.

Indications of a Pseudogap in the Nambu Jona-Lasinio model

Paolo Castorina*, Giuseppe Nardulli[†] and Dario Zappalà**

*Department of Physics, University of Catania, Italy and INFN-Catania, Italy.
[†]Department of Physics and TIRES Center, University of Bari, Italy and INFN-Bari, Italy
and PH Department, TH Unit, CERN, 1211 Geneva 23, Switzerland.
**INFN- Catania, Italy and Department of Physics, University of Catania, Italy.

Abstract.
The survival of $\bar{q}q$ bound states at temperatures higher than the chiral restoration temperature, T_c, recently observed in lattice QCD, is discussed in the framework of the Nambu Jona-Lasinio model. The perturbative determination of the spectral function provides an indication of a pseudogap phase above T_c.

Keywords: Chiral symmetry, pseudogap transition, bound states
PACS: 12.39.Ki, 11.30.Rd

INTRODUCTION

Recent lattice studies about the Quantum Chromodynamics transition at finite temperature from the hadronic to the deconfined phase, show that, heavy, and possibly also light mesonic bound states survive up to twice the deconfinement temperature (for a recent review see [1] and references therein), in contrast with the former suggestion that a strong suppression of heavy bound states, such as the J/ψ, would occur just above the critical temperature. This unexpected feature indicates some not yet understood mechanism at the deconfinement transition. In addition, the first experimental results collected at RHIC support the presence of a strongly interacting phase above the critical temperature.

This picture has some analogies with the physics of high temperature superconductors where the coherence length is much smaller than in ordinary superconductors and the temperature corresponding to the onset of superconductivity turns out to be much lower than the temperature related to the Cooper pairs formation. Between these two temperatures a pseudogap is observed, which consists in a depletion of the single particle density of states around the Fermi level. These features could be regarded as a phenomenological manifestation of a crossover from the ordinary Bardeen-Cooper-Schrieffer superconducting behavior to a Bose-Einstein Condensate behavior.

An explanation of the high temperature superconductivity has been provided by Emery and Kivelson (see [2]) who suggested that phase fluctuations of the condensate should be responsible for the spoiling of the long range coherence, the appearance of the pseudogap and the consequent lowering of the temperature associated to the onset of superconductivity . The relation between the pseudogap transition and field fluctuation has been explored within the framework of the Nambu Jona-Lasinio (NJL) model [3, 4, 5], and also indicated as a possible explanation of the observed lattice bound states above the critical temperature[6].

Following [6], after a brief analysis of some results in mean field theory for the NJL model, we briefly discuss the effects of the fluctuations and the consequent appearance of two characteristic temperatures: T_c, corresponding to the restoration of chiral symmetry for the NJL model, and T^* associated to the decoupling of the mesons $\bar{q}q$ from the spectrum. Finally we present a calculation of the pseudogap between T_c and T^*.

NJL MODEL IN THE MEAN FIELD APPROXIMATION

We consider the NJL model [7, 8] with an isospin doublet of massless quarks with three colors ($N_f = 2$, $N_c = 3$) at finite chemical potential μ

$$\mathcal{L} = \bar{\psi}(i\partial_\nu \gamma^\nu + \mu\gamma_0)\psi + \frac{G_0}{2N_c}\left[(\bar{\psi}\psi)^2 + (\bar{\psi}i\gamma_5\tau\psi)^2\right]. \quad (1)$$

As it is well known (for reviews on the NJL model see e.g. [9],[10]), the mean field approximation provides the self-consistent equation (at $\mu = 0$) for the mass gap m_q:

$$m_q = 4N_f N_c \frac{G_0}{2N_c} \int_\Lambda \frac{d^3 p}{(2\pi)^3} \frac{m_q}{\sqrt{p^2 + m_q^2}} \quad (2)$$

where Λ is a 3D cutoff. Moreover the pion decay constant f_π, again at $\mu = 0$, is given by

$$f_\pi^2 = -4im_q^2 N_c \int_\Lambda \frac{d^4 p}{(2\pi)^4} \frac{1}{(p^2 - m_q^2 + i\varepsilon)^2} \quad (3)$$

and one can use Eqs. (2) and (3) to get Λ by fixing m_q. In particular we use $m_q = 300$ MeV and $f_\pi = 93.3$ MeV as input, which gives $\Lambda = 675$ MeV.

The approach is then easily generalized at finite temperature and density to obtain a new gap equation for $m_q(T, \mu)$ which, after getting rid of the coupling G_0 by means of Eq. (2), reads

$$0 = \int_\Lambda \frac{d^3 p}{(2\pi)^3} \left[\frac{1}{\sqrt{p^2 + m_q^2}} - \frac{\sinh y}{\varepsilon(\cosh y + \cosh x)}\right] \quad (4)$$

where $\varepsilon = \sqrt{p^2 + m_q^2(T,\mu)}$ and $x = \mu/T$ and $y = \varepsilon/T$.

The gap equation at finite temperature and density naturally provides a critical line signaling a phase transition. In fact, the vanishing of the solution of Eq. (4): $m_q(T,\mu) = 0$ defines the critical temperature $T^*(\mu)$ and its plot in the $T - \mu$ plane is displayed in fig. 1, with the input for the 3D cutoff Λ obtained from the values of m_q and f_π at $T = \mu = 0$ given above. Our next step is to check to what extent the fluctuations can modify the results of the mean field analysis.

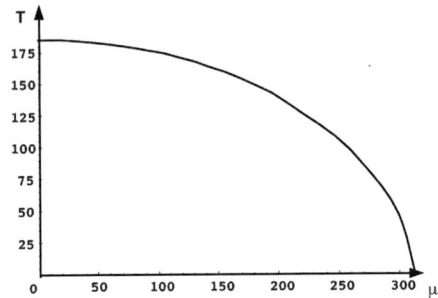

FIGURE 1. Plot of $T^*(\mu)$ in MeV, as obtained from Eq. (4) for $\Lambda = 675$ MeV.

FLUCTUATION EFFECTS

In order to study the modifications to the mean field analysis it is necessary to introduce a method that can properly account for the fluctuation effects. Such approach has been developed [4, 5] and its essential features will be briefly illustrated in the following.

The four fermion interaction in the model in Eq. (1) can be cast in the form of Yukawa coupling, by introducing the auxiliary scalar and pseudoscalar fields σ and $\vec{\pi}$:

$$\mathscr{L} = \bar{\psi}(i\partial_\nu \gamma^\nu + \mu \gamma_0 - g_0(\sigma + i\gamma_5 \vec{\tau} \cdot \vec{\pi}))\psi - \frac{g_0^2 N_c}{2G_0}\left[\sigma^2 + \vec{\pi}^2\right] \quad (5)$$

with the quark-meson coupling constant given by $g_0 = m_q/f_\pi$, which is the analogous of the Goldberger-Treiman relation. By functional integration of the fermionic degrees of freedom and derivative expansion, Eq. (5) becomes equivalent to the σ model with lagrangian

$$\mathscr{L}_\sigma = \frac{\beta}{2}\left((\partial \sigma)^2 + (\partial \vec{\pi})^2 - \frac{\kappa^2}{4}\left(\sigma^2 + \vec{\pi}^2 - f_\pi^2\right)^2\right) \quad (6)$$

where the stiffness parameter, β, which is relevant for our purposes, is given by [4, 11]

$$\beta = 4g_0^2 N_c N_f \left(\int_\Lambda \frac{d^4 p_E}{(2\pi)^4} \frac{1}{(p_E^2 + m_q^2)^2} - \frac{1}{2}\int_\Lambda \frac{d^4 p_E}{(2\pi)^4} \frac{p_E^2}{(p_E^2 + m_q^2)^3}\right) \quad (7)$$

and p_E is the Euclidean four-momentum.

The key point is that an equivalent description of the problem should be obtained by resorting to an effective theory, defined by a non-linear σ model for the fields σ and $\vec{\pi}$, satisfying the constraint $\sigma^2 + \vec{\pi}^2 = f_\pi^2$ [4, 5]. This constraint can be introduced into the functional generator Z of the non-linear σ model, by means of a functional Lagrange multiplier $\lambda(x)$

$$Z = \int [d\sigma][d\vec{\pi}][d\lambda]\exp\left\{i\frac{\beta}{2}\int d^4x\left((\partial \sigma)^2 + (\partial \vec{\pi})^2 - \lambda\left(\sigma^2 + \vec{\pi}^2 - f_\pi^2\right)\right)\right\} \quad (8)$$

The parameter β in Eq. (8) indicates the stiffness of the non-linear σ model.

The functional integration of the $\vec{\pi}$ fields in Eq. (8) can be performed and, afterward, one can use the saddle point approximation and search for x-independent solutions for σ and λ. As it is evident from Eq. (8), the constant λ plays the role of a square mass both for σ and $\vec{\pi}$ fields. Since we are looking for phase transitions at finite T and μ, we consider Matsubara frequencies $\tilde{\omega}_n = 2\pi n T$ and non-vanishing baryonic chemical potential and the two saddle point conditions read

$$0 = \lambda \sigma; \quad 0 = \beta(f_\pi^2 - \sigma^2) - (N_f^2 - 1) \sum_{n=-\infty}^{\infty} \int_{\Lambda_\pi} \frac{d^3k}{(2\pi)^3} \frac{T}{(\tilde{\omega}_n - i\mu)^2 + k^2 + \lambda} \quad (9)$$

where Λ_π is a suitable cutoff for the non-linear σ model.

The first condition $0 = \lambda \sigma$ indicates that at least one of the two variables must vanish and the other variable is implicitly determined from the second condition in Eq. (9), in terms of T, μ and of the stiffness β. The equivalence of the non-linear σ model with the lagrangian in Eq. (6) implies that the stiffness evaluated in the two cases must coincide. Therefore one can insert the expression of β of Eq. (7), properly modified to account for the finite T and μ effects, i.e.

$$\beta = 2g_0^2 N_f N_c T \sum_{n=-\infty}^{+\infty} \int_\Lambda \frac{d^3p}{(2\pi)^3} \left[\frac{1}{[\varepsilon^2 + (\omega_n - i\mu)^2]^2} + \frac{m_q^2(T,\mu)}{[\varepsilon^2 + (\omega_n - i\mu)^2]^3} \right] \quad (10)$$

into Eq. (9) and determine the corresponding solution for σ or λ.

As an example, in fig. 2 we plot the solutions of Eq. (9): σ and the mass of the pionic mode $m_\pi = \sqrt{\lambda}$ versus the temperature, at $\mu = 0$ and for two different values of the number of colors: $N_c = 3$ (solid lines) and $N_c = 10$ (dashed lines). The values of the two cutoffs entering Eqs. (9) and (10) are $\Lambda_\pi = 200$ MeV and $\Lambda = 675$ MeV. In fact Λ_π and Λ are related to different degrees of freedom and do not need to have the same value and we take a value of Λ_π close to f_π, which fixes the scale of the non-linear σ model. (The choice $\Lambda_\pi \sim \Lambda$ made in [3] has been criticized in [4]. For other references on this point see [6].)

We observe in fig. 2 that at low temperatures the solution of Eq. (9) corresponding to $\lambda = 0$ (not plotted) and non-vanishing σ is realized. Then σ decreases with T and eventually vanishes. By further increasing the temperature, the solution $\sigma = 0$ (not plotted) and $m_\pi = \sqrt{\lambda} \neq 0$ is realized.

The two regions meet at $T = T_c$ where $\sigma = \lambda = 0$. This corresponds in Eq. (9) to the critical stiffness

$$\beta_c = \frac{N_f^2 - 1}{f_\pi^2(T_c, \mu)} \sum_{n=-\infty}^{\infty} \int_{\Lambda_\pi} \frac{d^3k}{(2\pi)^3} \frac{T_c}{(\tilde{\omega}_n - i\mu)^2 + k^2} \quad (11)$$

and T_c is determined by the comparison of Eqs. (10) and (11) In the particular case considered in fig. 2 we get $T_c \approx 146$ MeV for $N_c = 3$ and $T_c \approx 172$ MeV for $N_c = 10$.

One should notice that in the non-linear σ model the region below T_c shows a non-vanishing expectation value of the field σ which implies a dynamical breaking of the chiral symmetry. Above T_c this expectation value vanishes and at the same time a finite

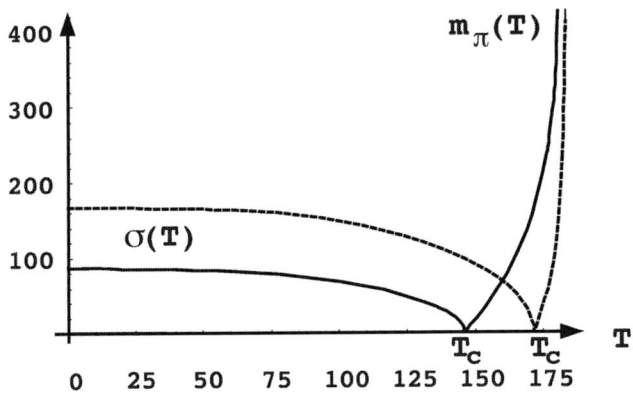

FIGURE 2. σ and m_π vs. T at $\mu = 0$ (in MeV) as obtained from Eq. (9). See text for the specific values of the other parameters.

mass for the pions is generated. The absence from the spectrum of massless Goldstone bosons is a signal of chiral symmetry restoration. Therefore, within this framework, T_c is the critical temperature related to the chiral symmetry.

In fig. 2, when the temperature is increased, m_π grows and, at some temperature, T_{pair}, eventually diverges and the $\vec{\pi}$ mesons decouple and disappear from the spectrum. Reasonably, T_{pair} could be interpreted as the temperature of dissociation of the $\bar{q}q$ pair. A remarkable feature which appears in fig.2 is that T_{pair} does not depend on N_c and, in practice, it numerically coincides with the critical temperature T^* determined by the mean field analysis: $T_{pair} \sim T^* \sim 185$ MeV.

Therefore, the comparison of the NJL model with the non-linear σ model indicates that fluctuations indeed modify the mean field conclusions and the new critical temperature T_c, associated with the chiral symmetry restoration, is generated. T_c turns out to be smaller than T^* so that we find a temperature interval $T_c < T < T^*$ where the gap equation still provides $m_q \neq 0$ but the chiral symmetry is restored as an effect of the fluctuations. Only above $T^* \sim T_{pair}$ the mesons disappear and m_q vanishes. Consistently with this picture we note that T_c depends on N_c and $T_c \to T^*$ when $N_c \to \infty$, i.e. when the fluctuation effects become suppressed and the mean field regime holds.

A comment about m_q is in order. In fact the common belief is that a non-vanishing m_q is directly related to the breakdown of chiral symmetry. This is true only if the gap m_q appears as a pole in the two point function of the fermion ψ. In our specific case m_q corresponds to the pole mass of ψ only in the mean field approximation. This property has been first observed in 2D models in [12], and, for a detailed treatment of the problem, we refer to [13, 14]. With a suitable change of variables, corresponding to the introduction of polar 'coordinates': $\sigma + i\gamma^5 \pi = \rho \exp(i\gamma^5 \theta)$ and of the fermion $\chi = \exp(i\gamma^5 \theta/2)$, where ρ and χ are chirally neutral, it is possible to show that m_q, as obtained from the gap equations (2) and (4) is a pole in the χ two point function (of course in our specific case the definition of the new variables should be modified to account for the flavor degrees of freedom). Due to the chiral neutrality of χ, this excitation does not break chiral symmetry. Conversely, it does not appear in the ψ propagator as a pole but rather, due to the θ field fluctuations, as a branch cut. In this sense, a finite m_q is still compatible with chiral symmetry.

The presence of a region between two characteristic temperatures where the fluctuations cancel the long range order associated to the quark mass, restoring the symmetry, shows a clear resemblance with the pseudogap phase of high temperature superconductors [3, 4, 14] and this motivates the following analysis of the spectral function.

EVALUATION OF THE PSEUDOGAP

After having determined the two characteristic temperatures T_c and T^*, we conclude by showing that a simple approximation in the calculation of the spectral function provides clear indications of a pseudogap [6]. The spectral function of the fermionic quasiparticles, $N(\omega)$, is obtained from the imaginary part of $G^R(\vec{k}, \omega)$:

$$N(\omega) = -\frac{1}{\pi} \int \frac{d\vec{k}}{(2\pi)^3} \text{Tr}_{c,f,d} \; \gamma^0 \; \text{Im} \; G^R(\vec{k}, \omega) \tag{12}$$

where $\text{Tr}_{c,f,d}$ is the trace over Dirac, color and flavor degrees of freedom. $G^R(\vec{k}, \omega)$ is the analytical continuation of the imaginary time retarded Green function: $G^R(\vec{k}, \omega) = \left[G^{-1}(\vec{k}, \omega + i\varepsilon) - \Sigma_R(\vec{k}, \omega) \right]^{-1}$ where $G^{-1}(\vec{k}, \omega + i\varepsilon)$ corresponds to the free term contribution and $\Sigma_R(\vec{k}, \omega)$ includes higher order corrections. The starting point to calculate the retarded Green function is the Lagrangian

$$\mathscr{L}_{eff} = \bar{\psi}[i\partial_\nu \gamma^\nu + \mu\gamma_0 - g_0(\sigma + i\gamma_5 \vec{\tau} \cdot \vec{\pi})]\psi - \frac{g_0^2 N_c}{2G_0} \left[\sigma^2 + \vec{\pi}^2 \right] \tag{13}$$

where the σ field is replaced by expanding the constraint : $\sigma = \left(f_\pi^2 - \vec{\pi}^2 \right)^{1/2} \sim f_\pi - \vec{\pi}^2/(2f_\pi)$. The correction $\Sigma_R(\vec{k}, \omega)$ to the free fermion propagator is computed to the one loop level from Eq. (13), [11], and its imaginary part generates the correction, N_{pert}, to the free fermion contribution in the spectral function

$$N(\omega) = \frac{N_c N_f}{\pi^2}(\omega + \mu)\sqrt{(\omega + \mu)^2 - m_q^2} + N_{pert} \; . \tag{14}$$

In fact we have included the mass m_q, which is dynamically generated, in the free fermion contribution and we have also included in the definition of g_0 the form factor $g_0^2(\vec{q}^2) = g_0^2 m_P^4 / \left(m_P^2 + (2\pi n T)^2 + q^2 \right)^2$ with $m_P = 100$ MeV, which acts as a regulator that corrects the large momentum behavior of the loop.

We normalize $N(\omega)$ to the free massless case : $R(\omega) = N(\omega)/(\frac{N_c N_f}{\pi^2}(\omega + \mu)^2)$ and plot $R(\omega)$ for $(T - T_c)/T_c = 0.02, 0.15, 0.20$ ($\mu = 10$ MeV) in fig. 3. The solid straight line corresponds to $T = T^*$. In fig. 3 we observe the pseudogap for values of ω just above

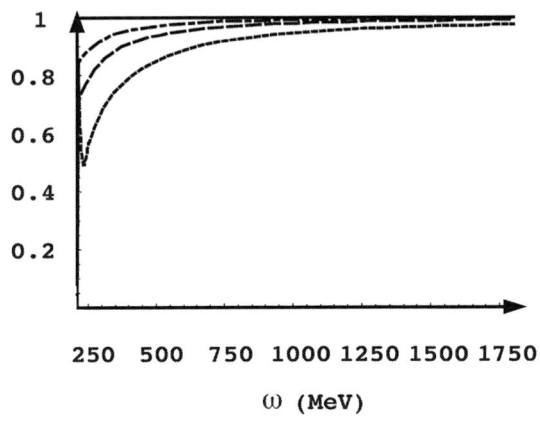

FIGURE 3. The ratio $N(\omega)/N_{free}(\omega)$ as a function of the energy ω, at $\mu = 10$ MeV and for various temperatures between T_c and T^*. See text.

the value $\omega = m_q - \mu$, where the free part of $N(\omega)$ vanishes, whereas it disappears for large values of ω. The effect of the pseudogap is larger for temperatures just greater than T_c and vanishes at $T = T^*$. Therefore, even in this simple approximation, it is possible to realize the presence of the depletion of states in the spectral function and this calculation supports the picture of a pseudogap appearing between the two characteristic temperatures. Since the NJL model effectively contains many ingredients of QCD, this analysis is a starting point to check the existence of a pseudogap in QCD and its possible role in the explanation of the mesonic bound states observed in lattice calculations.

REFERENCES

1. F. Karsch (2005), hep-lat/0502014.
2. V. J. Emery, and S. A. Kivelson, *Nature* **374**, 434 (1995).
3. T. Hatsuda, and T. Kunihiro, *Phys. Rev. Lett.* **55**, 158–161 (1985).
4. H. Kleinert, and B. van den Bossche, *Phys. Lett.* **B474**, 336–346 (2000), hep-ph/9907274.
5. E. Babaev, *Phys. Rev.* **D62**, 074020 (2000), hep-ph/0006087.
6. P. Castorina, G. Nardulli, and D. Zappala (2005), hep-ph/0505089.
7. Y. Nambu, and G. Jona-Lasinio, *Phys. Rev.* **122**, 345–358 (1961).
8. Y. Nambu, and G. Jona-Lasinio, *Phys. Rev.* **124**, 246–254 (1961).
9. S. P. Klevansky, *Rev. Mod. Phys.* **64**, 649–708 (1992).
10. T. Hatsuda, and T. Kunihiro, *Phys. Rept.* **247**, 221–367 (1994), hep-ph/9401310.
11. T. Eguchi, *Phys. Rev.* **D14**, 2755 (1976).
12. E. Witten, *Nucl. Phys.* **B145**, 110 (1978).
13. E. N. Nikolov, W. Broniowski, C. V. Christov, G. Ripka, and K. Goeke, *Nucl. Phys.* **A608**, 411–436 (1996), hep-ph/9602274.
14. V. Gusynin, V. Loktev, and S. Sharapov, *JETP* **88**, 685 (1999).

Color superconductivity and the strange quark

Mark Alford

Physics Department, Washington University, Saint Louis, MO 63130, USA

Abstract. At ultra-high density, matter is expected to form a degenerate Fermi gas of quarks in which there is a condensate of Cooper pairs of quarks near the Fermi surface: color superconductivity. In these proceedings I review some of the underlying physics, and discuss outstanding questions about the phase structure of ultra-dense quark matter.

Keywords: dense quark matter, color superconductivity
PACS: 12.38.-t, 26.60.+c, 25.75.Nq

INTRODUCTION

The exploration of the phase diagram of matter at ultra-high temperature or density is an area of great interest and activity, both on the experimental and theoretical fronts. Heavy-ion colliders such as the SPS at CERN and RHIC at Brookhaven have probed the high-temperature region, searching for the transition to deconfined quark matter. In this paper we discuss a different part of the phase diagram, the low-temperature high-density region. Here there are as yet no experimental constraints, but we expect to find phases characterized by Cooper pairing of quarks, i.e. color superconductivity, driven by the Bardeen-Cooper-Schrieffer (BCS) [1] mechanism. The BCS mechanism operates when there exists an attractive interaction between fermions at a Fermi surface. The QCD quark-quark interaction is strong, and is attractive in many channels, so we expect cold dense quark matter to *generically* exhibit color superconductivity. Moreover, quarks, unlike electrons, have color and flavor as well as spin degrees of freedom, so many different patterns of pairing are possible. This leads us to expect a rich phase structure in matter beyond nuclear density.

Calculations using a variety of methods agree that at sufficiently high density, the favored phase is color-flavor-locked (CFL) color-superconducting quark matter [2] (for reviews, see Ref. [3]). However, there is still uncertainty over the nature of the next phase down in density. Recent work [4] suggests that when the density drops low enough so that the mass of the strange quark can no longer be neglected, there is a continuous phase transition from the CFL phase to a new gapless CFL (gCFL) phase, which could lead to observable consequences if it occurred in the cores of neutron stars [5]. However, it now appears that some of the gluons in the gCFL phase have imaginary Meissner masses, indicating an instability towards an unknown lower-energy phase [6, 7, 8, 9]. The nature of this phase is still unclear, although the crystalline "LOFF" phase is a strong candidate [10, 8].

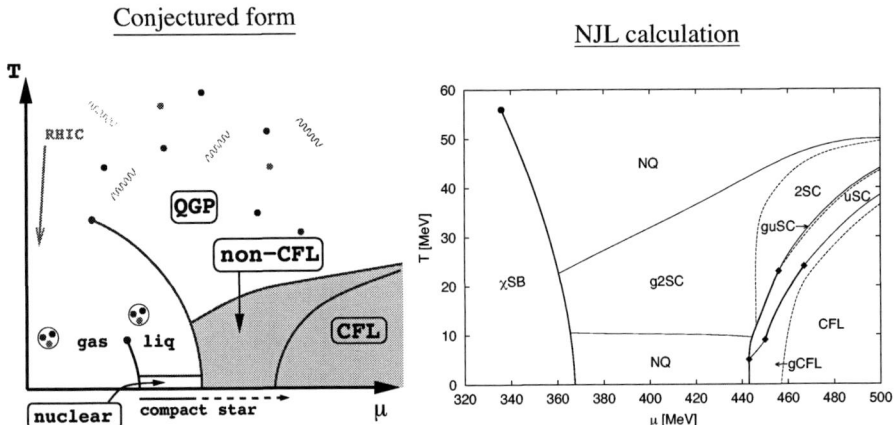

FIGURE 1. On the left, the conjectured form of the phase diagram for matter at ultra-high density and temperature. On the right, the result of a calculation using a NJL model [17]. At high density we find a rich structure of color-superconducting phases.

REVIEW OF COLOR SUPERCONDUCTIVITY

The phase diagram of quark matter

In the real world there are two light quark flavors, the up (u) and down (d), with masses $\lesssim 5$ MeV, and a medium-weight flavor, the strange (s) quark, with mass ~ 100 MeV. (Their effective "constituent" masses in dense matter may be much larger.) The strange quark therefore plays a crucial role in the phases of QCD. Fig. 1 shows a conjectured phase diagram for QCD, and also a calculated phase diagram obtained using a Nambu–Jona-Lasinio model of QCD. In both cases, along the horizontal axis the temperature is zero, and the density rises from the onset of nuclear matter through the transition to quark matter. Compact stars are in this region of the phase diagram, although it is not known whether their cores are dense enough to reach the quark matter phase. Along the vertical axis the temperature rises, taking us through the crossover from a hadronic gas to the quark gluon plasma. This is the regime explored by high-energy heavy-ion colliders.

At the highest densities we find the CFL phase, in which the strange quark participates symmetrically with the up and down quarks in Cooper pairing—this is described in more detail below. The phases that occur at intermediate density are still not well understood. The NJL calculation works with a limited set of possibilities, and the NJL calculation neglects more exotic possibilities such as kaon condensation [11], crystalline color superconductivity (LOFF) [10], and single-flavor pairing [12, 13, 14, 15, 16].

Color superconductivity

The fact that QCD is asymptotically free implies that at sufficiently high density and low temperature, there is a Fermi surface of weakly-interacting quarks. The interaction between these quarks is certainly attractive in some channels (quarks bind together to form baryons), so we expect the formation of a condensate of Cooper pairs. We can see this by considering the grand canonical potential $F = E - \mu N$, where E is the total energy of the system, μ is the chemical potential, and N is the number of quarks. The Fermi surface is defined by a Fermi energy $E_F = \mu$, at which the free energy is minimized, so adding or subtracting a single particle costs zero free energy. Now switch on a weak attractive interaction. It costs no free energy to add a pair of particles (or holes), and if they have the right quantum numbers then the attractive interaction between them will lower the free energy of the system. Many such pairs will therefore be created in the modes near the Fermi surface, and these pairs, being bosonic, will form a condensate. The ground state will be a superposition of states with all numbers of pairs, breaking the fermion number symmetry.

A pair of quarks cannot be a color singlet, so the resulting condensate will break the local color symmetry $SU(3)_{\text{color}}$. The formation of a condensate of Cooper pairs of quarks is therefore called "color superconductivity". The condensate plays the same role here as the Higgs particle does in the standard model: the color-superconducting phase can be thought of as the Higgs phase of QCD.

Highest density: Color-flavor locking (CFL)

At the highest densities, where the strange quark Fermi momentum is close to the up and down quark Fermi momenta, the favored phase is "color-flavor locking" (CFL) [2]. This has been confirmed by both NJL [2, 18] and gluon-mediated interaction calculations [19]. The CFL pairing pattern is

$$\langle q_i^\alpha C\gamma_5 q_j^\beta \rangle_{1PI} \propto (\kappa+1)\delta_i^\alpha \delta_j^\beta + (\kappa-1)\delta_j^\alpha \delta_i^\beta = \varepsilon^{\alpha\beta N}\varepsilon ijN + \kappa(\cdots)$$

$$[SU(3)_{\text{color}}] \times \underbrace{SU(3)_L \times SU(3)_R}_{\supset [U(1)_Q]} \times U(1)_B \to \underbrace{SU(3)_{C+L+R}}_{\supset [U(1)_{\tilde{Q}}]} \times \mathbb{Z}_2 \quad (1)$$

Color indices α, β and flavor indices i, j run from 1 to 3, Dirac indices are suppressed, and C is the Dirac charge-conjugation matrix. The term multiplied by κ corresponds to pairing in the $(\mathbf{6}_S, \mathbf{6}_S)$, which although not energetically favored breaks no additional symmetries and so κ is in general small but not zero [2, 19, 20, 21]. The Kronecker deltas connect color indices with flavor indices, so that the condensate is not invariant under color rotations, nor under flavor rotations, but only under simultaneous, equal and opposite, color and flavor rotations. Since color is only a vector symmetry, this condensate is only invariant under vector flavor+color rotations, and breaks chiral symmetry. The features of the CFL pattern of condensation are

- The color gauge group is completely broken. All eight gluons become massive. This ensures that there are no infrared divergences associated with gluon propagators.

- All the quark modes are gapped. The nine quasiquarks (three colors times three flavors) fall into an $\mathbf{8} \oplus \mathbf{1}$ of the unbroken global $SU(3)$, so there are two gap parameters. The singlet has a larger gap than the octet.
- A rotated electromagnetism ("$\tilde{Q}$") survives unbroken. It is a combination of the original photon and one of the gluons.
- Two global symmetries are broken, the chiral symmetry and baryon number, so there are two gauge-invariant order parameters that distinguish the CFL phase from the QGP, and corresponding Goldstone bosons which are long-wavelength disturbances of the order parameter. When the light quark mass is non-zero it explicitly breaks the chiral symmetry and gives a mass to the chiral Goldstone octet, but the CFL phase is still a superfluid, distinguished by its baryon number breaking.
- The symmetries of the 3-flavor CFL phase are the same as those one might expect for 3-flavor hypernuclear matter [18], so it is possible that there is no phase transition between them.

REAL-WORLD QUARK MATTER

Stresses on the CFL phase

The CFL phase is characterized by pairing between different flavors and different colors of quarks. We can easily understand why this is to be expected. Firstly, the QCD interaction between two quarks is most attractive in the channel that is antisymmetric in color (the $\bar{\mathbf{3}}$). Secondly, pairing tends to be stronger in channels that do not break rotational symmetry [12, 13, 14, 15, 16], so we expect the pairing to be a spin singlet, i.e. antisymmetric in spin. Finally, fermionic antisymmetry of the Cooper pair wavefunction then forces the Cooper pair to be antisymmetric in flavor.

Pairing between different colors/flavors can occur easily when they all have the same chemical potentials and Fermi momenta. This is the situation at very high density, where the strange quark mass is negligible. However, in a real compact star we must take into account the forces that try to split those Fermi momenta apart, imposing an energy cost on cross-species pairing. We must require electromagnetic and color neutrality [22, 23] (possibly via mixing of oppositely-charged phases), allow for equilibration under the weak interaction, and include a realistic mass for the strange quark. These factors cause the different colors and flavors to have different chemical potentials, and this imposes a stress on cross-species pairing such as occurs in the CFL pairing pattern. As we come down in density, we expect the CFL pairing pattern to be distorted, and then to be replaced by some other pattern.

In the next few subsections we give a quick overview of the expected phases of real-world quark matter. We restrict our discussion to zero temperature because the critical temperatures for most of the phases that we discuss are expected to be of order 10 MeV or higher, and the core temperature of a neutron star is believed to drop below this value within minutes (if not seconds) of its creation in a supernova.

Kaon condensation: the CFL-K^0 phase

Bedaque and Schäfer [11] showed that when the stress is not too large (high density), it may simply modify the CFL pairing pattern by inducing a flavor rotation of the condensate which can be interpreted as a condensate of "K^0" mesons, i.e. the neutral anti-strange Goldstone bosons associated with the chiral symmetry breaking. This is the "CFL-K0" phase, which breaks isospin. The K^0 condensate can easily be suppressed by instanton effects [24], but if these are ignored then the kaon condensation occurs for $M_s \gtrsim m^{1/3}\Delta^{2/3}$ for light (u and d) quarks of mass m. This was demonstrated using an effective theory of the Goldstone bosons, but with some effort can also be seen in an NJL calculation [25, 26].

The non-CFL region

The nature of the next significant transition has been studied in NJL model calculations which ignore the K0-condensation in the CFL phase [27, 28, 17]. It has been found that the phase structure depends on the strength of the pairing. If the pairing is very strong (so that $\Delta_{CFL} \sim 100$ MeV where Δ_{CFL} is what the CFL gap would be at $\mu \sim 500$ MeV if M_s were zero) then the CFL phase survives all the way down to the transition to nuclear matter. For less strong pairing, there may be a transition to a two-flavor pairing ("2SC") phase [29, 30] and/or a single-flavor pairing phase (see below), and then to nuclear matter.

For a wide range of parameter values, however, we find something more interesting. We can make a rough quantitative analysis by expanding in powers of M_s/μ and Δ/μ, and ignoring the fact that the effective strange quark mass may be different in different phases [23]. Such an analysis shows that as we come down in density we find a transition at $\mu \approx \frac{1}{2}M_s^2/\Delta_{CFL}$ from CFL to another phase, the gapless CFL phase (gCFL) [4]. The gCFL phase is also found in more complete NJL calculations that do not use the assumptions of Ref [4]. This is seen in the right-hand panel of Fig. 1 (the "gCFL" region) and is discussed in detail below.

Single-flavor pairing

If M_s is sufficiently large at densities where quark matter is favored over nuclear matter, then (via the neutrality requirement) it splits the chemical potentials of the different flavors so far apart that no cross-species pairing can occur at all. There is no CFL or 2SC pairing. In most NJL studies this is described loosely as "unpaired" quark matter. However, it is well known that there are attractive channels for a single flavor, although they are much weaker than the 2SC and CFL channels. Thus in these regions we expect some form of single-flavor pairing. There are various "1SC" phases, many of which break rotational invariance, and a very interesting color-spin-locked (CSL) phase which is rotationally invariant [12, 13, 14, 15, 16]. These phases have much lower critical temperatures than the others (from a few MeV down to eV).

The gapless CFL phase

As mentioned above, an expansion in M_s/μ for pairing strength $\Delta_{CFL} \lesssim 25$ MeV shows that, for $M_s^2/\mu \gtrsim 2\Delta$, the CFL phase has higher free energy than an alternative phase called gapless CFL ("gCFL"). This follows from the energetic balance, mentioned above, between the cost of keeping the Fermi surfaces together and the benefit of the pairing that can then occur. The leading effect of M_s is like a shift in the chemical potential of the strange quarks, so the bd and gs quarks feel "effective chemical potentials" $\mu_{bd}^{\rm eff} = \mu - \frac{2}{3}\mu_8$ and $\mu_{gs}^{\rm eff} = \mu + \frac{1}{3}\mu_8 - \frac{M_s^2}{2\mu}$. In the CFL phase $\mu_8 = -M_s^2/(2\mu)$ [23], so $\mu_{bd}^{\rm eff} - \mu_{gs}^{\rm eff} = M_s^2/\mu$. The CFL phase will be stable as long as the pairing makes it energetically favorable to maintain equality of the bd and gs Fermi momenta, despite their differing chemical potentials [31]. It becomes unstable when the energy gained from turning a gs quark near the common Fermi momentum into a bd quark (namely M_s^2/μ) exceeds the cost in lost pairing energy $2\Delta_1$. So the CFL phase is stable when

$$\frac{M_s^2}{\mu} < 2\Delta_{CFL}, \qquad (2)$$

For larger M_s^2/μ, the CFL phase is replaced by some new phase with unpaired bd quarks, which cannot be neutral unpaired or 2SC quark matter because the new phase and the CFL phase must have the same free energy at the critical $M_s^2/\mu = 2\Delta_{CFL}$.

The obvious approach to finding this phase is to perform a NJL model calculation with a general ansatz for the pairing that includes differences between the flavors, for example by allowing different pairing strengths Δ_{ud}, Δ_{ds}, Δ_{us}. This was done in Ref. [4], and the resultant "gCFL" phase was described in detail. In Fig 2 we show the results. The gCFL phase takes over from CFL at $M_s^2/\mu \approx 2\Delta_{CFL}$, and remains favored beyond the value $M_s^2/\mu \approx 4\Delta_{CFL}$ at which the CFL phase would become unfavored.

Crystalline pairing

The pairing patterns discussed so far have been translationally invariant. But in the region of parameter space where cross-species pairing is just barely excluded by stresses that pull apart the Fermi surfaces, one expects a position-dependent pairing known as the "LOFF" phase [33, 10, 34, 35]. This arises because one way to achieve pairing between different flavors while accomodating the tendency for the Fermi momenta to separate is to only pair over part of the Fermi surface. As we will discuss below, the LOFF phase competes with the gCFL phase, and may resolve that phase's stability problems.

Mixed Phases

Another way for a system to deal with a stress on its pairing pattern is phase separation. In the context of quark matter this corresponds to relaxing the requirement of local charge neutrality, and requiring neutrality only over long distances, so we allow a mixture of a positively charged and a negatively charged phase, with a common pressure and a common value of the electron chemical potential μ_e that is not equal to the

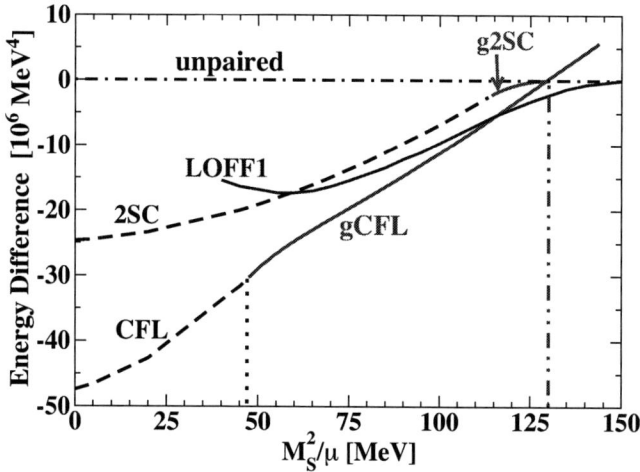

FIGURE 2. Free energy of various phases in an NJL model, allowing different pairing strengths Δ_{ud}, Δ_{ds}, Δ_{us} for the different flavors. The CFL pairing strength is $\Delta_{CFL} = 25$ MeV. Note that the gCFL phase takes over from CFL at $M_s^2/mu \approx 2\Delta_{CFL}$, and remains favored beyond the value $M_s^2/\mu \approx 4\Delta_{CFL}$ at which the CFL phase would become unfavored. The "LOFF1" curve is the single-plane-wave LOFF ansatz of [32].

neutrality value for either phase. Such a mixture of nuclear and CFL quark matter was studied in Ref. [36]. In quark matter it has been found that as long as we require local color neutrality such mixed phases are not the favored response to the stress imposed by the strange quark mass [4, 37]. Phases involving color charge separation have been studied [38] but it seems likely that the energy cost of the color-electric fields will disfavor them.

Beyond gapless CFL

The arguments above led us to the conclusion that the favored phase of quark matter at the highest densities is the CFL phase, and that as the density is decreased there is a transition to another color superconducting phase, the gapless CFL phase. However, it turns out that the gCFL phase is itself unstable, and that there is therefore another phase of even lower free energy, that occurs below the gCFL phase in the phase diagram. The nature of that phase remains uncertain at present.

The instability of the gCFL phase appears to be related to one of its most interesting features, namely the presence of gapless fermionic excitations around the ground state. These are illustrated in Fig. 3, which shows that there is one mode (the bu-rs quasiparticle) with an unusual quadratic dispersion relation, which is expected to give rise to exotic transport properties [5]. The instability of the gCFL phase was established in Refs. [7, 9] after an analogous instability in the gapless 2SC phase had been discovered [6, 8]. The instability manifests itself in imaginary Meissner masses M_M for some of the gluons. M_M^2 is the low-momentum current-current two-point function, and $M_M^2/(e^2\Delta^2)$

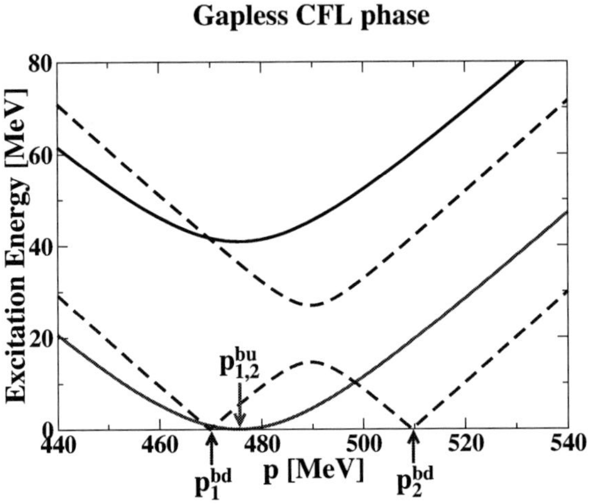

Gapless CFL phase

FIGURE 3. Dispersion relations of the lightest quasiquark excitations in the gCFL phase, at $\mu = 500$ MeV, with $m_s = 200$ MeV and $\Delta_{CFL} = 25$ MeV. Note that in there is a gapless mode with a *quadratic* dispersion relation (energy reaching zero at momentum $p_{1,2}^{bu}$) as well as two gapless modes with more conventional linear dispersion relations.

is the coefficient of the gradient term in the effective theory of small fluctuations around the ground-state condensate. The fact that we find a negative value when the quasiparticles are gapless indicates an instability towards spontaneous breaking of translational invariance. Calculations in a simple two-species model [39] show that imaginary M_M is generically associated with the presence of gapless charged fermionic modes.

The nature of the true ground state in the intermediate density regime remains unclear. It could be a mixed phase [40], a crystalline (LOFF) phase [8], or a p-wave meson condensate [41, 42]. Recent calculations [32] for the 3-flavor case show that even a very simple LOFF ansatz yields a state that has lower free energy than gCFL in the region where the gCFL→unpaired transition occurs (see Fig. 2). Based on what was found in the two-flavor case [34], it is reasonable to expect that when the full space of crystal structures is explored, the LOFF state will be preferred to gCFL over a much wider range of the stress parameter $M_s^2\Delta/(2\mu)$, and it might turn out that the whole gCFL region is actually a LOFF region.

An alternative explanation was advanced by Hong [43] (see also Ref. [44]): since the instability is generically associated with the presence of gapless fermionic modes, and the BCS mechanism implies that any gapless fermionic mode is unstable to Cooper pairing in the most attractive channel, one might expect that the instability will simply be resolved by "secondary pairing". This means the formation of a $\langle qq \rangle$ condensate where q is either one of the gapless quasiparticles whose dispersion relation is shown in Fig. 3. After the formation of such a secondary condensate, the linear gapless dispersion relations would be modified by "rounding out" of the corner where the energy falls to zero, leaving a secondary energy gap Δ_s, which renders the mode gapped, and removes the instability. In the case of the quadratically gapless mode there is a greatly increased density of states at low energy (in fact, the density of states diverges as $E^{-1/2}$), so Hong calculated that the secondary pairing should be much stronger than would be predicted by BCS theory, and he specifically predicted $\Delta_s \propto G_s^2$ for coupling strength G_s, as compared with the standard BCS result $\Delta \propto \exp(-\text{const}/G)$.

This possibility was worked out in an NJL model in Ref. [45], using a two-species model. This allowed a detailed exploration of the strength of secondary pairing. The calculation confirmed Hong's prediction that in typical secondary channels $\Delta_s \propto G_s^2$. However, in all the secondary channels that were analyzed it was found that the secondary gap, even with this enhancement, is from ten to hundreds of times smaller than the primary gap at reasonable values of the secondary coupling. This shows that that secondary pairing does not generically resolve the magnetic instability of the gapless phase, since it indicates that there is a temperature range $\Delta_s \ll T \ll \Delta_p$ in which there is primary pairing (of strength Δ_p) but no secondary pairing, and at those temperatures the instability problem would arise again.

CONCLUSION

As I have described, the project of delineating a plausible phase diagram for high-density quark matter is still not complete. I have discussed some ideas for the "non-CFL" region, but there are others such as a suggested gluon condensation in two-flavor quark matter [46], and deformation of the Fermi surfaces (discussed so far only in non-beta-equilibrated nuclear matter [47]). It is very interesting to note that the problem of how a system with pairing responds to a stress that separates the chemical potentials of the pairing species is a very generic one, arising in condensed matter systems and cold atom systems as well as in quark matter. Recent work by Son and Stephanov [48] on a two-species model characterized by a diluteness parameter and a splitting potential shows that between the BCS-paired region and the unpaired region in the phase diagram one should expect a translationally-broken region. In QCD this could correspond to a p-wave meson condensate or a LOFF state (see above). What is particularly exciting is that the technology of cold atom traps has advanced to the point where fermion superfluidity can now be seen in conditions where many of the important parameters can be manipulated, and it may soon be possible to investigate the response of the pairing to external stress under controlled experimental conditions.

REFERENCES

1. J. Bardeen, L. Cooper, J. Schrieffer, Phys. Rev. **106**, 162 (1957); Phys. Rev. **108**, 1175 (1957)
2. M. Alford, K. Rajagopal and F. Wilczek, Nucl. Phys. **B537**, 443 (1999) [hep-ph/9804403].
3. K. Rajagopal and F. Wilczek, hep-ph/0011333. M. G. Alford, Ann. Rev. Nucl. Part. Sci. **51** (2001) 131 [hep-ph/0102047]. D. K. Hong, Acta Phys. Polon. B **32**, 1253 (2001) [hep-ph/0101025]. D. H. Rischke, Prog. Part. Nucl. Phys. **52**, 197 (2004) [nucl-th/0305030]. T. Schäfer, hep-ph/0304281. S. Reddy, Acta Phys. Polon. B **33**, 4101 (2002) [arXiv:nucl-th/0211045].
4. M. Alford, C. Kouvaris and K. Rajagopal, Phys. Rev. Lett. **92**, 222001 (2004) [arXiv:hep-ph/0311286]; M. Alford, C. Kouvaris and K. Rajagopal, arXiv:hep-ph/0406137.
5. M. Alford, P. Jotwani, C. Kouvaris, J. Kundu and K. Rajagopal, arXiv:astro-ph/0411560.
6. M. Huang and I. A. Shovkovy, Phys. Rev. D **70**, 094030 (2004) [arXiv:hep-ph/0408268].

7. R. Casalbuoni, R. Gatto, M. Mannarelli, G. Nardulli and M. Ruggieri, arXiv:hep-ph/0410401.
8. I. Giannakis and H. C. Ren, arXiv:hep-ph/0412015.
9. K. Fukushima, arXiv:hep-ph/0506080.
10. M. G. Alford, J. A. Bowers and K. Rajagopal, Phys. Rev. D **63**, 074016 (2001) [arXiv:hep-ph/0008208].
11. P. F. Bedaque and T. Schäfer, Nucl. Phys. A **697** (2002) 802 [hep-ph/0105150].
12. M. Iwasaki, T. Iwado, Phys. Lett. **B350**, 163 (1995); M. Iwasaki, Prog. Theor. Phys. Suppl. **120**, 187 (1995)
13. T. Schafer, Phys. Rev. D **62**, 094007 (2000) [arXiv:hep-ph/0006034].
14. M. Buballa, J. Hosek and M. Oertel, Phys. Rev. Lett. **90**, 182002 (2003) [arXiv:hep-ph/0204275].
15. M. G. Alford, J. A. Bowers, J. M. Cheyne and G. A. Cowan, Phys. Rev. D **67**, 054018 (2003) [arXiv:hep-ph/0210106].
16. A. Schmitt, Q. Wang and D. H. Rischke, Phys. Rev. D **66**, 114010 (2002) [arXiv:nucl-th/0209050].
17. S. B. Ruster, V. Werth, M. Buballa, I. A. Shovkovy and D. H. Rischke, Phys. Rev. D **72**, 034004 (2005) [arXiv:hep-ph/0503184].
18. T. Schafer and F. Wilczek, Phys. Rev. D **60**, 074014 (1999) [arXiv:hep-ph/9903503].
19. T. Schafer, Nucl. Phys. B **575**, 269 (2000) [arXiv:hep-ph/9909574].
20. I. A. Shovkovy and L. C. R. Wijewardhana, Phys. Lett. B **470**, 189 (1999) [arXiv:hep-ph/9910225].
21. R. D. Pisarski and D. H. Rischke, "Why color-flavor locking is just like chiral symmetry breaking". To be published in, *Proceedings of the Judah Eisenberg Memorial Symposium, "Nuclear Matter, Hot and Cold"*, Tel Aviv, April 14 - 16, 1999 [nucl-th/9907094].
22. K. Iida and G. Baym, Phys. Rev. D **63**, 074018 (2001) [Erratum-ibid. D **66**, 059903 (2002)] [arXiv:hep-ph/0011229].
23. M. Alford and K. Rajagopal, JHEP **0206**, 031 (2002) [arXiv:hep-ph/0204001].
24. T. Schafer, Phys. Rev. D **65**, 094033 (2002) [arXiv:hep-ph/0201189].
25. M. Buballa, Phys. Lett. B **609**, 57 (2005) [arXiv:hep-ph/0410397].
26. M. M. Forbes, arXiv:hep-ph/0411001.
27. K. Fukushima, C. Kouvaris and K. Rajagopal, Phys. Rev. D **71**, 034002 (2005) [arXiv:hep-ph/0408322].
28. D. Blaschke, S. Fredriksson, H. Grigorian, A. M. Oztas and F. Sandin, arXiv:hep-ph/0503194.
29. M. Alford, K. Rajagopal and F. Wilczek, Phys. Lett. **B422**, 247 (1998) [hep-ph/9711395].
30. R. Rapp, T. Schäfer, E. V. Shuryak and M. Velkovsky, Phys. Rev. Lett. **81**, 53 (1998) [hep-ph/9711396].
31. K. Rajagopal and F. Wilczek, Phys. Rev. Lett. **86**, 3492 (2001) [hep-ph/0012039].
32. R. Casalbuoni, R. Gatto, N. Ippolito, G. Nardulli and M. Ruggieri, arXiv:hep-ph/0507247.
33. A. I. Larkin and Yu. N. Ovchinnikov, Zh. Eksp. Teor. Fiz. **47**, 1136 (1964) [Sov. Phys. JETP **20**, 762 (1965)]; P. Fulde and R. A. Ferrell, Phys. Rev. **135**, A550 (1964).
34. J. A. Bowers and K. Rajagopal, Phys. Rev. D **66**, 065002 (2002) [arXiv:hep-ph/0204079].
35. R. Casalbuoni and G. Nardulli, Rev. Mod. Phys. **76**, 263 (2004) [arXiv:hep-ph/0305069].
36. M. G. Alford, K. Rajagopal, S. Reddy and F. Wilczek, Phys. Rev. D **64**, 074017 (2001) [arXiv:hep-ph/0105009].
37. M. Alford, C. Kouvaris and K. Rajagopal, arXiv:hep-ph/0407257.
38. F. Neumann, M. Buballa and M. Oertel, Nucl. Phys. A **714**, 481 (2003) [arXiv:hep-ph/0210078].
39. M. Alford and Q. h. Wang, J. Phys. G **31**, 719 (2005) [arXiv:hep-ph/0501078].
40. S. Reddy and G. Rupak, Phys. Rev. C **71**, 025201 (2005) [arXiv:nucl-th/0405054].
41. T. Schafer, arXiv:hep-ph/0508190.
42. A. Kryjevski, arXiv:hep-ph/0508180.
43. D. K. Hong, arXiv:hep-ph/0506097.
44. M. Huang and I. Shovkovy, Nucl. Phys. A **729**, 835 (2003) [arXiv:hep-ph/0307273].
45. M. Alford and Q. h. Wang, arXiv:hep-ph/0507269.
46. E. V. Gorbar, M. Hashimoto and V. A. Miransky, arXiv:hep-ph/0507303.
47. A. Sedrakian, arXiv:nucl-th/0312053.
48. D. T. Son and M. A. Stephanov, arXiv:cond-mat/0507586.

Smeared gap equations in crystalline color superconductivity

M. Ruggieri

Universita di Bari, I-70126 Bari, Italy
and
I.N.F.N., Sezione di Bari, I-70126 Bari, Italy

Abstract. In the framework of HDET, we discuss an averaging procedure of the NJL quark-quark interaction lagrangian, treated in the mean field approximation, for the two flavor LOFF phase of QCD. This procedure gives results which are valid in domains where Ginzburg-Landau results may be questionable. We compute and compare the free energy for different LOFF crystalline structures.

Keywords: High density QCD; Color superconductivity; Smeared gap equations; LOFF phase.
PACS: 12.38.Aw, 12.38.t

INTRODUCTION AND THE LOFF STATE

The behaviour of QCD at very high baryon density and low temperature has recently attracted a lot of interest. In these conditions, quarks are expected to deconfine [1] and occupy (in momentum space) large Fermi spheres. At very high densities the relevant interaction is the one gluon exchange, which is attractive in the antisymmetric $\bar{3}$ color channel. Thus, a Cooper-like pairing phenomenon is expected to occur. This is color superconductivity [2, 3, 4, 5, 6, 7, 8, 9] (see [10, 11, 12, 13, 14] for reviews).

Regimes of high baryon densities and low temperatures are expected to be realized in the core of compact stars; therefore these superdense objects could be the places where color superconductivity is realized in nature. Apart the astrophysical speculations, the study of color-superconductive QCD is an intriguing challenge itself because its understanding implies a deeper understanding of the QCD phase diagram.

In nature, as a consequence of electrical and color neutrality and of the different masses, the Fermi momenta of the quarks should depend on their color and their flavor. When the difference of the Fermi momenta of the pairing quarks are too different, a superconductive ground state in which the Cooper pairs have a net momentum is energetically favored with respect to the usual, zero momentum state. The resulting state is known as LOFF phase, and has been studied in the sixties in the context of condensed matter physics [15, 16]. Possible realizations of the LOFF state in the frame of color superconductivity have been considered for the first time in Refs. [17, 18, 19] (see [20] for a review and Refs.[21, 22, 23] for recent developments).

A color-superconductive phase is characterized by a nonzero expectation value of a bilinear quark operator, namely a condensate. In the LOFF phase with two flavors (u and d) one has

$$\langle \psi_{\alpha i} C \gamma_5 \psi_{\beta j} \rangle \propto \Delta(\mathbf{r}) \varepsilon^{\alpha\beta 3} \varepsilon_{ij3} , \qquad (1)$$

where α, β are color indices and i, j denote the flavor. A three flavor case (u, d and s) has been recently considered in Ref. [24]. As for the space dependence of the condensate one usually decomposes $\Delta(r)$ as a sum of plane waves. The most simple ansatz is [16]:

$$\Delta(r) = \Delta\, e^{2i\mathbf{q}\cdot\mathbf{r}}, \qquad (2)$$

and the corresponding superconductive phase is known as FF state. The net momentum of the Cooper pair is $2\mathbf{q}$, which is the same for all the pairs while its direction is chosen spontaneously.

One of the most important problem is to calculate the gap parameter Δ in Eq. (2). For the FF state this problem can be solved exactly: one has simply to shift the momenta of the paired quarks, $\mathbf{p}_u \to \mathbf{p}_u + \mathbf{q}$, $\mathbf{p}_d \to \mathbf{p}_d - \mathbf{q}$ so the $\mathbf{q}$ dependence of the gap parameter disappears. Then one can write a self-consistency and exactly soluble equation for Δ. In the context of QCD this procedure has been applied in Ref. [18] where a diagrammatic approach to the FF color superconductive state is considered. Unfortunately, this procedure cannot be applied to the general case of the linear combination of N plane waves. To overcome this difficulty, one can employ a Ginzburg-Landau (GL) expansion of the general LOFF free energy functional Ω. GL expansion works where Δ is small when compared to the typical mass scales of the model, in this case q and $\delta\mu$. In the framework of the GL expansion, in Ref. [19] it is conjectured that a face centered cube is a good candidate for the crystalline color superconductive state.

In this paper we wish to define a procedure that allows to study the crystalline color superconductor with a generic crystalline structure in a space of parameters which is complementary to the GL one. This procedure is based on a weighted average of the interaction lagrangian over the lattice cell. The paper is organized as follows: in Sec. we discuss the smearing procedure used to obtain the effective gap equations, in the simple case of the FF state. Sec. is devoted to the generalization of this procedure to a generic crystalline structure. In Sec. we show one result of the solution of the smeared gap equations, namely the free energy plots for the different crystalline structures. Finally, in Sec. we summarize the leading results of our work.

SMEARED LAGRANGIAN IN THE FF PHASE

We shall consider Cooper pairing of the massless quarks up u and down d, with chemical potential μ_u, μ_d. We define $\mu = (\mu_u + \mu_d)/2$ and $\delta\mu = |\mu_u - \mu_d|/2 \ll \mu$. We begin with a review of the FF state. Although this case can be solved exactly, it is useful to consider it here in order to fix the notations and introduce some definitions to be used later on.

We work in the framework of High Density Effective Theory (HDET) [12, 25, 26, 27, 28, 29, 30]. The lagrangian for free quarks can be written as

$$\mathscr{L}_0 = \sum_{\vec{v}} \left[\psi_+^\dagger iV \cdot \partial \psi_+ + \psi_-^\dagger i\tilde{V} \cdot \partial \psi_- \right] + (L \to R). \qquad (3)$$

Here the sum represents an average over velocities; $\psi_\pm \equiv \psi_{\pm \mathbf{v}}$ are velocity dependent, positive energy left handed fields (the negative energy part has been integrated out). $\psi_{\mathbf{v}}$ depends on the residual momentum ℓ, corresponding to the decomposition of the quark momentum $p = \mu v + \ell$, with $v^\mu = (0, \mathbf{v})$ and $\ell_\parallel = \boldsymbol{\ell}\cdot\mathbf{v} = \xi$. We also introduce $V^\mu = (1, \mathbf{v})$ and $\tilde{V}^\mu = (1, -\mathbf{v})$.

Next we turn to the interaction term. We consider a Nambu-Jona Lasinio (NJL) inspired four fermion interaction to mimic the one gluon exchange of QCD, namely [17, 31]

$$\mathscr{L}_I = -\frac{3}{8} G \bar{\psi} \gamma^\mu \lambda_a \psi \, \bar{\psi} \gamma^\mu \lambda_a \psi . \tag{4}$$

Here G is a coupling constant, with dimension $mass^{-2}$; λ_a are color matrices and a sum over flavors is understood (the FF state with quark-quark interaction mediated by one gluon exchange has been considered in Ref.[32]). In the mean field approximation, after Fierzing, we get

$$\mathscr{L}_I = -\frac{1}{2}\varepsilon_{\alpha\beta 3}\varepsilon^{ij}(\psi_i^\alpha \psi_j^\beta \Delta(r) + \text{c.c.}) + (L \to R) - \frac{1}{G}\Delta(r)\Delta^*(r), \tag{5}$$

where i, j are flavor indices and α, β are color indices. In the FF state the total momentum of the Cooper pair is $2\mathbf{q}$ and the condensate has the space-dependence of a single plane wave, see Eq. (2). In the HDET formalism Eq. (5) can be recast in the form

$$\begin{aligned}\mathscr{L}_I &= -\frac{\Delta}{2}\sum_{\mathbf{v_i},\mathbf{v_j}} \exp\{i\mathbf{r}\cdot\boldsymbol{\alpha}(\mathbf{v_i}, \mathbf{v_j}, \mathbf{q})\}\varepsilon_{ij}\varepsilon_{\alpha\beta 3}\psi^T_{\mathbf{v_i};i\alpha}(x) C \psi_{-\mathbf{v_j};j\beta}(x) \\ &\quad -(L \to R) + \text{h.c.} - \frac{1}{g}\Delta(r)\Delta^*(r) ,\end{aligned} \tag{6}$$

where

$$\boldsymbol{\alpha}(\mathbf{v_i}, \mathbf{v_j}, \mathbf{q}) = 2\mathbf{q} - \mu_i \mathbf{v_i} - \mu_j \mathbf{v_j} . \tag{7}$$

Eqs. (3) and (7) are the HDET lagrangian of paired quarks in the FF state. Armed with them one can write a gap equation, namely a self-consistence (Schwinger-Dyson) equation for the gap parameter Δ.

Now we have the ingredients necessary to define our smearing procedure. To this end we recall the exact FF gap equation [31],

$$\Delta = i\frac{g\rho}{2}\int\frac{d\mathbf{v}}{4\pi}\int_0^\delta \frac{d\xi}{2\pi}\int d\ell_0 \frac{\Delta_{eff}}{\ell_0^2 - \xi^2 - \Delta_{eff}^2} = \frac{g\rho}{2}\int\frac{d\mathbf{v}}{4\pi}\int_0^\delta d\xi \frac{\Delta_{eff}}{\sqrt{\xi^2 + \Delta_{eff}^2}}, \tag{8}$$

where we have defined an effective gap parameter,

$$\Delta_{eff} = \Delta\theta(E_u)\theta(E_d) = \begin{cases} \Delta & \text{for } (\xi, \mathbf{v}) \in PR \\ 0 & \text{elsewhere} \end{cases}, \tag{9}$$

and $E_{u,d}$ are the dispersion laws for u and d quarks respectively [17, 33]

$$E_{d,u} = \pm\delta\mu \mp \mathbf{q}\cdot\mathbf{v} + \sqrt{\xi^2 + \Delta^2} ; \tag{10}$$

PR denotes the pairing region,

$$PR = \{(\xi, \mathbf{v}) | E_u > 0 \text{ and } E_d > 0\}. \tag{11}$$

We stress that Eq. (8) is exact. The key observation is that we can obtain the *same* gap equation (8) in the framework of HDET by defining a weighted smearing procedure of the gap lagrangian (6) over the lattice cell. First of all, we note that in the gap equation the relevant momenta are small with respect to the gap which is of the order of q. Therefore we may assume that the velocity dependent fields are slowly varying over regions of the order of the lattice size. This means that in the average we can treat them as constant, and in conclusion the average is made only on the coefficient $\exp\{i\mathbf{r}\cdot\boldsymbol{\alpha}\}$. Therefore what we are computing is

$$I(\boldsymbol{\alpha}) = \left\langle \exp\{i\mathbf{r}\cdot\boldsymbol{\alpha}\} g_R(\mathbf{r}) \right\rangle \tag{12}$$

where the bracket means average over the cell, and the weight function $g_R(\mathbf{r})$ can be chosen in such a way that

$$I(\boldsymbol{\alpha}) = \delta_R^3\left(\frac{\boldsymbol{\alpha}}{2q}\right), \quad \text{where} \quad \delta_R^3(\mathbf{x}) = \begin{cases} 1 & \text{for } |\mathbf{x}| < \frac{\pi}{2R}, \\ 0 & \text{elsewhere} \end{cases} \tag{13}$$

and $R/\pi \approx 1$. Independently of the exact form of $g_R(\mathbf{r})$, we will assume that the average procedure gives as a result the brick-shaped function δ_R defined in (13). As shown in Ref. [31], choosing

$$R = \frac{\pi|\delta\mu - \mathbf{v}\cdot\mathbf{q}|}{2\sqrt{\xi^2 + \Delta^2}|h(\mathbf{v}\cdot\hat{\mathbf{q}})|}, \quad h(\mathbf{v}\cdot\hat{\mathbf{q}}) = 1 - \frac{z_q}{\mathbf{v}\cdot\hat{\mathbf{q}}}, \tag{14}$$

one obtains from Eq. (6) the smeared lagrangian

$$\mathscr{L}_I = -\frac{1}{2}\sum_{\vec{v}} \Delta_{eff}\, \varepsilon_{ij}\varepsilon_{\alpha\beta 3}\, \psi_{\mathbf{v};i\alpha}^T(\ell) C \psi_{-\mathbf{v};j\beta}(-\ell) - (L \to R) + \text{h.c.} - \frac{1}{g}\Delta\Delta^*, \tag{15}$$

from which the desired gap equation (8) is obtained (the fermion propagator can be read in Eqs. (22) and (23) of Ref. [31]). Since $R/\pi \approx 1$, then Δ has not to be small, meaning that we should be far from a second order phase transition.

Let us finally notice that in Eq. (8) one can perform the ℓ_0 integration by substituting the previous expression for R with

$$R = \frac{\pi|\delta\mu - \mathbf{v}\cdot\mathbf{q}|}{2|\ell_0|\cdot|h(\mathbf{v}\cdot\hat{\mathbf{q}})|}. \tag{16}$$

In fact, observing that in any case Δ_{eff} is equal to 0 or Δ, at the pole we get back the expression (14). Then written as in (16), R and analogously Δ_{eff} become functions of the velocity and the energy; therefore our average should be better taken in the momentum space.

SMEARING FOR GENERIC CRYSTALLINE STRUCTURES

We have defined in the previous section a smearing procedure which, in the case of the FF state, allows to write in the HDET formalism a gap equation which coincides with the exact gap equation in Eq. (8). Apart the technical details for the definition of the weighting function, the smearing procedure can be viewed as a recipe which allows to replace an **r**-dependent gap function by an **r**-independent one. All the details of the crystalline structure are embodied into the definition of the pairing regions.

We now use this recipe to smear the interaction lagrangian for generic crystalline structures, defined by the pairing ansatz

$$\Delta(r) = \Delta \sum_{m=1}^{P} e^{2i\mathbf{q_m}\cdot\mathbf{r}}, \quad \mathbf{q_m} = q\mathbf{n_m}. \tag{17}$$

This procedure, although artificial, allows to study generic crystalline structures in a domain where Ginzburg-Landau expansion may not work well. Generalizing the results of the previous equations we substitute in the Lagrangian (15) $\Delta_{eff}(\mathbf{v}\cdot\mathbf{n},\ell_0)$ with

$$\Delta_E(\mathbf{v},\ell_0) = \sum_{m=1}^{P} \Delta_{eff}(\mathbf{v}\cdot\mathbf{n_m},\ell_0), \tag{18}$$

and the arguments of the θ functions in Eq. (9) will be the appropriate fermion dispersion laws. After smearing the gap equation reads

$$P\Delta = i\frac{g\rho}{2}\int\frac{d\mathbf{v}}{4\pi}\int\frac{d^2\ell}{2\pi}\frac{\Delta_E(\mathbf{v},\ell_0)}{\ell_0^2 - \ell_\parallel^2 - \Delta_E^2(\mathbf{v},\ell_0)} \tag{19}$$

which generalizes Eq. (8). The energy integration is performed by the residue theorem and the phase space is divided into different regions according to the pole positions. We get

$$P\Delta\ln\frac{2\delta}{\Delta_0} = \sum_{k=1}^{P}\int\int_{P_k}\frac{d\mathbf{v}}{4\pi}d\xi\frac{\Delta_E(\mathbf{v},\varepsilon)}{\sqrt{\xi^2+\Delta_E^2(\mathbf{v},\varepsilon)}} = \sum_{k=1}^{P}\int\int_{P_k}\frac{d\mathbf{v}}{4\pi}d\xi\frac{k\Delta}{\sqrt{\xi^2+k^2\Delta^2}} \tag{20}$$

where the pairing regions P_k are defined as follows

$$P_k = \{(\mathbf{v},\xi)\,|\,\Delta_E(\mathbf{v},\varepsilon) = k\Delta\} \tag{21}$$

and we have ruled out the coupling constant G by means of the BCS gap Δ_0. The first term in the sum, corresponding to the region P_1, has P equal contributions with a dispersion rule equal to the Fulde and Ferrel case. This can be interpreted as a contribution from P non interacting plane waves. In the other regions the different plane waves have an overlap.

Thus for N plane waves the smearing procedure does not simply leads to a lagrangian which is the sum of N plane waves (FF) lagrangians. This "would be nice" scenario is complicated by the presence of the pairing regions with two or more overlaps $(P_2,\ldots,P_N)$.

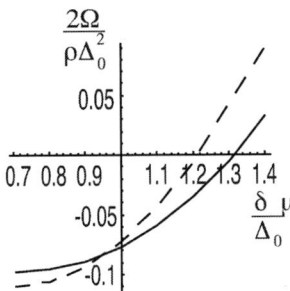

FIGURE 1. The values of the free energies of the bcc (dashed line) and of the fcc (full line) crystalline LOFF structures as a function of $\delta\mu/\Delta_0$. The bcc is the favored structure up to $\delta\mu \approx .95\Delta_0$; for $.95\Delta_0 < \delta\mu < 1.32\Delta_0$ the fcc is favored. Here, for each value of $\delta\mu$, the values of z_q and Δ are those that minimize the free energy.

NUMERICAL RESULTS FOR THE FREE ENERGY

In this section we present results for the free energy Ω of the structures that we have considered in Ref. [31], computed as integral of the gap equation (20). In particular, in Fig. 1 we show $\Omega(\Delta, \delta\mu) - \Omega(0, \delta\mu)$ against $\delta\mu$ for the face centered (fcc) and the body centered (bcc) cubic structures (the other structures have a higher free energy and therefore they are not shown here). Our central result is that the bcc is a good candidate for the ground state for $\delta\mu \leq 0.95\Delta_0$. Above this value of $\delta\mu$ the fcc is the good ground state for $\delta\mu \leq 1.32\Delta_0$. For higher values of $\delta\mu$, the fermion condensation is energetically disfavored. The transition to the normal state is first order.

CONCLUSIONS

In the framework of HDET, we have discussed an averaging procedure of the NJL quark-quark interaction lagrangian, treated in the mean field approximation, for the LOFF phase of QCD. This procedure gives results which are valid in domains where Ginzburg-Landau results may be questionable. Among the several structures considered, we find that a body centered cube is the favorite ground state for $\delta\mu \leq 0.95\Delta_0$, while for higher values of $\delta\mu$ the face centered cube is favoured. A first order transition to the normal state is found for $\delta\mu \approx 1.32\Delta_0$. The method exposed here can be applied also to the LOFF phase in the electromagnetic superconductors (see for example Ref. [34]).

ACKNOWLEDGMENTS

I am in debt with R. Casalbuoni, R. Gatto, N. Ippolito, G. Nardulli and M. Ciminale for fruitful collaboration; moreover I thank M. Alford, A. Gerhold, M. Mannarelli and I. Shovkovy for clarifying discussions, comments and helpful correspondence.

REFERENCES

1. J. C. Collins, and M. J. Perry, *Phys. Rev. Lett.* **34**, 1353 (1975).
2. B. C. Barrois, *Nucl. Phys.* **B129**, 390 (1977).
3. D. Bailin, and A. Love, *Phys. Rept.* **107**, 325 (1984).
4. M. G. Alford, K. Rajagopal, and F. Wilczek, *Phys. Lett.* **B422**, 247–256 (1998), hep-ph/9711395.
5. R. Rapp, T. Schafer, E. V. Shuryak, and M. Velkovsky, *Phys. Rev. Lett.* **81**, 53–56 (1998), hep-ph/9711396.
6. T. Schafer, and F. Wilczek, *Phys. Rev.* **D60**, 114033 (1999), hep-ph/9906512.
7. M. G. Alford, K. Rajagopal, and F. Wilczek, *Nucl. Phys.* **B537**, 443–458 (1999), hep-ph/9804403.
8. I. Shovkovy, and M. Huang, *Phys. Lett.* **B564**, 205 (2003), hep-ph/0302142.
9. M. Alford, C. Kouvaris, and K. Rajagopal, *Phys. Rev. Lett.* **92**, 222001 (2004), hep-ph/0311286.
10. K. Rajagopal, and F. Wilczek (2000), hep-ph/0011333.
11. M. G. Alford, *Ann. Rev. Nucl. Part. Sci.* **51**, 131–160 (2001), hep-ph/0102047.
12. G. Nardulli, *Riv. Nuovo Cim.* **25N3**, 1–80 (2002), hep-ph/0202037.
13. T. Schafer (2003), hep-ph/0304281.
14. T. Schaefer (2005), hep-ph/0509068.
15. A. I. Larkin, and Y. N. Ovchinnikov, *Zh. Eksp. Teor. Fiz.* **47**, 1136 (1964).
16. P. Fulde, and R. A. Ferrel, *Phys. Rev.* **135**, A550 (1964).
17. M. G. Alford, J. A. Bowers, and K. Rajagopal, *Phys. Rev.* **D63**, 074016 (2001), hep-ph/0008208.
18. J. A. Bowers, J. Kundu, K. Rajagopal, and E. Shuster, *Phys. Rev.* **D64**, 014024 (2001), hep-ph/0101067.
19. J. A. Bowers, and K. Rajagopal, *Phys. Rev.* **D66**, 065002 (2002), hep-ph/0204079.
20. R. Casalbuoni, and G. Nardulli, *Rev. Mod. Phys.* **76**, 263–320 (2004), hep-ph/0305069.
21. I. Giannakis, and H.-C. Ren, *Phys. Lett.* **B611**, 137–146 (2005), hep-ph/0412015.
22. I. Giannakis, D.-f. Hou, and H.-C. Ren (2005), hep-ph/0507306.
23. I. Giannakis, and H.-C. Ren, *Nucl. Phys.* **B723**, 255–280 (2005), hep-th/0504053.
24. R. Casalbuoni, R. Gatto, N. Ippolito, G. Nardulli, and M. Ruggieri (2005), hep-ph/0507247.
25. D. K. Hong, *Phys. Lett.* **B473**, 118–125 (2000), hep-ph/9812510.
26. D. K. Hong, *Nucl. Phys.* **B582**, 451–476 (2000), hep-ph/9905523.
27. S. R. Beane, P. F. Bedaque, and M. J. Savage, *Phys. Lett.* **B483**, 131–138 (2000), hep-ph/0002209.
28. T. Schafer, *Nucl. Phys.* **A728**, 251–271 (2003), hep-ph/0307074.
29. R. Casalbuoni, *AIP Conf. Proc.* **602**, 358–367 (2001), hep-th/0108195.
30. T. Schafer, *ECONF* **C030614**, 038 (2003), hep-ph/0310176.
31. R. Casalbuoni, et al., *Phys. Rev.* **D70**, 054004 (2004), hep-ph/0404090.
32. A. K. Leibovich, K. Rajagopal, and E. Shuster, *Phys. Rev.* **D64**, 094005 (2001), hep-ph/0104073.
33. R. Casalbuoni, et al., *Phys. Lett.* **B575**, 181–189 (2003), hep-ph/0307335.
34. R. Casalbuoni, R. Gatto, M. Mannarelli, G. Nardulli, and M. Ruggieri, *Phys. Lett.* **B600**, 48–56 (2004), hep-ph/0407210.

Asymmetric neutrino emission from spin-1 color superconductor

Andreas Schmitt*, Igor A. Shovkovy[†,**] and Qun Wang[‡]

*Center for Theoretical Physics, Massachusetts Institute of Technology, Cambridge, MA 02139, USA
[†]Frankfurt Institute for Advanced Studies, J.W. Goethe-Universität, D-60054 Frankfurt am Main, Germany
[**]Bogolyubov Institute for Theoretical Physics, 03143, Kiev, Ukraine
[‡]Department of Modern Physics, University of Science and Technology of China, Hefei, Anhui 230026, People's Republic of China

Abstract. We discuss the spatial asymmetry in the neutrino emission from spin-1 color-superconducting quark matter in the A phase. The asymmetry could potentially lead to a "neutrino rocket" mechanism for neutron stars whose cores are made of quark matter. However, an estimate of the resulting velocities, using the specific heat and the total emissivity of the A phase, shows that no observable "kicks" due this mechanism is expected.

Keywords: color superconductivity, quark matter, compact stars
PACS: 12.39.-x, 11.15.Ex, 21.65.+f

Introduction. Matter at large baryon density is expected to be deconfined and color-superconducting (for reviews see, e.g., Ref. [1]). Color-superconducting quark matter could exist inside neutron stars, whose central densities are highest in Nature. Therefore, it is important to study the physical implications of such a possibility in detail. At present, this is not easy because our current knowledge regarding the ground state of neutral, β-equilibrated dense matter is very limited [2, 3]. Many color superconducting phases were proposed [4, 5, 6, 7, 8, 9, 10, 11], but it is not clear which of them can be realized inside stars.

Here we discuss an unusual physical property of the transverse spin-1 color-superconducting A phase of dense quark matter [12]. In this phase, the neutrino emission is not symmetric in space.

Neutrino emission. Let us start by outlining the main steps in the derivation of a general expression for the neutrino emissivity in spin-1 color superconducting phases. We use the Kadanoff-Baym formalism [13, 14] to derive the following differential expression for the emissivity [15]:

$$\frac{d\varepsilon_\nu}{dp_\nu d\Omega_\nu} = \frac{G_F^2}{8(2\pi)^6} \int p_e dp_e \int d\Omega_e p_\nu^2 \, n_B(p_\nu - p_e + \mu_e)$$
$$\times \, n_F(p_e - \mu_e) L_{\lambda\sigma}(P_e, P_\nu) \, \mathrm{Im}\Pi_R^{\lambda\sigma}(\delta P_e - P_\nu), \qquad (1)$$

where G_F is the Fermi coupling constant, μ_e is the electron chemical potential, and $\delta P_e^\lambda \equiv P_e^\lambda - \delta_0^\lambda \mu_e$. Here, particle four-momenta are denoted by capital Latin letters, while the absolute values of the three-momenta are denoted by lowercase letters. The metric tensor is $g_{\lambda\sigma} = \mathrm{diag}(1,-1,-1,-1)$. The Bose and the Fermi distribution functions are denoted by $n_B(\omega) \equiv [\exp(\omega/T)-1]^{-1}$ and $n_F(\omega) \equiv [\exp(\omega/T)+1]^{-1}$, respectively. The lepton tensor $L_{\lambda\sigma}(P_e, P_\nu)$ is defined as follows:

$$L_{\lambda\sigma}(P_e,P_\nu) = \mathrm{Tr}\left[P_e^\kappa \gamma_\kappa \gamma_\sigma (1-\gamma^5) P_\nu^\rho \gamma_\rho \gamma_\lambda (1-\gamma^5)\right]. \tag{2}$$

Finally, the last factor in the integrand on the right hand side of Eq. (1) is the imaginary part of the retarded polarization tensor of the W-boson,

$$\Pi^{\lambda\sigma}(Q) = T\sum_n \int \frac{d^3\mathbf{k}}{(2\pi)^3} \mathrm{Tr}\left[\Gamma_-^\lambda S(K)\Gamma_+^\sigma S(K+Q)\right], \tag{3}$$

with the trace running over flavor, color, Dirac and Nambu-Gorkov indices. The quark propagator $S(K)$ is diagonal in flavor space, and its components have the following Nambu-Gorkov structure:

$$S_f = \begin{pmatrix} G_f^+(K) & \Xi_f^-(K) \\ \Xi_f^+(K) & G_f^-(K) \end{pmatrix}, \quad \text{for} \quad f = u,d. \tag{4}$$

We consider the ultrarelativistic limit. The explicit color and Dirac structure of $G_f^\pm$ and $\Xi_f^\pm$ can be found in Ref. [12]. Here, we note only that the poles of the quark propagators appear at $k_0 = k + \mu_f$ (antiquarks) and at

$$k_0 = \varepsilon_{k,r,f} \equiv \sqrt{(k-\mu_f)^2 + \lambda_{k,r}|\phi_f|^2}, \quad r = 1,2,3, \tag{5}$$

where ϕ_f is the gap parameter, and the functions $\lambda_{k,r}$ are specified by the choice of the phase.

The order parameter of the A phase has a special direction in color space: quarks of one color do not pair. Also, it has a special direction in momentum space, say, the z-direction. If $\theta_\mathbf{k}$ denotes the angle between the three-momentum of a quasiparticle and the z-axis, the three low-energy quasiparticle modes in the transverse A phase are defined by $\lambda_{k,1} = (1+|\cos\theta_\mathbf{k}|)^2$, $\lambda_{k,2} = (1-|\cos\theta_\mathbf{k}|)^2$, and $\lambda_{k,3} = 0$. Here, "transverse" refers to the fact that quarks of opposite chirality form Cooper pairs.

The explicit expressions of the vertices in Eq. (3) read

$$\Gamma_\pm^\lambda = \begin{pmatrix} \gamma^\lambda(1-\gamma^5)\tau_\pm & 0 \\ 0 & -\gamma^\lambda(1+\gamma^5)\tau_\mp \end{pmatrix}, \tag{6}$$

where the flavor matrix $\tau_\pm \equiv (\tau_1 \pm i\tau_2)/2$ is constructed from Pauli matrices.

By substituting the quark propagators (4) and the vertices (6) into Eq. (3), we calculate the imaginary part of the retarded polarization tensor. Then, by making use of the result, we derive an expression for the emissivity in the following approximate form [15]:

$$\begin{aligned} \varepsilon_\nu &\approx 2\left(\varepsilon_\nu^{(11)} + \varepsilon_\nu^{(22)} + \varepsilon_\nu^{(33)}\right) \\ &\approx \frac{457}{630}\left[\frac{2}{3}G\left(\frac{\phi_u}{T},\frac{\phi_d}{T}\right) + \frac{1}{3}\right]\alpha_s G_F^2 \mu_e \mu_u \mu_d T^6, \end{aligned} \tag{7}$$

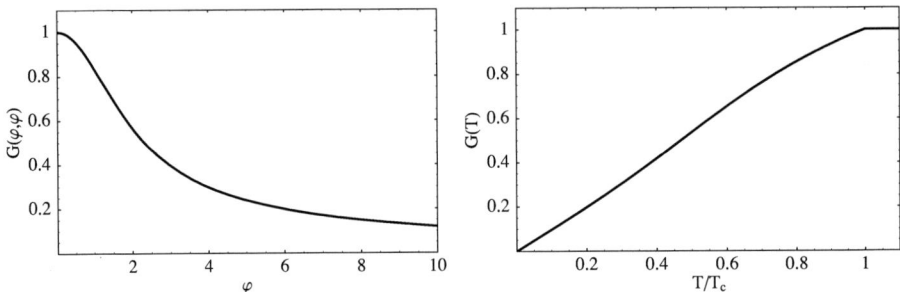

FIGURE 1. The suppression functions $G(\varphi, \varphi)$ (left panel) and $G(T)$ (right panel) in the A phase.

where $\varepsilon_\nu^{(rr)}$ is the partial contribution involving the rth type quasiparticle modes, see Eq. (5). The factor 2 comes from taking into account both the neutrino and the antineutrino emissivities. In the final result, the contribution of the ungapped modes $\varepsilon_\nu^{(33)}$ is, up to a factor $1/3$, the same as in the normal phase of quark matter [16, 17]. The contribution of the other two modes is suppressed by the following function:

$$G(\varphi_u, \varphi_d) \approx \frac{2520}{457\pi^6} \int_0^\infty dv v^3 \int_{-1}^1 d\xi f(v,\xi), \qquad (8)$$

where

$$f(v,\xi) = \sum_{e_1,e_2=\pm} \int_0^\infty \int_0^\infty \frac{\left(e^{v+e_1\tilde{\varepsilon}_u - e_2\tilde{\varepsilon}_d} + 1\right)^{-1} dx_u dx_d}{\left(e^{-e_1\tilde{\varepsilon}_u} + 1\right)\left(e^{e_2\tilde{\varepsilon}_d} + 1\right)}, \qquad (9)$$

and $\tilde{\varepsilon}_f = \sqrt{x_f^2 + (1+\xi)^2 \varphi_f^2}$ with $f = u, d$. Note that $0 \le G(\varphi_u, \varphi_d) \le 1$ and $G(0,0) = 1$.

The function $G(\varphi_u, \varphi_d)$ can be easily evaluated numerically. For the sake of simplicity, we set $\varphi_u = \varphi_d \equiv \varphi$ in the calculation. The results are plotted in the left panel of Fig. 1. By definition, the function $G(\varphi_u, \varphi_d)$ describes the suppression of the emissivity due to the presence of the gap in the quasiparticle spectrum. It is interesting to note that, at large values of φ, the function is power suppressed, $G(\varphi, \varphi) \sim 1/\varphi$. This property is connected with the fact that the gap function in the A phase has nodes at the north and south poles of the Fermi sphere.

It might be also instructive to calculate the temperature dependence of the suppression function. This is shown in the right panel of Fig. 1. To make the plot, we made use of the following model temperature dependence of the gap parameter:

$$\phi = \phi_0 \sqrt{1 - \left(\frac{T}{T_c}\right)^2}, \qquad (10)$$

where ϕ_0 is the value of the gap parameter at $T=0$, and T_c is the value of the critical temperature. It is evident that the function $G(T/T_c)$ behaves almost linearly all the way down to zero temperature. Note that the numerical data for $G(T/T_c)$ is well approxi-

metric tensor is $g_{\lambda\sigma} = \text{diag}(1,-1,-1,-1)$. The Bose and the Fermi distribution functions are denoted by $n_B(\omega) \equiv [\exp(\omega/T)-1]^{-1}$ and $n_F(\omega) \equiv [\exp(\omega/T)+1]^{-1}$, respectively. The lepton tensor $L_{\lambda\sigma}(P_e, P_\nu)$ is defined as follows:

$$L_{\lambda\sigma}(P_e, P_\nu) = \text{Tr}\left[P_e^\kappa \gamma_\kappa \gamma_\sigma (1-\gamma^5) P_\nu^\rho \gamma_\rho \gamma_\lambda (1-\gamma^5) \right]. \tag{2}$$

Finally, the last factor in the integrand on the right hand side of Eq. (1) is the imaginary part of the retarded polarization tensor of the W-boson,

$$\Pi^{\lambda\sigma}(Q) = T \sum_n \int \frac{d^3\mathbf{k}}{(2\pi)^3} \text{Tr}\left[\Gamma_-^\lambda S(K) \Gamma_+^\sigma S(K+Q) \right], \tag{3}$$

with the trace running over flavor, color, Dirac and Nambu-Gorkov indices. The quark propagator $S(K)$ is diagonal in flavor space, and its components have the following Nambu-Gorkov structure:

$$S_f = \begin{pmatrix} G_f^+(K) & \Xi_f^-(K) \\ \Xi_f^+(K) & G_f^-(K) \end{pmatrix}, \quad \text{for} \quad f = u, d. \tag{4}$$

We consider the ultrarelativistic limit. The explicit color and Dirac structure of $G_f^\pm$ and $\Xi_f^\pm$ can be found in Ref. [12]. Here, we note only that the poles of the quark propagators appear at $k_0 = k + \mu_f$ (antiquarks) and at

$$k_0 = \varepsilon_{k,r,f} \equiv \sqrt{(k-\mu_f)^2 + \lambda_{k,r}|\phi_f|^2}, \quad r=1,2,3, \tag{5}$$

where ϕ_f is the gap parameter, and the functions $\lambda_{k,r}$ are specified by the choice of the phase.

The order parameter of the A phase has a special direction in color space: quarks of one color do not pair. Also, it has a special direction in momentum space, say, the z-direction. If $\theta_\mathbf{k}$ denotes the angle between the three-momentum of a quasiparticle and the z-axis, the three low-energy quasiparticle modes in the transverse A phase are defined by $\lambda_{k,1} = (1+|\cos\theta_\mathbf{k}|)^2$, $\lambda_{k,2} = (1-|\cos\theta_\mathbf{k}|)^2$, and $\lambda_{k,3} = 0$. Here, "transverse" refers to the fact that quarks of opposite chirality form Cooper pairs.

The explicit expressions of the vertices in Eq. (3) read

$$\Gamma_\pm^\lambda = \begin{pmatrix} \gamma^\lambda(1-\gamma^5)\tau_\pm & 0 \\ 0 & -\gamma^\lambda(1+\gamma^5)\tau_\mp \end{pmatrix}, \tag{6}$$

where the flavor matrix $\tau_\pm \equiv (\tau_1 \pm i\tau_2)/2$ is constructed from Pauli matrices.

By substituting the quark propagators (4) and the vertices (6) into Eq. (3), we calculate the imaginary part of the retarded polarization tensor. Then, by making use of the result, we derive an expression for the emissivity in the following approximate form [15]:

$$\begin{aligned}\varepsilon_\nu &\approx 2\left(\varepsilon_\nu^{(11)} + \varepsilon_\nu^{(22)} + \varepsilon_\nu^{(33)}\right) \\ &\approx \frac{457}{630}\left[\frac{2}{3}G\left(\frac{\phi_u}{T}, \frac{\phi_d}{T}\right) + \frac{1}{3}\right]\alpha_s G_F^2 \mu_e \mu_u \mu_d T^6,\end{aligned} \tag{7}$$

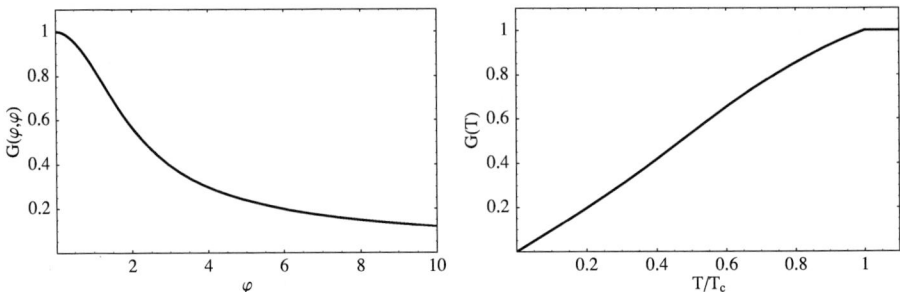

FIGURE 1. The suppression functions $G(\varphi,\varphi)$ (left panel) and $G(T)$ (right panel) in the A phase.

where $\varepsilon_\nu^{(rr)}$ is the partial contribution involving the rth type quasiparticle modes, see Eq. (5). The factor 2 comes from taking into account both the neutrino and the antineutrino emissivities. In the final result, the contribution of the ungapped modes $\varepsilon_\nu^{(33)}$ is, up to a factor $1/3$, the same as in the normal phase of quark matter [16, 17]. The contribution of the other two modes is suppressed by the following function:

$$G(\varphi_u, \varphi_d) \approx \frac{2520}{457\pi^6} \int_0^\infty dv v^3 \int_{-1}^1 d\xi\, f(v,\xi), \qquad (8)$$

where

$$f(v,\xi) = \sum_{e_1,e_2=\pm} \int_0^\infty \int_0^\infty \frac{\left(e^{v+e_1\tilde{\varepsilon}_u - e_2\tilde{\varepsilon}_d}+1\right)^{-1} dx_u dx_d}{\left(e^{-e_1\tilde{\varepsilon}_u}+1\right)\left(e^{e_2\tilde{\varepsilon}_d}+1\right)}, \qquad (9)$$

and $\tilde{\varepsilon}_f = \sqrt{x_f^2 + (1+\xi)^2 \varphi_f^2}$ with $f = u,d$. Note that $0 \leq G(\varphi_u, \varphi_d) \leq 1$ and $G(0,0) = 1$.

The function $G(\varphi_u, \varphi_d)$ can be easily evaluated numerically. For the sake of simplicity, we set $\varphi_u = \varphi_d \equiv \varphi$ in the calculation. The results are plotted in the left panel of Fig. 1. By definition, the function $G(\varphi_u, \varphi_d)$ describes the suppression of the emissivity due to the presence of the gap in the quasiparticle spectrum. It is interesting to note that, at large values of φ, the function is power suppressed, $G(\varphi,\varphi) \sim 1/\varphi$. This property is connected with the fact that the gap function in the A phase has nodes at the north and south poles of the Fermi sphere.

It might be also instructive to calculate the temperature dependence of the suppression function. This is shown in the right panel of Fig. 1. To make the plot, we made use of the following model temperature dependence of the gap parameter:

$$\phi = \phi_0 \sqrt{1 - \left(\frac{T}{T_c}\right)^2}, \qquad (10)$$

where ϕ_0 is the value of the gap parameter at $T=0$, and T_c is the value of the critical temperature. It is evident that the function $G(T/T_c)$ behaves almost linearly all the way down to zero temperature. Note that the numerical data for $G(T/T_c)$ is well approximated by the following fit:

$$G(\varphi,\varphi) \approx \sum_{n=1}^{5} \frac{g_n}{[1+(r_0\varphi)^2]^{n/2}} = \sum_{n=1}^{5} g_n \left(\frac{T}{T_c}\right)^n, \quad (11)$$

with $g_1 \approx 0.9707$, $g_2 \approx -0.2498$, $g_3 \approx 2.0946$, $g_4 \approx -2.9288$, $g_5 \approx 1.1133$, and $r_0 = e^{\gamma+\bar{\zeta}}/\pi \approx 0.8125$ is the ratio of the critical temperature to the value of the gap at $T=0$ in the A phase, i.e., $r_0 = T_c/\phi_0$ which is expressed in terms of the Euler constant $\gamma \approx 0.577$ and $\bar{\zeta} = \ln 2 - 1/3$ [12].

In passing, we note that the pair breaking processes [18] do not play any significant role in the case of spin-1 color-superconducting quark matter under consideration. The corresponding contribution to the emissivity [19] is parametrically suppressed by factor $T/\mu_e \sim 10^{-3}$.

From Eq. (7), we see that the (anti-)neutrino emissivity in the A phase differs only by a factor of order 1 from the corresponding result in the normal phase of quark matter [16, 17]. Therefore, the A phase should have qualitatively the same effect on cooling of stars as the normal phase.

Nevertheless, the neutrino emission from the A phase is very unusual. It is not symmetric with respect to reversing the direction of the z-axis. To quantify the asymmetry, we calculate the value of the z-component of the momentum carried away by neutrinos per unit volume of quark matter, per unit time. This is obtained by replacing one power of p_ν on the right hand side of Eq. (1) by $p_\nu \cos\theta_{\mathbf{p}_\nu}$. Taking into account that the neutrino and the antineutrino emissions give the same contributions, we arrive at the final result [15]

$$\frac{dP_z^{(\text{tot})}}{dVdt} \approx \frac{2}{3} H\left(\frac{\phi_u}{T},\frac{\phi_d}{T}\right) \frac{457}{630} \alpha_s G_F^2 \mu_e \mu_u \mu_d T^6, \quad (12)$$

where

$$H(\varphi_u,\varphi_d) \approx -\frac{840}{457\pi^6} \int_0^\infty d\nu \nu^3 \int_{-1}^1 d\xi \xi f(\nu,\xi). \quad (13)$$

Note that $H(0,0) = 0$ which is consistent with the fact that the momentum kick is vanishing in the normal phase of quark matter. The numerical result for the function H at equal values of its two arguments is well approximated by the following expression:

$$H(\varphi,\varphi) \approx \sum_{n=1}^{5} \frac{h_n}{[1+(r_0\varphi)^2]^{n/2}} = \sum_{n=1}^{5} h_n \left(\frac{T}{T_c}\right)^n, \quad (14)$$

with $h_1 = 0.3068$, $h_2 = -0.1977$, $h_3 = -0.7838$, $h_4 = 1.0286$, $h_5 = -0.3539$. Here we used again the notation $r_0 \equiv T_c/\phi_0 = e^{\gamma+\bar{\zeta}}/\pi \approx 0.8125$ [12]. Note that $\sum_{n=1}^5 h_n = 0$.

The physical reason for the breakdown of the reflection symmetry lies in the pairing pattern of the transverse A phase. Quasiquarks of the first branch, $r=1$, have helicity $+1$ if the projection of their momentum onto the z-axis is negative, $\cos\theta_k < 0$, and helicity -1 if $\cos\theta_k > 0$. Quasiquarks of the second branch, $r=2$, have opposite helicities. Only left-handed quarks (in the ultrarelativistic limit, quarks with negative helicity) participate in the Urca processes. Thus, the quasiquarks of the first (second) branch contribute

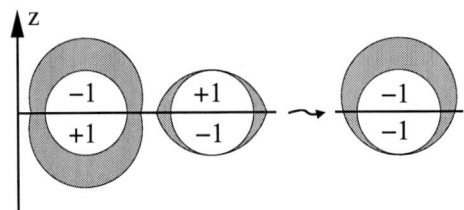

FIGURE 2. Gap functions for the first (left) and the second (middle) excitation branch with specified helicities of quasiparticles in the upper and the lower half-spaces. The "effective" gap relevant for the neutrino emission is shown on the right.

only if their momenta are in the upper (lower) half-space. Taking this into account, the effective branch relevant for the emission has gap $\phi_{\text{eff}} \sim 1 + \cos\theta_k$, which discriminates between $+z$ and $-z$ directions, see Fig. 2. Since neutrinos are emitted preferably in the direction opposite to the quark momenta, this asymmetry manifests itself in the neutrino emission.

Estimate of the maximum velocity kick. In order to make a simple estimate of the velocity kick due to the neutrino emission from a color-superconducting quark matter core in the A phase, we need to evaluate cooling rate of bulk matter in spin-1 color superconducting A phases. When the cooling is only due to the neutrino emissivity, one has the following relation,

$$\varepsilon_V(T) = -c_V(T)\frac{dT}{dt}. \tag{15}$$

In order to derive the change of temperature in time, one has to integrate the above equation,

$$t - t_0 = -\int_{T_0}^{T} dT' \frac{c_V(T')}{\varepsilon_V(T')}, \tag{16}$$

where T_0 is the temperature at time t_0. The specific heat of the A phase of quark matter can be given approximately by the corresponding expression in the normal phase, $c_V = (\mu_u^2 + \mu_d^2)T$ (see, for example, Ref. [17]). Also, the neutrino emissivity of the A phase can be replaced by the normal phase expression derived by Iwamoto [16]. As we argued earlier, both results are the same up to a factor of order 1. Thus, we arrive at

$$t - t_0 = \frac{315}{914} \frac{\mu_u^2 + \mu_d^2}{\alpha_s G_F^2 \mu_e \mu_u \mu_d} \left(\frac{1}{T^4} - \frac{1}{T_0^4}\right), \tag{17}$$

By inverting this relation, we arrive at the following model time dependence of the core temperature:

$$T(t) = T_0 \frac{\tau^{1/4}}{(t - t_0 + \tau)^{1/4}}, \tag{18}$$

where the initial condition is fixed as follows: $T_0 = 100$ keV at $t_0 = 100$ yr. This model time dependence is justified for the long-term stellar cooling after the thermal relaxation of crust is completed ($t \gtrsim 10^2$ yr) and before the surface cooling by photons starts to dominate ($t \lesssim 10^5$ yr), e.g., see Ref. [20]. In the above expression, the following shorthand notation was introduced

$$\tau \equiv \frac{315}{914} \frac{\mu_u^2 + \mu_d^2}{\alpha_s G_F^2 \mu_e \mu_u \mu_d} \frac{1}{T_0^4} \approx 10^{-5} \text{ yr.} \quad (19)$$

To make the estimate, we used $\alpha_s = 1$, as well as the following values of the chemical potentials: $\mu_u = 400$ MeV, $\mu_d = 500$ MeV, $\mu_e = 100$ MeV.

In general, we assume that $T_c < T_0$, i.e., the system is too hot for spin-1 pairing initially, and then it cools through the transition point. The velocity kick for a star of mass $1.4 M_\odot$ with the quark core of radius R_c is given by the expression:

$$\delta v \equiv \frac{\Delta P_z^{(\text{tot})}}{1.4 M_\odot} = \frac{457 \alpha_s}{945} G_F^2 \mu_e \mu_u \mu_d \frac{4\pi}{3} \frac{R_c^3}{1.4 M_\odot} T_0^4 T_c^2 \tau \theta(t - t_c)$$
$$\times \sum_{n=1}^{5} \frac{4 h_n}{2+n} \left[1 - \left(\frac{t_c - t_0 + \tau}{t - t_0 + \tau} \right)^{(2+n)/4} \right], \quad (20)$$

where we used the approximate expression in Eq. (14), as well as the simplified temperature dependence for the gap parameter in Eq. (10). The notation $t_c \equiv \tau(T_0^4/T_c^4 - 1) + t_0$ stands for the time when the temperature of the quark core drops below the critical value.

From Eq. (20), we can also derive the expression for the maximum velocity kick ($t = \infty$):

$$\delta v_{\text{max}} \approx 0.033 \, \alpha_s G_F^2 \mu_e \mu_u \mu_d \frac{4\pi}{3} \frac{R_c^3}{1.4 M_\odot} T_0^4 T_c^2 \tau. \quad (21)$$

The results for the maximum velocity kicks differ by a factor $\tau/t_0 \approx 10^{-7}$ from the original prediction in Ref. [21]. Consequently, the predicted kick velocities become extremely small. For example, instead of velocities of the order of $\delta v_{\text{max}} \sim 1000$ km/s, one gets $\delta v_{\text{max}} \sim 10^{-4}$ km/s.

Discussion. The main observation of this study is that the neutrino emission from the spin-1 color-superconducting transverse A phase is not symmetric with respect to reversing one spatial direction. This is a special property of the A phase which, as a more detailed analysis shows [15], is based on a helicity order of its quasiparticles and which is not shared by other known spin-1 color-superconducting phases.

One may suggest that the unusual neutrino emission from the color-superconducting A phase may have a rather spectacular implication for the physics of neutron (hybrid) stars. For example, the spatial asymmetry in the emission could result in a non-vanishing velocity kick transfered to a star [21]. If realized in nature, such a phenomenon would have specific predictions, related directly to the properties of the A phase. Our estimates show, however, that the maximum velocity kicks due to the asymmetric neutrino emission are many orders of magnitude smaller than those observed in stars.

In principle, the acceleration of stars due to asymmetric neutrino emission could also be detected through high precision timing of compact stars. Because of the Doppler effect, the acceleration should modify the measured spin period of a star, as well as other related observables. The difficulty in interpreting such measurements may exist, however, because other sources of the acceleration exist.

In conclusion, we find that the spin-1 color-superconducting A phase has very special physical properties that, in principle, could be detected in the observation data from stars, containing such a phase of quark matter. In future studies, it would be of great interest, therefore, to address the issue of possible signature-type observables in detail.

ACKNOWLEDGMENTS

This work was supported in part by the Virtual Institute of the Helmholtz Association under grant No. VH-VI-041 and by Gesellschaft für Schwerionenforschung (GSI), Bundesministerium für Bildung und Forschung (BMBF). A.S. thanks the German Academic Exchange Service (DAAD) for financial support.

REFERENCES

1. K. Rajagopal and F. Wilczek, hep-ph/0011333; M. Alford, Ann. Rev. Nucl. Part. Sci. **51**, 131 (2001); S. Reddy, Acta Phys. Polon. B **33**, 4101 (2002); T. Schäfer, hep-ph/0304281; D. H. Rischke, Prog. Part. Nucl. Phys. **52**, 197 (2004); M. Buballa, Phys. Rep. **407**, 205 (2005); H.-C. Ren, hep-ph/0404074; M. Huang, Int. J. Mod. Phys. E **14**, 675 (2005); I. A. Shovkovy, nucl-th/0410091.
2. S. B. Rüster, I. A. Shovkovy and D. H. Rischke, Nucl. Phys. A **743**, 127 (2004); I. A. Shovkovy, S. B. Rüster and D. H. Rischke, J. Phys. G **31**, S849 (2005).
3. K. Fukushima, C. Kouvaris and K. Rajagopal, Phys. Rev. D **71**, 034002 (2005).
4. M. Alford, K. Rajagopal, and F. Wilczek, Phys. Lett. B **422**, 247 (1998); R. Rapp, T. Schäfer, E. V. Shuryak, and M. Velkovsky, Phys. Rev. Lett. **81**, 53 (1998).
5. M. G. Alford, K. Rajagopal, and F. Wilczek, Nucl. Phys. B**537**, 443 (1999).
6. I. Shovkovy and M. Huang, Phys. Lett. B **564**, 205 (2003); M. Huang and I. Shovkovy, Nucl. Phys. A **729**, 835 (2003); M. Alford, C. Kouvaris, and K. Rajagopal, Phys. Rev. Lett. **92**, 222001 (2004); M. Alford, C. Kouvaris, and K. Rajagopal, Phys. Rev. D **71**, 054009 (2005).
7. M. Iwasaki and T. Iwado, *Phys. Lett. B* **350**, 163 (1995).
8. T. Schäfer, Phys. Rev. D **62**, 094007 (2000).
9. M. Buballa, J. Hošek and M. Oertel, Phys. Rev. Lett. **90**, 182002 (2003); A. Schmitt, Q. Wang and D. H. Rischke, Phys. Rev. D **66**, 114010 (2002); A. Schmitt, Q. Wang and D. H. Rischke, Phys. Rev. Lett. **91**, 242301 (2003); A. Schmitt, nucl-th/0405076.
10. M. G. Alford, J. A. Bowers, and K. Rajagopal, Phys. Rev. D **63**, 074016 (2001); R. Casalbuoni, R. Gatto, M. Mannarelli, and G. Nardulli, Phys. Rev. D **66**, 014006 (2002); I. Giannakis, J. T. Liu, and H.-C. Ren, Phys. Rev. D **66**, 031501(R) (2002); R. Casalbuoni, R. Gatto, N. Ippolito, G. Nardulli and M. Ruggieri, hep-ph/0507247.
11. E. V. Gorbar, M. Hashimoto, and V. A. Miransky, hep-ph/0507303.
12. A. Schmitt, Phys. Rev. D **71**, 054016 (2005).
13. L. P. Kadanoff and G. Baym, *Quantum Statistical Mechanics* (Benjamin, New York, 1962).
14. A. Sedrakian and A. Dieperink, Phys. Lett. B **463**, 145 (1999); A. Sedrakian and A. Dieperink, Phys. Rev. D **62**, 083002 (2000).
15. A. Schmitt, I. A. Shovkovy, and Q. Wang, in preparation.
16. N. Iwamoto, Phys. Rev. Lett. **44**, 1637 (1980).
17. T. Schäfer and K. Schwenzer, Phys. Rev. D **70**, 114037 (2004).
18. E. Flowers, M. Ruderman, and P. Sutherland, Astrophys. J. **205**, 541 (1976); D. N. Voskresensky and A. V. Senatorov, Sov. Phys. JETP **63**, 885 (1986).
19. P. Jaikumar and M. Prakash, Phys. Lett. B **516**, 345 (2001).
20. D. G. Yakovlev, A. D. Kaminker, O. Y. Gnedin and P. Haensel, Phys. Rept. **354**, 1 (2001).
21. A. Schmitt, I. A. Shovkovy and Q. Wang, Phys. Rev. Lett. **94**, 211101 (2005); Erratum *ibid.* **94**, 159902(E) (2005).

Participants

Mark Alford
Physics Department - Washington University
CB 1105, 1 Brookings Dr,
St Louis, MO 63130
USA
e-mail: alford@wuphys.wustl.edu

Charalampos Anastasiou
Institute for Theoretical Physics - ETH
Rämistrasse 101
CH - 8092 Zürich
SWITZERLAND
e-mail: babis@phys.ethz.ch

Antonella Antonelli
Istituto Nazionale di Fisica Nucleare
Laboratori Nazionali di Frascati
Via Fermi 40
I-00044 Frascati
ITALY
e-mail: antonella.antonelli@lnf.infn.it

Marco Battaglieri
Istituto Nazionale di Fisica Nucleare
Sezione di Genova
Via Dodecaneso 33
I-16139 Genova
ITALY
e-mail: battaglieri@ge.infn.it

Francesco Becattini
Department of Physics - University of Florence & INFN
Via G. Sansone 1
I-50019 Sesto Fiorentino (Firenze)
ITALY
e-mail: becattini@fi.infn.it

Nora Brambilla
Department of Physics - University of Milan & INFN
Via Celoria 16
I-20133 Milano
ITALY
e-mail: nora.brambilla@mi.infn.it

Giuseppe Eugenio Bruno
Department of Physics - University of Bari & INFN
Via Orabona 4
I-70126 Bari
ITALY
e-mail: giuseppe.bruno@ba.infn.it

Roberto Casalbuoni
Department of Physics - University of Florence & INFN
Via G. Sansone 1
I-50019 Sesto Fiorentino (Firenze)
ITALY
e-mail: roberto.casalbuoni@fi.infn.it

Stefano Catani
Istituto Nazionale di Fisica Nucleare
Sezione di Firenze
via G. Sansone 1
I-50019 Sesto Fiorentino (Firenze)
ITALY
e-mail: stefano.catani@fi.infn.it

Marco Ciminale
Department of Physics - University of Bari & INFN
Via Orabona 4
I-70126 Bari
ITALY
e-mail: marco.ciminale@ba.infn.it

Pietro Colangelo
Istituto Nazionale di Fisica Nucleare
Sezione di Bari
Via Orabona 4
I-70126 Bari
ITALY
e-mail: pietro.colangelo@ba.infn.it

Leonardo Cosmai
Istituto Nazionale di Fisica Nucleare
Sezione di Bari
Via Orabona 4
I-70126 Bari
ITALY
e-mail: leonardo.cosmai@ba.infn.it

Donato Creanza
Department of Physics - Politecnico di Bari & INFN
Via Orabona 4
I-70126 Bari
ITALY
e-mail: donato.creanza@ba.infn.it

Fulvia De Fazio
Istituto Nazionale di Fisica Nucleare
Sezione di Bari
Via Orabona 4
I-70126 Bari
ITALY
e-mail: fulvia.defazio@ba.infn.it

Nicola De Filippis
Department of Physics - University of Bari & INFN
Via Orabona 4
I-70126 Bari
ITALY
e-mail: nicola.defilippis@ba.infn.it

Luigi Del Debbio
CERN
CH-1211 Geneve 23
SWITZERLAND
e-mail: luigi.del.debbio@cern.ch

David d'Enterria
Nevis Laboratories - Columbia University
538 W. 120th St., NY 10027
USA
e-mail: denterria@nevis.columbia.edu

Adriano Di Giacomo
Department of Physics - University of Pisa & INFN
L.go Pontecorvo, 3
I-56127 Pisa
ITALY
e-mail: adriano.digiacomo@df.unipi.it

Pasquale Di Nezza
Istituto Nazionale di Fisica Nucleare
Laboratori Nazionali di Frascati
via E.Fermi 40
I-00044 Frascati (Roma)
ITALY
e-mail: Pasquale.DiNezza@lnf.infn.it

Elvio Di Salvo
Department of Physics - University of Genoa & INFN
Via Dodecaneso 33
I-16146 Genova
ITALY
e-mail: disalvo@ge.infn.it

Gerhard Ecker
Inst. Theor. Physik - University of Wien
Boltzmanng. 5
A-1090 Wien
AUSTRIA
e-mail: Gerhard.Ecker@univie.ac.at

Jan Olav Eeg
Department of Physics - University of Oslo
P.O.Box 1048
Blindern, N-Oslo 0316
NORWAY
e-mail: j.o.eeg@fys.uio.no

Pietro Faccioli
Department of Physics - University of Trento and ECT*
Via Sommarive 15
I-38050 Povo (Trento)
ITALY
e-mail: faccioli@ect.it

Svjetlana Fajfer
Department of Physics - University of Ljubljana
Jadranska 19
SI-1000 Ljubljana
SLOVENIA
e-mail: svjetlana.fajfer@ijs.si

Angelo Raffaele Fazio
University of the Witwatersrand
Johannesburg, 2050 Wits
SOUTH AFRICA
e-mail: fazioA@physics.wits.ac.za

Rossella Ferrandes
Department of Physics - University of Bari & INFN
Via Orabona 4
I-70126 Bari
ITALY
e-mail: rossella.ferrandes@ba.infn.it

Fernando Ferroni
Department of Physics - University of Rome "La Sapienza" & INFN
Piazza A. Moro 1
I-00185 Roma
ITALY
e-mail: fernando.ferroni@roma1.infn.it

Ferruccio Feruglio
Department of Physics - University of Padova & INFN
Via Marzolo 8
I-35131 Padova
ITALY
e-mail: ferruccio.feruglio@pd.infn.it

Rosa Anna Fini
Istituto Nazionale di Fisica Nucleare
Sezione di Bari
Via Orabona 4
I-70126 Bari
ITALY
e-mail: rosanna.fini@ba.infn.it

Domenico Giordano
Department of Physics - University of Bari & INFN
Via Orabona 4
I-70126 Bari
ITALY
e-mail: domenico.giordano@ba.infn.it

Tobias Hurth
CERN
CH-1211 Geneve 23
SWITZERLAND
e-mail: Tobias.Hurth@cern.ch

Nicola Ippolito
Department of Physics - University f Bari &INFN
Via Orabona 4
I-70126 Bari
ITALY
e-mail: nicola.ippolito@ba.infn.it

Yu Jia
Department of Physics - University of Milan & INFN
via Celoria 16
I-20133 Milano
ITALY
e-mail: Yu.Jia@mi.infn.it

Vincenzo Laporta
Department of Physics - University of Bari & INFN
Via Orabona 4
I-70126 Bari
ITALY
e-mail: vincenzo.laporta@ba.infn.it

Maria Paola Lombardo
Istituto Nazionale di Fisica Nucleare
Laboratori Nazionali di Frascati
via E.Fermi 40
I-00044 Frascati (Roma)
ITALY
e-mail: lombardo@lnf.infn.it

Luciano Maiani
CERN & Department of Physics - University of Rome "La Sapienza" & INFN
Piazza A. Moro 1
I-00185 Roma
ITALY
e-mail: luciano.maiani@roma1.infn.it

Sandra Malvezzi
Istituto Nazionale di Fisica Nucleare
Sezione di Milano
Via Celoria 16
I-20133 Milano
ITALY
e-mail: sandra.malvezzi@mi.infn.it

Massimo Mannarelli
Center for Theoretical Physics - MIT
285 Harvard Street 310
Cambridge MA 02139
USA
e-mail: massimo@lns.mit.edu

Pierpaolo Mastrolia
Dept. of Physics and Astronomy - UCLA
Los Angeles, CA 90095-1547
USA
e-mail: mastrolia@physics.ucla.edu

Annalisa Mastroserio
Department of Physics - University of Bari & INFN
Via Orabona 4
I-70126 Bari
ITALY
e-mail: annalisa.mastroserio@ba.infn.it

Salvatore My
Department of Physics - Politecnico di Bari & INFN
Via Orabona 4
I-70126 Bari
ITALY
e-mail: salvatore.my@ba.infn.it

Eugenio Nappi
Istituto Nazionale di Fisica Nucleare
Sezione di Bari
Via Orabona 4
I-70126 Bari
ITALY
e-mail: eugenio.nappi@ba.infn.it

Giuseppe Nardulli
Department of Physics - University of Bari & INFN
Via Orabona 4
I-70126 Bari
ITALY
e-mail: giuseppe.nardulli@ba.infn.it

Antti Niemi
Department of Theoretical Physics - Uppsala University
Box 803, S-75108 Uppsala
SWEDEN
e-mail: antti.niemi@teorfys.uu.se

Luis Oliver
Laboratoire de Physique Thèorique
Universitè Paris-Sud XI
Batiment 210
F-91405 Orsay
FRANCE
e-mail: luis.oliver@th.u-psud.fr

Altug Ozpineci
Istituto Nazionale di Fisica Nucleare
Sezione di Bari
Via Orabona 4
I-70126 Bari
ITALY
e-mail: altug.ozpineci@ba.infn.it

Davide Perrino
Department of Physics - University of Bari & INFN
Via Orabona 4
I-70126 Bari
ITALY
e-mail: davide.perrino@ba.infn.it

T.N. Pham
CPhT - Ecole Polytechnique
91128 Palaiseau Cedex
FRANCE
e-mail: tri-nang.pham@cpht.polytechnique.fr

Philippe Pillot
Institut de Physique Nucleaire de Lyon
43, Bd du 11 Novembre 1918
69622 Villeurbanne Cedex
FRANCE
e-mail: pillot@ipnl.in2p3.fr

Antonio Polosa
Department of Physics - University of Bari & INFN
Via Orabona 4
I-70126 Bari
ITALY
e-mail: antonio.polosa@cern.ch

Alexis Pompili
Department of Physics - University of Bari & INFN
Via Orabona 4
I-70126 Bari
ITALY
e-mail: pompili@ba.infn.it

Jorge Portoles
Instituto de Fisica Corpuscular – IFIC
Ed. Institutos de Investigacion , Apt. 22085
E-46071 Valencia
SPAIN
e-mail: Jorge.Portoles@ific.uv.es

Giovanni M. Prosperi
Department of Physics - University of Milan & INFN
Via Celoria 16
I-20133 Milano
ITALY
e-mail: prosperi@mi.infn.it

Marco Ruggieri
Department of Physics - University of Bari & INFN
Via Orabona 4
I-70126 Bari
ITALY
e-mail: marco.ruggieri@ba.infn.it

Pietro Santorelli
Department of Physics - University of Naples & INFN
Via Cinthia
I-80126 Napoli
ITALY
e-mail: pietro.santorelli@na.infn.it

Hagop Sazdjian
IPN, Universite Paris XI
91406 Orsay Cedex
FRANCE
e-mail: sazdjian@ipno.in2p3.fr

Dmitri Shirkov
Joint Institute for Nuclear Research
Joliot-Curie 6, Dubna, 141980
RUSSIA
e-mail: shirkovd@theor.jinr.ru

Igor Shovkovy
Frankfurt Institute for Advanced Studies - J. W. Goethe-University
Senckenberganlage 31
D-60054 Frankfurt am Main
GERMANY
e-mail: shovkovy@th.physik.uni-frankfurt.de

Claudia Simolo
Department of Physics - University of Milan & INFN
Via Celoria 16
I-20133 Milano
ITALY
e-mail: Claudia.Simolo@mi.infn.it

Luca Trentadue
Department of Physics - University of Parma & INFN
Parco Area delle Scienze 7/A
I-43100 Parma
ITALY
e-mail: luca.trentadue@pr.infn.it

Antonio Vairo
Department of Physics - University of Milan & INFN
Via Celoria 16
I-20133 Milano
ITALY
e-mail: antonio.vairo@mi.infn.it

Giacomo Volpe
Department of Physics - University of Bari & INFN
Via Orabona 4
I-70126 Bari
ITALY
e-mail: giacomo.volpe@ba.infn.it

Andrzej Wereszczynski
Institute of Physics - Jagiellonian University
Reymonta 4
30-059 Krakow
POLAND
e-mail: wereszczynski@th.if.uj.edu.pl

Valeriy Zamiralov
Scobeltsyn Institute of Nuclear Physics, Moscow State University
SINP , MGU, Leninskie Gory
119 992, Moscow
RUSSIA
e-mail: zamir@depni.sinp.msu.ru

Dario Zappalà
Istituto Nazionale di Fisica Nucleare
Sezione di Catania
Via S. Sofia 65
I-95123 Catania
ITALY
e-mail: dario.zappala@ct.infn.it

Author Index

A

Alford, M., 293
Aliev, T., 40
Anastasiou, C., 75
Antonelli, A., 18
Arnaldi, R., 279
Averbeck, R., 279

B

Banicz, K., 279
Battaglieri, M., 48
Becattini, F., 266
Bruno, G. E., 272

C

Casalbuoni, R., 104
Castor, J., 279
Castorina, P., 286
Catani, S., 67
Chaurand, B., 279
Cicalo, C., 279
Colangelo, P., 217
Colla, A., 279
Cortese, P., 279
Cosmai, L., 144

D

Damjanovic, S., 279
David, A., 279
Deandrea, A., 197
de Falco, A., 279
De Fazio, F., 217
Del Debbio, L., 137
D'Elia, M., 130, 245
d'Enterria, D., 252
Devaux, A., 279
De Vita, R., 48
Di Giacomo, A., 130
Di Nezza, P., 57
Di Renzo, F., 245

Di Salvo, E., 33
Drees, A., 279
Ducroux, L., 279

E

Ecker, G., 1
Eeg, J. O., 183
En'yo, H., 279

F

Faccioli, P., 26
Fajfer, S., 183, 203
Ferretti, A., 279
Ferroni, F., 151
Floris, M., 279
Force, P., 279

G

Gómez Dumm, D., 11
Guettet, N., 279
Guichard, A., 279
Gulkanian, H., 279

H

Heuser, J., 279
Hurth, T., 164

J

Jia, Y., 231
Jugeau, F., 173, 238

K

Kamenik, J., 203
Keil, M., 279
Kluberg, L., 279

Kubarovsky, V., 48

L

Ladisa, M., 197
Laporta, V., 197
Le Yaouanc, A., 173
Lombardo, M. P., 245
Lourenço, C., 279
Lozano, J., 279

M

Malvezzi, S., 210
Mannarelli, M., 259
Manso, F., 279
Martins, P., 279
Masoni, A., 279
Mastrolia, P., 89

N

Nardulli, G., 197, 286
Neves, A., 279
Niemi, A. J., 114

O

Ohnishi, H., 279
Oliver, L., 173
Oppedisano, C., 279
Ozpineci, A., 40, 217

P

Parracho, P., 279
Pham, T. N., 190
Pica, C., 130
Pich, A., 11
Pillot, P., 279
Portolés, J., 11
Prapotnik Brdnik, A., 183
Puddu, G., 279

R

Radermacher, E., 279
Ramalhete, P., 279
Rapp, R., 259
Raynal, J.-C., 173
Rosinsky, P., 279
Ruggieri, M., 303

S

Santorelli, P., 197
Sazdjian, H., 238
Schmitt, A., 310
Scomparin, E., 279
Seixas, J., 279
Serci, S., 279
Shahoyan, R., 279
Shirkov, D. V., 97
Shovkovy, I. A., 310
Sonderegger, P., 279
Specht, H. J., 279

T

Tieulent, R., 279
Trentadue, L., 80

U

Usai, G., 279

V

Vairo, A., 224
Veenhof, R., 279

W

Wang, Q., 310
Wereszczyński, A., 124
Wöhri, H. K., 279

Y

Yakovlev, S. B., 40

Z

Zamiralov, V. S., 40
Zappalà, D., 286